Environmental Sec
Use - with special r

NATO Security through Science Series

This Series presents the results of scientific meetings supported under the NATO Programme for Security through Science (STS).

Meetings supported by the NATO STS Programme are in security-related priority areas of Defence Against Terrorism or Countering Other Threats to Security. The types of meeting supported are generally "Advanced Study Institutes" and "Advanced Research Workshops". The NATO STS Series collects together the results of these meetings. The meetings are co-organized by scientists from NATO countries and scientists from NATO's "Partner" or "Mediterranean Dialogue" countries. The observations and recommendations made at the meetings, as well as the contents of the volumes in the Series, reflect those of participants and contributors only; they should not necessarily be regarded as reflecting NATO views or policy.

Advanced Study Institutes (ASI) are high-level tutorial courses to convey the latest developments in a subject to an advanced-level audience

Advanced Research Workshops (ARW) are expert meetings where an intense but informal exchange of views at the frontiers of a subject aims at identifying directions for future action

Following a transformation of the programme in 2004 the Series has been re-named and re-organised. Recent volumes on topics not related to security, which result from meetings supported under the programme earlier, may be found in the NATO Science Series.

The Series is published by IOS Press, Amsterdam, Dordrecht, in conjunction with t he NATO Public Diplomacy Division.

Sub-Series

A. Chemistry and Biology	Springer
B. Physics and Biophysics	Springer
C. Environmental Security	Springer
D. Information and Communication Security	IOS Press
E. Human and Societal Dynamics	IOS Press

http://www.nato.int/science
http://www.springer.com
http://www.iospress.nl

Series C: Environmental Security – Vol. 7

Environmental Security and Sustainable Land Use - with special reference to Central Asia

edited by

Hartmut Vogtmann
President, Federal Agency for Nature Conservation,
Bonn, Germany

and

Nikolai Dobretsov
President, Siberian Branch of Russian Academy of Sciences,
Novosibirsk, Russia

With the collaboration of Astrid Mittelstaedt

Published in cooperation with NATO Public Diplomacy Division

Proceedings of the NATO Advanced Research Workshop on
Environmental Security and Sustainable Land Use of Mountain and Steppe
Territories of Mongolia and Altai
Barnaul, Russia
25-27 October 2004

A C.I.P. Catalogue record for this book is available from the Library of Congress.

ISBN-10 1-4020-4492-5 (PB)
ISBN-13 978-1-4020-4492-2 (PB)
ISBN-10 1-4020-4491-7 (HB)
ISBN-13 978-1-4020-4491-5 (HB)
ISBN-10 1-4020-4493-3 (e-book)
ISBN-13 978-1-4020-4493-9 (e-book)

Published by Springer,
P.O. Box 17, 3300 AA Dordrecht, The Netherlands.

www.springer.com

Printed on acid-free paper

All Rights Reserved
© 2006 Springer
No part of this work may be reproduced, stored in a retrieval system, or transmitted in
any form or by any means, electronic, mechanical, photocopying, microfilming, recording
or otherwise, without written permission from the Publisher, with the exception of any
material supplied specifically for the purpose of being entered and executed on a
computer system, for exclusive use by the purchaser of the work.

Printed in the Netherlands.

CONTENTS

Preface ix

Chapter 1: Environmental Consequences of Climate Change

Desertification of mid-latitude Northern Asia and global change
periodicity in the Quaternary
    N.L. Dobretsov, V.S. Zykin & V.S. Zykina 3

Regional climate and environmental change in Central Asia
    N.F. Kharlamova & V.S. Revyakin 19

The reasons and consequences of climate changes
    V.F. Loginov 27

Climate change consequences in steppe-forest transition zone in Moravia
    M. Klimanek 47

Chapter 2: Sustainable Land Use and Regional Development

Assessment of sustainability of ecological-economic systems by indicators of sustainable development
    E.M. Rodina 59

Overview of NATO CCMS Pilot Study on environmental
decision making for sustainable development in Central Asia
    M. Khankhasayev, R. Herndon, J. Moerlins & C. Teaf 65

Experiences in the study of land cover transformation on Mediterranean
islands caused by change in land tenure
    B. Cyffka 85

Study of soils modified with structure forming agents
    P. Petranka 105

Problems of instability in agrarian nature management and food safety
in large countries of Central Asia
    B. Krasnoyarova & I. Orlova 115

Sustainable land use as a basis for a healthy nutrition and a corner
stone for regional development
    A. Meier – Ploeger 129

Chapter 3: Sustainable Development in Mountain and Steppe Regions

Problems of sustainable development of mountainous regions of Tajikistan
  H.M. Muhabbatov & H.U. Umarov 141

Implication of environmental law for mountain protection in China
  J. Zhang 149

Actual ecological situation in the territory of mountain regions and biodiversity problems – the case of Georgia
  Z. Tatashidze, I. Bondyrev & E. Tsereteli 159

New ways and new forms of limited and controlled nature management in the steppe region of northern Eurasia
  A.A. Chibilyov 175

Chapter 4: Beyond Boarders: Transboundary Biosphere Reserve "Altai" as an Approach for Regional Development

Sustainable land use convergence in border area in Central Europe
  J. Kolejka & D. Marek 183

Sustainable development beyond administrative boundaries
Case study: Rhön Biosphere Reserve, Germany
  DR. D. Pokorny 199

The links between poverty and environment: the rationale for environmentally sustainable resource use, with application to land management in the Altai region
  D.H. Smith 215

Conditions and trends in natural systems of the Altai-Sayan ecoregion
  A. Mandych 231

Transboundary Biosphere Territory "Altai": Expert evaluation for the establishment
  YU. Vinokurov, B. Krasnoyarova & S. Surazakova 277

Chapter 5: Landscape Planning as an Integrative Tool for Sustainable Development

Landscape planning as a tool for sustainable development of the territory
  D. Gruehn 297

The Russian school of landscape planning
　　A.N. Antipov & YU.M. Semenov 309

Chapter 6: Challenges and Threats for Environmental Stability in Central Asia

Historical experience and estimation of modern land tenure of the Inner Asia
　　A. Tulokhonov 323

No Man's Land – Environmental influences in Central Asian security
　　P.H. Liotta 335

Physical - geographical characteristics of the Altai region
　　DR.D. Enkhtaivan 349

New challenge of the modern civilization and global economy in the north of Western Siberia
　　G.N. Grebenuk & F.N. Rjansky 353

Summary of other Workshop Lectures
　　N. Dobretsov 365

List of Workshop Participants 375

PREFACE

Climate change, non-sustainable land management and the insufficient participation of the local population leads to land degradation problems in parts of the Altai Region, what includes the territory of China, Kazakhstan, Mongolia and the Russian Federation. This territory is characterized by mountain and steppe landscapes, high biological diversity and a high dependency on natural resources. An environmentally sustainable use of natural resources remains a fundamental foundation for sustainable economic and social development. The achievement of ecologically sound land management practices, especially in vulnerable regions, was identified as a particular important goal for the economic significance of agricultural production and a predominantly rural population. So far, in some parts of the region, land use has been unsustainable over the long term.

The workshop offered important impulses for the solution of the land-use problems in this region: strategic approaches for further actions as well as proposals for concrete measures. First achievements of the workshop are the continuation of the work towards the establishment of the Transboundary Biosphere Territory "Altai" between China, Kazakhstan, Mongolia and the Russian Federation as a model region for sustainable regional development. Further, the three days of intensive scientific exchange and informal discussion had contributed to improve the development of an international program on environmental monitoring and resource management practices, and the establishment of landscape planning as an instrument for the implementation of sustainable land management.

This book presents important aspects of environmental security and sustainable land use in general and in particular for Central Asia, i.e. the environmental consequences of climate change; sustainable land use and regional development; sustainable development in mountain and steppe regions; beyond boarders: Transboundary Biosphere Reserve "Altai" as an approach for regional development; landscape planning as an integrative tool for sustainable development; challenges and threats for environmental stability in Central Asia arisen during the workshop.

A final recommendation in the area of sustainable development and land use as well as the proceeding efforts towards the establishment of the Transboundary Biosphere Reserve "Altai" and the implementation of landscape planning were set in the book as some of the main results of the meeting.

The Editors.

CHAPTER 1
ENVIRONMENTAL CONSEQUENCES
OF CLIMATE CHANGE

DESERTIFICATION OF MID-LATITUDE NORTHERN ASIA AND GLOBAL CHANGE PERIODICITY IN THE QUATERNARY

N.L. DOBRETSOV[1], V.S. ZYKIN[2] AND V.S. ZYKINA[2]
1 – Siberian Branch of Russian Academy of Sciences, Russia
2 – Institute of Geology SB RAS, Russia

Abstract: The knowledge of the history of regional geosystems and prediction of their future evolution trends are indispensable for nature conservation. Global warming threatens to become catastrophic and is thus an urgent scientific and social problem. The last century of the past millennium was marked by an exceptional growth of global air temperature which became 0.6° C higher than at the end of the Little Ice Age (1550-1850). Warming was especially rapid after the 1960s, with a linear trend of 0.20° C per decade (global) and 0.29° C per decade in the Northern Hemisphere (Grusa et al., 2001). The past decade was the warmest over the millennium, and 1998 was the globally warmest year. Arctic ice sheets in warm season have reduced in surface area for 10-15% and have become 40% thinner for the past 50 yr. Mountain glaciers in Asia have been reducing and permafrost has been degrading. Scientists are not unanimous about the prospects, some believing that warming-related global change can speed up and cause regional- and global-scale socioeconomic ill effects, and others considering the problem ambiguous and poorly understood; the latter opinion is that prediction has even increased in uncertainty lately instead of being resolved (Boehmer-Christiansen, 2000). Prediction for global change and its short-term consequences is difficult because the changes are driven by sophisticated interplay of numerous climate controls and feedback mechanisms, while the available field and modeling data remain insufficient. The relative contributions of natural and cultural effects to the ongoing warming have not been so far constrained unambiguously.

Changes in air humidity on continents and desertification processes constitute a key warming-related problem (Drozdov, 1981; Fairbridge, 1989; Kovda, 1977; Shnitnikov, 1957; Jones, Reid, 2001). The UN Convention to Combat Desertification (UNCCD), signed by over 150 countries and ratified in 1994, puts special emphasis on desertification studies. The humidity regime in the 21st century is expected to undergo strong changes accompanying global warming (Climate Change..., 2001). The amount of precipitation in Asian Russia has reduced notably since the 1960s, especially in the warm season (Gruza et al., 2001), which, together with the air temperature growth, is fraught with drought hazard. The humidity/temperature relationship is intricate and nonlinear, and remains an under explored problem associated with climate events. The mechanisms that control mid-latitude aridization are especially poorly understood (Manabe, Broccoli, 1990).

Reconstructions of past global and regional climates, along with numerical modelling, make an effective prediction tool for climate change and for the consequences of global warming. The vulnerability of the modern geosystems, which have lived through a long history, to long-lasting cultural and natural impacts, including climate events, cannot be assessed without the knowledge of their past evolution. Studies of climate change in space and time and revealing its periodicity of different scales provide clues to the regularities of the changes, to the present state and dynamics of climates, and to the natural evolution trends. The impact of many climate-forming factors, their interplay, and cause-effect relations can be inferred from their long-lasting action in the geological and historic past. Investigation into past climate events can make basis for empirical estimates of the vulnerability of the global thermal regime to changes in air chemistry, reveal the dependence of separate environment components on global temperature trends and their feedback, and allow specific predictions for different regions. Documenting and interpretation of global and regional climate records are of special importance for gaining insight into the temperature/humidity relationship. Paleoclimate evidence is practically the only available source of this knowledge as climate models cannot give an accurate account of regional precipitation dynamics (Borzenkova, 1987, 1992).

The Cenozoic history of environments in continental Asia is recorded in sedimentary land and limnic sections making unique archives that store yet undeciphered evidence of climate change. The great amount of paleoclimate data collected through the recent decade shows that the bottom fill of Lake Baikal, loess-soil sequences in southern West Siberia, the sediments of closed (not overflowing) lakes, and peatbogs offer the most complete record of Quaternary climates in mid-latitude continental Asia. Analyzed at high-resolution, it can display the history of global and regional climate events, possible responses of natural systems, and the trends of humidity variations in inner Asia associated with warming progress.

1. CLIMATE RECORD IN THE BOTTOM FILL OF LAKE BAIKAL

Wetting/drying cycles caused strong influence on sedimentation in Lake Baikal. Baikal bottom sediments were stripped by numerous shallow boreholes and by deep drilling in four sites (BDP-93-1, BDP-93-2; BDP-96-1, BDP-96-2; BDP-98; and BDP-99) as part of the international Baikal Drilling Project (Karabanov et al., 2001; Kuzmin et al., 2001). The cores show a well pronounced cyclic structure produced by alternated layers of diatomaceous mud and diatom-barren clay. Climate events in the > 10 myr continuous sedimentary archive of Lake Baikal are recorded by geochemical and biological proxies. Diatom abundances and related biogenic silica contents reflect biological production in the lake being a highly sensitive climate proxy: diatomaceous mud was deposited during interglacials and diatom-barren clay accumulated during glacials (Bezrukova et al., 1991).

The Baikal BiSi paleoclimate record for the past 800 kyr correlates well with the marine oxygen-isotope stratigraphy (Colman et al., 1995; Prokopenko et al., 2001; Williams et al., 1997) and includes 19 stages (Figure 1). BiSi peaks corresponding to warm times are equated to the warm odd stages and the dips to the cold even stages of the $\delta^{18}O$ curve.

Multi-element SR XFA analysis revealed subtle geochemical responses in bottom sediments and several types of terrigenous markers of cold and warm climates (Goldberg et al., 2000). Sr/Ba ratios (Rb, Cs, Ti), U/Th, Zn/Nb, and high contents of U, Mo, Br, Eu, Tb, Yb, and Lu show positive correlation with BiSi and indicate warm intervals. Cold periods are marked by high contents of Th, Ba, Rb, Cs, La, Ce, and Nd and high ratios of La (Ce, Ba)/Yb(Y, Zr).

Spectral analysis of the Baikal BiSi records (Colman et al., 1995; Williams et al., 1997) and geochemical proxies (Goldberg et al., 2000) for the past 800 kyr detected orbitally-forced cycles of 100, 42 and 23-19 kyr. The agreement of the Baikal climate record with the marine oxygen-isotope curve in number, amplitude, and geometry of peaks (Bassinot et al., 1994) allows an inference that climate change in Siberia has been controlled by Earth's orbital parameters and followed the global climate trends (Kuzmin et al., 2001).

Although diatom abundance in the alternated layers of diatomaceous mud and diatom-barren clay is the key climate proxy, the cause of its correlation with global climate remains disputable. Temperature being not critical to diatom production, several factors were suggested to explain the observed abundance variations through the Pleistocene (Grachev et al., 2002), including water turbidity and supply of nutrients (dissolved Si and P) (Gavshin et al., 2001; Lisitsin, 1966). Abrupt turbidity decrease in Lake Baikal after thawing of mountain glaciers provoked dramatic change to the lake ecosystem and diatom production growth (Bezrukova et al., 1991). The change in Si and P supply is attributed to abrupt slowing of chemical weathering associated with a 6°C fall of mean annual air temperature (Grachev et al., 2002). Reduced input of plainland rivers into the lake as a result of precipitation decrease associated with climate drying during glacial peaks was another cause of break in Si supply and periodic extinction of diatoms (Goldberg et al., 2005). One more factor of diatom fall in the lake is related to considerable increase in atmospheric dust during glacials (Broecker, 2000) recorded in loess deposition.

Abundant dust in the air deteriorated its transparence, increased water turbidity when precipitated into the lake, and exerted strong control on sedimentation. Moreover, dust screened light penetration when settled on the ice, while the duration of the ice period can have been two months or more longer than at present (Shimaraev et al., 1995). Diatoms in Lake Baikal now bloom in spring under the ice (Verkhozina et al., 1997), and the dust screen must have impeded photosynthesis and reduced diatom production in the periods of dry and cold climate. Climate drying during glacials in the Baikal catchment is indicated by loess occurrences in the Baikal region and the presence of ventifacts in the Selenga terraces (Bazarov, 1986) which are indicative of eolian activity and a runoff break in the valley. Therefore, the Baikal climate record provides evidence of global temperature change as well as climate drying variations which approximately correlate with the temperature dynamics.

2. LOESS-SOIL SEQUENCE IN SOUTHERN WEST SIBERIA

Loess deposition and soil formation are informative of the periodicity and regularities of wetting and drying (desertification) processes. Loess-soil sequences in Middle and Late Pleistocene continental deposits make up the most complete and detailed natural account of climate and environment events. The composition and structure of the loess-soil sequence reflects the general intensity of atmospheric circulation (Broecker, 2000): Biogenic sedimentation and soil formation mostly occurred in periods of weak circulation, whereas loess deposition was associated with active wind transport when the air was thickly saturated with dust.

The loess-soil sequence of West Siberia consists of rhythmically alternated thick layers of loess and complexes of fossil soils intervened by thin loess horizons. The stratigraphic record *(Table 1)* and evolution of the loess-soil sequence was constrained by paleopedological, paleomagnetic, and biostratigraphic evidence and by ^{14}C and TL dating (Zykina et al., 1981, 2000; Volkov, Zykina, 1991; Zykina, 1999; Zykin et al., 2000; Zander et al., 2003).

Fossil soils in soil complexes formed during warm Pleistocene stages as indicated by their age constraints from different methods and their morphology that reflects formation in a temperate climate. The rhythmic pattern of fossil soils in the West Siberian loess stratigraphy and the structure of the soil

Table 1: Neopleistocene loess-soil sequence of Siberia

complexes correlate well with warm odd MIS stages (Bassinot et al., 1994), warm stages of the Baikal sedimentary record (Kuz'min et al., 2001; Goldberg et al., 2000), temperature and dust curves from Vostok ice cores in Antarctica (Petit et al, 1999), and with magnetic susceptibility variations in the loess-soil sequences of China (Kukla et al., 1990) *(Figure 1)*.

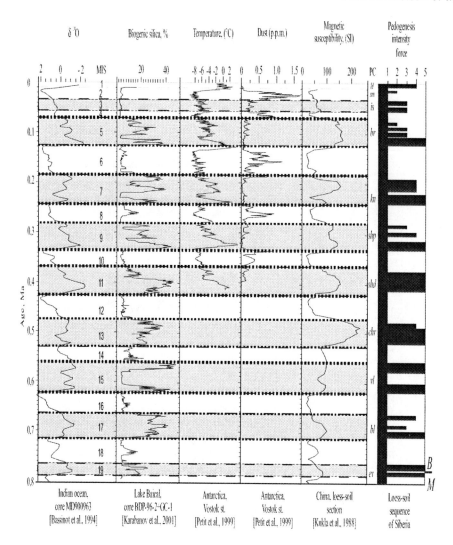

Fig. 1: Comparison of loess-soil of Siberia and records of Lake Baical biogenic silica, Vostok ice core in Antarctic, China loess-soil magnetic susceptibility and marine isotope stratigraphy

Fossil soils within the West Siberian loess-soil sequence *(Figure 2,3)* faithfully record warm odd stages of the continuous global archive (Dobretsov et al., 2003) which include closely positioned warm events intervened by relatively brief cold excursions *(Figure 2, 3)*.

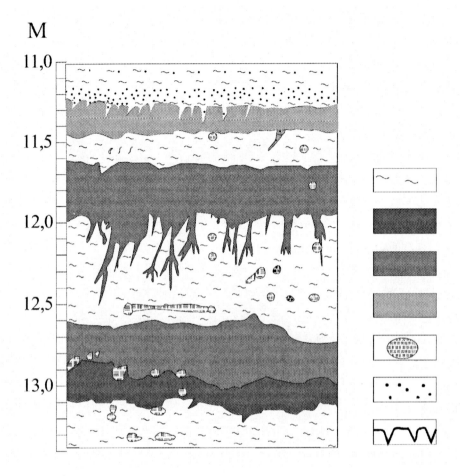

Fig. 2: Shipunovo pedocomlex in West Siberia (MIS 9) (Iskitim region, near Novosibirsk)
1 – loess, 2 – high-humus loam, 3 – medium-humus loam, 5 – burrows, 6 – sandy loam
7 – frost wedges

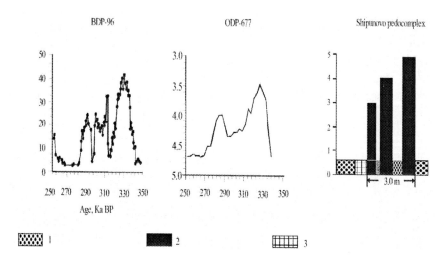

Fig. 3: Climate change during MIS 9 recorded in bottom sediments of Lake Baikal (ODP-677, Chackleton et al, 1995), and in loess-soil sequence of Siberia.
1 – loess, 2 – chernozem, 3 – cryogenic stage

The thick loess layers match the cold stages of the global record, and high dust contents in the cold intervals of the Antarctic and Greenland ice cores (Petit et al., 1999; 2000; Biscaye et al., 1997) suggests that loess was deposited in cold and dry periods (*Figure 1*). Dust contents in air during glacial peaks were up to 30 times those during the peaks of interglacial (Broecker, 2000).

The good agreement of the West Siberian loess-soil sequence and the continuous global climate record (Dobretsov et al., 2003) *(Figure 1)* attest to the global scale of loess deposition and confirms the common periodicity, trends, and mechanisms of Pleistocene climate change both in mid-latitude Central Asia and on the Earth as a whole.

The West Siberian loess-soil sequence that matches the global record (Bassinot et al., 1994; Kuzmin et al., 2001; Goldberg, 2000; Petit et al., 1999; Kukla et al., 1990), where spectral analysis revealed 20, 40 and 100 kyr orbital forcing cycles, can be expected to bear the same periodicity. The Brunhes chron is dominated by 100 kyr cycles which control the large-scale glacial-interglacial periodicity. 40 kyr cycles govern the alternation of thick loess layers correlated with the even stages of the marine oxygen-isotope stratigraphy and soil complexes corresponding to the odd stages. Deposition during most of warm stages followed 20 kyr orbital cycles which is reflected in the alternated soils and thin loess interbeds in soil complexes. The 20 kyr orbital cycles are absent from the Middle Pleistocene cold stages in West Siberia associated with the deposition of thick loess layers but are well pronounced in the Late Pleistocene loess sequence with intermittent six poorly developed soils. These cycles may have been then of a lower amplitude and remained obscure in the loess record of mid-latitude Siberia.

The loess deposition in the western West Siberian Plain was accompanied by the formation of large deflation surfaces and closed deflation basins. The latter are widespread in southern West Siberia and are often filled with closed (not overflowing) lakes. Their eolian origin in arid climate is indicated by desert pavement, ventifacts, carbonate crusts, and desert varnish on bedrock pebble and debris. More evidence is found in cracked cobble on the floor of the Lake Aksor deflation basin in the Pavlodar Irtysh region, which subsided during the Ermakovian glacial equated to MIS 4 (Zykin et al., 2003), and in mud crack wedges on the bottom of Lake Chany whose basin dates back to the Sartan glacial or stage 2 of the $\delta^{18}O$ stratigraphy. The >70 m deep closed deflation basins of lakes Kyzyl-Kak, Teke, Kishi-Karoi, and Ul'ken-Karoi appear to be older. The wind-borne sediment transport from the deflation basins occurred repeatedly during the stages of cold and dry climate.

In addition to the erosional etch features, other wind-built landforms widespread in mid-latitude Inner Asia are likewise genetically related to loess deposition. In West Siberia they form the well preserved sand-ridge topography of the Last Glacial and long-lasting wide low ridges in eastern Kulunda [Volkov, 1976]. The distribution and orientation of the eolian landforms produced during glacials suggest their origin by transport with mostly western winds.

The temperature and humidity regime of the soil formation stages in West Siberia were similar to those of Holocene environments or of the previous interglacials and interstadials (Zykin et al., 2000a,b; Zykina et al., 1981). The humidity conditions in northern Kazakhstan and Central Asia during the last interglacial warming when the global mean annual air temperature reached 1.5-2.0° C in the optimum (Velichko et al., 1984) were much better than now (Zykin et al., 1995, 2000b). The inception of migrant species from Central Asia that belong to the genera *Corbicula, Corbiculina, Odhneripisidium,* and *Allocinma* in almost all large interglacials of Quaternary West Siberia indicates the presence of a developed river network and sufficient humidity in Central Asia and western Kazakhstan in those times.

3. BOTTOM FILL OF CLOSED LAKE BASINS

The Quaternary warm wet and cold dry stages were intervened by shorter millennium-scale temperature and humidity spells. These abrupt brief excursions were revealed in the upper bottom fill section of the closed basin of Lake Aksor (Pavlodar Irtysh region). The section of lacustrine sediments that was deposited during the Sartan Last glacial displays prominent cyclicity of limnic sand, sand-wedge polygons, and horizons of desert weathering and selective wind erosion *(figure 4)* which records abrupt changes in temperature and humidity. Climate reconstructions reveal at least eight 1100 – 1300 kyr long repeated cycles between 24 and 16 kyr BP *(Figure 5)*. Within that time span, strongly cold and extremely dry climate alternated with moderately cold wetter periods. The colder and drier climate caused lake drying and its floor freezing, formation of sand-wedge polygons, and active deflation. The cold episodes were followed by warmer and wetter spells when permafrost degraded and the basin became filled with water. Sand-wedge polygons which are diagnostic of extremely cold and dry climates can develop under -12 to $-20°$ C and a mean annual precipitation below 100 mm (Karte, 1983). Deposition of limnic sand without sand-wedge polygons can have occurred at mean annual air temperature no lower than $-3°$ C. Well preserved permafrost features, with prominent bulging parts of the polygons, evidence of rapidly changing deposition environments in the closed basin. The mean annual air temperature during the formation of frozen ground must have been 13-21° C colder than now, which is in line with paleoclimate modeling (Kutzbach et al., 1998) which predicts 10-15° C lower temperatures. Temperature contrasts between the intervals of stronger and weaker cooling in the Pavlodar Irtysh region reached from 9 to 17° C.

4. THE HOLOCENE

The temperature and humidity spells within warm and relatively wet stages in the Quaternary are best documented for the Holocene. That was the coldest time compared to the previous interglacials with quasi-periodic (every 1000 in the average) fluctuations. The early Holocene (10.2-5.3 kyr BP) climate in the Northern Hemisphere was warm and relatively stable, mean annual air temperature increased and reached an optimum that lasted from 6.2 to 5.3 kyr BP (Borzenkova, 1992; Velichko, 1989). Climate in the second half of the Holocene followed a cooling trend and was becoming ever less stable. The mean global temperature over the Northern Hemisphere during the Holocene optimum (6.2-5.3 kyr BP) was 1° C lower than in the optimum of the

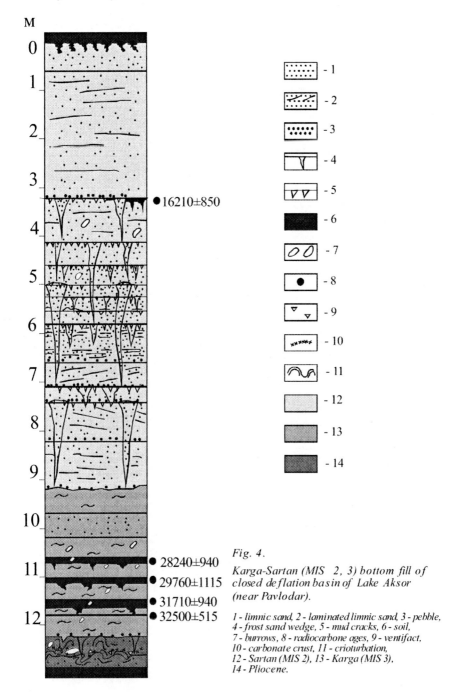

Fig. 4.

Karga-Sartan (MIS 2, 3) bottom fill of closed deflation basin of Lake Aksor (near Pavlodar).

1 - limnic sand, 2 - laminated limnic sand, 3 - pebble, 4 - frost sand wedge, 5 - mud cracks, 6 - soil, 7 - burrows, 8 - radiocarbone ages, 9 - ventifact, 10 - carbonate crust, 11 - crioturbation, 12 - Sartan (MIS 2), 13 - Karga (MIS 3), 14 - Pliocene.

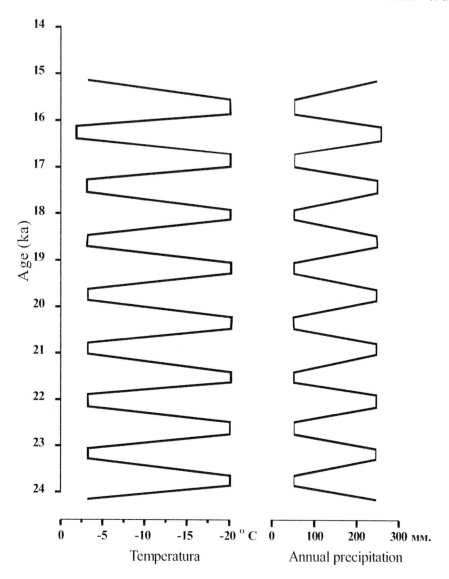

Fig. 5: Reconstruction of air temperature and annual precipitation in Sartan time, from bottom fill of Lake Aksor

last interglacial and 1° C higher than at present (Velichko, 1989). Numerous data indicate a complex relationship between wetting-drying and warming-cooling trends in mid-latitude Eurasia. Warming during the Holocene optimum in West Siberia was first accompanied by wetting and then by drying (Volkova et al., 1989). The same period in Mongolia was the time of cooling and precipitation increase (Khotinsky, 1989). The humidity conditions in mid-latitude Eurasia improved during the cooling of the Little Ice Age that lasted from the middle 15th to the latest 19th century (Monin, Shishkov, 1998). In the latest 18th century, the water level in the closed basin of Lake Chany was 3 m higher than now and its surface area was 1.5 – 2 times as large (Pulsating..., 1982). The climate of eastern Mongolia in the latest 17th - earliest 18th centuries was extremely dry, warm seasons were the shortest, winter seasons were the most severe, with the strongest winds and storms (Chichagov, 1996).

The vegetation composition in the Baraba Plain (West Siberia) in the Holocene changed in 1 kyr cycles (Levina, Orlova, 1993; Zykin et al., 2000a, b). Each of ten distinguished rhythms consists of a warm and a cold interval, about 500-600 yr long. These intervals, in turn, include two 250-300 yr subintervals of warm-dry/warm-wet and cold-wet/ cold-dry climates. Each rhythm began with a warm-dry stage followed by wetting under the same warm conditions; the second half of the rhythms was cold: first wet and then dry *(Table 2)*.

Thus, high-resolution analysis of climate events in the Holocene shows a weak correlation of temperature and humidity trends and even contrary contrasting changes in neighboring regions.

5. CONCLUSION

The reported Quaternary climate data from several regions of mid-latitude Asia allow empirical estimates of global temperature dependence of humidity and related aridization. Significant changes in the global temperature regime caused by Earth's orbital forcing were accompanied by humidity changes: humidity conditions improved during stages of global warming and global cooling stages were marked by active air circulation, drying, deflation, dust increase in the air, and loess deposition. Temperature and humidity conditions in cold times changed abruptly in periodic spells. Extremely cold stages were drier and less cold intervals were relatively wetter. Humidity in warm stages notably depended on temperature variations. A significant temperature increase of above 1º C was associated with higher humidity. The more subtle temperature-humidity pattern is quite intricate: moderate warming of < 1ºC was accompanied by weak drying and moderate cooling by relatively weak wetting.

Therefore, prediction of natural aridization and desertification processes for the nearest 1000 yr requires independent estimates of humidity trends controlled by mountain growth, volcanism, dynamics of Gulf Stream, and other factors.

This research is supported by RFBR grant 04-05-64486.

REFERENCES

1. Bassinot F.C., Labeyrie L.D., Vincent E., Quidelleur X., Shackleton N.J., Lancelot Y. (1994) The astronomical theory of climate and the age of the Brunhes-Matuyama magnetic reversal. Earth and Planetary Science Letters, vol. 126, p. 91-108.
2. Bazarov D.D.B. (1986) The Cenozoic of the Baikal and western Trans-Baikal regions. Nauka, Novosibirsk, 180 pp. (in Russian).
3. Bezrukova E.V., Bogdanov Y.A., Williams D.F., Granina L.Z., Grachev M.A., Ignatova N.V., Karabanov E.B., Kuptsov V.M., Kurylev A.V., Letunova P.P., Likhoshvay E.V., Chernyaeva G.P., Shimaraeva M.K. and Yakushin A.O. (1991) A dramatic change in the ecosystem of Lake Baikal in the Holocene. Doklady Akademii Nauk SSSR, 321, 1032-1037. (in Russian).
4. Biscaye P.I., Crousset F.E., Revel M., Van der Gaast S., Zielinski G.A., Vaars A., Kukla G. (1997) Asian provenance of glacial dust (stage 2) in the Greenland Ice Sheet Project 2 Ice Core, Summit, Greenland. J. Geophys. Res., 102, pp. 26765-26781.
5. Boehmer-Christiansen S. (2000) Who and how makes the policy concerning global change. Izvestia Russian Geographical Society, 132, 3, pp. 6-22. (in Russian).
6. Borzenkova I.I. (1987) Land wetting in the Northern Hemisphere in the geological past. Meteorology and hydrology, 10, pp. 53-61. (in Russian).
7. Borzenkova I.I. (1992) Climate change in the Cenozoic. Hydrometeoizdat, St. Petersburg, 248 pp. (in Russian).
8. Broecker W.S. (2000) Abrupt climate change: causal constraints provided by the paleoclimate record. Earth-Science Reviews, 51, pp. 137-154.
9. Chichagov V.P. (1996) Arid peneplain of Central Asia and its formation in the Little Ice Age in Eastern Mongolia. Izvestia Russian Geographical Society, 6, pp. 28-38. (in Russian).
10. Climate Change 2001: The Science of Climate Change. Intergovernmental Panel on Climate Change. J.T.Houghton et al. (Eds). Cambridge University Press. Cambridge. 2001. 881 pp.
11. Colman S.M., Peck J.A., Karabanov E.B., Carter S.J., Bradbury J.P., King J.W. and Williams D.F. (1995) Continental climate response to orbital forcing from biogenic silica records in Lake Baikal. Nature, 378, 769-771.
12. Dobretsov N.L., Zykin V.S., Zykina V.S. (2003) Structure of the Pleistocene Loess-Soil Sequence of Western Siberia and Its Correlation with the Baikalian and Global Records of Climatic Change. Doklady Earth Sciences, 391A, pp. 921-924.
13. Drozdov O.A. (1981) Land moistening and climate change. Meteorology and hydrology, 4, pp. 17-28. (in Russian).
14. Fairbridge Rh.W. (1989) Water deficiency versus water excess: global management potential. In Paepe R., Fairbrdge Rh.W., Jelgersma S. (eds.) Greenhouse Effect, Sea Level and Drought. NATO ASI Series. Series C. Vol. 325. Kluwer Academic Publishers, Dordrecht, Boston, London, pp. 185-197.
15. Gavshin V.M., Bobrov V.A., Khlystov O.K. (2001) Periodicity in diatom sedimentation and geochemistry of diatomaceous mud in Lake Baikal: global aspect. Russian Geology and Geophysics, 42, pp. 317-325.
16. Goldberg E.L., Phedorin M.A., Grachev M.A., Bobrov V.A., Dolbnya I.P., Khlystov O.M., Levina O.V., Ziborova G.A. (2000) Geochemical signals of orbital forcing in the records of paleoclimates found in the sediments of Lake Baikal. Nucl. Instr. and Meth. in Phys. Res. A., 448, 1-2, pp. 384-393.
17. Goldberg E.L., Chebykin E.P., Vorobyeva S.S., Grachev M.A. (2005) Uranium signals of paleoclimate humidity recorded in sediments of Lake Baikal. Doklady Earth Sciences, 400, 1, pp. 52-56.
18. Grachev M.A., Gorshkov A.G., Azarova I.N., Goldberg E.L., Vorob'eva S.S., Zheleznyakova T.O., Bezrukova E.V., Krapivina S.M., Letunova P.P., Khlystov O.K., Levina O.V., Chebykin E.P. (2002) Regular climate oscillations in millennial-scale and

speciation in Lake Baikal. Major regularities of global and regional climatic and environmental changes in the Late Cenozoic of Siberia. E.A.Vaganov et al. (Eds). Institute of Archaeology and Ethnography SB RAS Press, Novosibirsk, pp. 107-121. (in Russian).
19. Gruza G.V., Bardin M.Yu., Ran'kova E.Ya., Rocheva E.V., Sokolov Yu.Yu., Samokhina O.F., Platova T.V. (2001) On air temperature and precipitation changes in the territory of Russia during the 20th century. In Integrated monitoring of the environment and climate. Limits of change. Yu.A.Izrael (Ed.), Nauka, Moscow, pp. 18-39. (in Russian).
20. Jones P.D., Reid P.A. (2001) Temperature trends in regions affected by increasing aridity/humidity // Geophysical Research Letters, 28, 20, pp. 3919-3922.
21. Karabanov E.B., Prokopenko A.A., Kuzmin M.I., Williams D., Gvozdkov A.N. and Kerber E.V. (2001) Glacial and interglacial periods of Siberia: the Lake Baikal paleoclimate record and correlation with West Siberian stratigraphic scheme (*the Brunhes Chron*). Russian Geology and Geophysics, 42, 1, pp. 41-54.
22. Karte J. (1987) Periglacial Phenomena and their Significance as Climatic and Edaphic Indicators. Geojournal, 7, 4, pp. 329-340.
23. Kovda V.A. (1977) Aridization of land and drought mitigation. Nauka, Moscow, 270. (in Russian).
24. Khotinsky N.A. (1989) Problems of reconstruction and correlation of Holocene paleoclimates. In Khotinsky N.A. (Ed.) Paleoclimates of the Late Glacial and Holocene. Nauka, Moscow, pp. 12-17. (in Russian).
25. Kukla G., An Z.S., Melice J.L., Gavin J., Xiao J.L. (1990) Magnetic susceptibility record of Chinese Loess. Transactions of the Royal Society of Edinburg. Earth Sci., 81, pp. 263-288.
26. Kutzbach J., Gallimore R., Harrison S., Behling P., Selin R., Laarif T. (1998) Climate and biome simulations for the past 21,000 years. Quaternary Science Reviews, 17, pp. 473-506.
27. Kuzmin M.I., Karabanov E.B., Kawai T., Williams D., Bychinsky V.A., Kerber E.V., Kravchinsky V.A., Bezrukova E.V., Prokopenko A.A., Geletii V.F., Kalmychkov G.V., Goreglyad A.V., Antipin V.S., Khomutova M.Yu., Soshina N.M., Ivanov E.V., Khursevich G.K., Tkachenko L.L., Solotchina E.P., Ioshida N. and Gvozdkov A.N. (2001) Deep drilling on Lake Baikal: main results. Russian Geology and Geophysics, 42, 1, pp. 3-28.
28. Levina T.P., Orlova L.A. (1993) Holocene climatic rhythms of southern West Siberia. Russian Geology and Geophysics, 34, 3, pp. 36-51.
29. Lisitzin A.P. (1966) Main regularities in the distribution of recent siliceous sediments and their relations with climatic zonality. Geochemistry of Silica. N.M.Strakhov (Ed.). Nauka, Moscow, pp. 90-191. (in Russian).
30. Manabe S., Broccoli A.J. (1990) Mountains and Arid Climates of Middle Latitudes. Science, 247, pp. 192-195.
31. Monin A.S., Shishkov Yu.A. (1998) On Statistical Characteristics of the Little Ice Age. Doklady Akademii Nauk SSSR, 358, 2, pp. 252-255. (in Russian).
32. Petit J. R., Jouzel J., Raynaud D., Barkov N.I., Barnola J.-M., Basile I., Bender M., Chappellaz J., Davis M., Delaygue G., Delmotte M., Kotlyakov V.M., Legrand M., Lipenkov V.Y., Lorius C., Pépin L., Ritz C., Saltzman E., Stievenard M. (1999) Climate and atmospheric history of the past 420,000 years from the Vostok ice core, Antarctica. Nature, 399, pp. 429-436.
33. Prokopenko A.A., Karabanov E.B., Williams D.F., Kuzmin M.I., Shackleton N.J., Crowhurst S.J., Peck J.A., Gvozdkov A.N., King J.W. (2001) Biogenic silica record of the Lake Baikal response to climatic forcing during the Brunhes. Quaternary Research, 55, pp. 123-132.
34. Pulsating Lake Chany (1982). N.P.Smirnova, A.V. Shnitnikov (Eds). Nauka, Leningrad, 304 pp. (in Russian).
35. Shimaraev M.N., Granin N.G., Kuimova L.N. (1995) Experience of reconstruction of Baikal hydrophysical conditions in the Late Pleistocene and Holocene. Russian Geology and Geophysics, 36, 8, pp. 94-99.

36. Shnitnikov A.V. (1957) Humidity changes in the continental Northern Hemisphere. Transactions, USSR Geographical Society, New Series. T. 16. Moscow, Leningrad, Academy Science Press, 337 pp. (in Russian).
37. Velichko A.A. (1989) Holocene as element of the planetary natural process. In Khotinsky N.A. (ed.) Paleoclimates of the Late Glacial and Holocene. Nauka, Moscow, pp. 5-12. (in Russian).
38. Velichko A.A., Barash M.S., Grichuk V.P., Gurtovaya E.E., Zelikson E.M. (1984) The climate of the northern hemisphere in the epoch of the last Mikulino interglacial. Proceedings of the USSR Academy of Sciences. Geographical series, 1, pp. 5-18. (in Russian).
39. Verkhozina V.A., Kozhova J.M., Kusner Y.S. (1997) Hydrodynamics as a limiting factor in Lake Baikal ecosystem. Ecovision, 6, pp. 73-83.
40. Volkov I.A. (1976) The role of eolian factor in the topography evolution. In Timofeev D.A. (ed.) Problems of exogenic topography formation. Book 1. Nauka, Moscow, pp. 264-269.
41. Volkov I.A., Zykina V.S. (1991) Cyclicity of subaerial deposits of West Siberia and the Pleistocene climate history. In Zakharov V.A. (Ed.) Climatic Evolution, Biota and Envinronments of Man in the Late Cenozoic of Siberia. Institute of Geology, Geophysics and Mineralogy SB AN USSR, Novosibirsk, pp. 40-51. (in Russian).
42. Volkova V.S., Bakhareva V.A., Levina T.P. (1989) Vegetation and climate of Holocene of West Siberia. In Khotinsky N.A. (Ed.) Paleoclimates of the Lateglacial and Holocene. Nauka, Moscow, pp. 90-95. (in Russian).
43. Williams D.F., Peck J., Karabanov E.B., Prokopenko A.A., Kravchinsky V., King J., Kuzmin M.I. (1997) Lake Baikal Record of Continental Climate Response to Orbital Insolation During the Past 5 Million Years. Science, 278, pp. 1114-1117.
44. Zander A., Frechen M., Zykina V., Boenigk W. (2003) Luminescence chronology of the Upper Pleistocene loess record at Kurtak in Middle Siberia. Quaternary Science Reviews, 22, pp. 999–1010.
45. Zykin V.S., Zazhigin V.S., Zykina V.S. (1995) Changes in Environment and Climate during Early Pliocene in the Southern West-Siberian Plain. Russian Geology and Geophysics, 36, 8, pp. 37–47.
46. Zykin V.S., Zykina V.S., Orlova L.A. (2000a) Stratigraphy and major regularities of environmental and climatic changes in the Pleistocene and Holocene of Western Siberia. Archaeology, ethnology and anthropology of Eurasia, 1, pp. 3-22.
47. Zykin V.S., Zykina V.S., Orlova L.A. (2000b) Quaternary warming stages in Southern West Siberia: environment and climate. Russian Geology and Geophysics, 41, 3, pp. 295-312.
48. Zykin V.S., Zykina V.S., Orlova L.A. (2003) Reconstruction of environmental and climatic change during the Late Pleistocene in southern West Siberia using data from the Lake Aksor Basin. Archaeology, ethnology and anthropology of Eurasia, 4, pp. 2-16.
49. Zykina V.S. (1999) Pedogenesis and climate change history during Pleistocene in Western Siberia. Anthropozoikum, 23, pp. 49-54.
50. Zykina V.S., Volkov I.A., Dergachtva M.I. (1987) Upper Quaternary deposits and fossil soils of Novosibirsk Priobie. Nauka, Moscow, 204 pp.
51. Zykina V.S., Volkov I.A., Semenov V.V. (2000) Reconstruction of Neopleistocene climates in West Siberia based on study of Belovo key section. In: Problems of reconstruction of Holocene and Pleistocene Climate and Environment in Siberia. E.A.Vaganov et al. (Eds). Institute of Archaeology and Ethnography SB RAS Press, Novosibirsk, pp. 229-249.

REGIONAL CLIMATE AND ENVIRONMENTAL CHANGE IN CENTRAL ASIA

N.F. KHARLAMOVA & V.S. REVYAKIN
Barnaul, Russia

Abstract: The solution of problems of sustainable development of agricultural production and biodiversity conservation calls for the consideration of regional climate change. According to the data given by weather station in Barnaul city mean annual air temperature rose by 2.8oC over the period from 1838 till 2004. Warming was observed mainly during the cold period (October – March) and constituted 3.4oC. Secular cycle of precipitation variations with its minimum in the mid-nineteenth century and maximum in the early twentieth century has been defined. New secular circle of humidity variations has started. However the rate of temperature rise exceeds precipitation rate resulting in drastic reduction of glaciers in Altai-Sayan Mountains that leads to desertification and other negative after effects.

Keywords: Regional climate change, global warming, glacier recession, desertification.

Thorough review of Russian and foreign publications given by K.Ya. Kondratiev and devoted to the outcomes of International Conference on Climate Change held in 2003 in Moscow reinforces the increased attention to the problems associated with climate. This is a sufficient proof of not only scientific but also practical importance of research carried out in global and regional climatology. Not touching upon the problem of revealing the anthropogenic reasons of global warming and the changes taking place, we would like to pay our attention to the other problems as well.

Experts engaged in climate study lack for objective information on the past, current and potential climate change, its specific environmental impact and the assessment of the possible response. Thus, the main objectives of climate study are as follows:

1. To reveal the peculiarities of the current change, particularly on the regional level, in order to work out the appropriate development scenarios for the nearest 10-25 years;
2. To determine the specific character of the change impact on other environmental components (water resources, longstanding frozen condition of ground, mountain glaciation, etc.) and different natural complexes (ecosystems, agrocoenosis and biocoenosis, etc.).

In our opinion the solution of these problems is a prime necessity for implementation of activities specified in the statement of International group G-8 of July 2, 2003, namely to provide the sustainable agricultural production and biodiversity conservation. Climate factor is the most important control factor of the influence on physical and energy resources essential to any ecosystem. Components of the complex multi-level biogeocoenosis show different climate change response resulting in their instability and disturbance of development. In this case any extra anthropogenic impact can act as a "trigger" for environmental disruption, degradation and severe ecological consequences. Therefore the assessment of the current climate change impact on the natural systems components, agricultural sectors' functioning (vulnerability, adaptivity and stability) is of crucial importance.

The extent to which natural components and sectors of national economy are affected by the current weather conditions predetermines the significance of such research in all regions of Russia. We would like to emphasize the specific features of intercontinental territories ignoring the coastal densely populated areas where weather anomalies are apparent. When defining the region under study one should take into consideration the following A. Voyeikov's opinion expressed in 1909: "It's high time to leave out of account the expression "Middle Asia" with reference to our Turkestan, to avoid any misunderstanding the central part of Asia should be called "central". This region should include the countries from Altai and Sayany in the north to the Himalayas in the south, from Pamirs and west Tien Shan in the west up to the proper China in the east". It is expected to establish the transboundary biosphere territory "Altai" within this region incorporating districts of Kazakhstan, Mongolia and China with the exception of Russia.

Low humidity and strong dependence of all natural components on precipitation are aggravated by wide variations of climatic parameters year after year due to continental character of climate in Central Asia. Increased amplitude of annual and daily temperature, irregular annual precipitation when the major part occurs during warm period (spring in desert areas, the second half of warm period in steppe and forest-steppe regions) determine the rhythm of development not only of natural processes but of economical activity of population engaged in agricultural production. Mountain regions are distinguished by heavy dependence on environment and natural resources due to the predominance of traditional nature management under the lack of transport facilities and low level of industrial development. As a result of the recent political and economic reforms the rural population has come back to the original methods of management, mainly to cattle-breeding. Aridization and the increase of livestock in Kazakhstan, Mongolia and Republic of Altai have caused land degradation and even its desertification. Biodiversity conservation in the regions mentioned is of special importance. Apartness of Central Asia favored the conservation of many species endangered in the adjacent regions. The problem of relationship between the area occupied by special protected territories where economical activity is limited including the rapidly developed recreational land tenure and the problems of living standard increase, poverty extermination, sustainable economic development calls for thorough consideration and acceptable solution. However this region is greatly affected by climate change. For instance, the migration of animals increases due to the lack of forage caused by heavy snow and variations in winter temperature followed by the formation of the thin crust of ice over snow that hampers the hoofed animals' feeding; some bird species that hibernate on the unfrozen lakes are endangered to poaching. Rapid extension of areas occupied by forests attests to the shift of altitudinal belts' boundaries. The rate and tendencies of this process should be covered since the possible deterioration of environment can have an adverse effect upon flora and fauna of high mountains. Recreational development of picturesque pristine sub mountain and mountain regions of Altai has demonstrated negative after-effects. If the development of economically sound and ecologically safe land tenure prevails, to what extent the most important components of economy as agricultural production and recreational activity should be advanced?

One way or the other the problems mentioned above are associated with climate change. According to scientific data in XX century mean annual surface air temperature in the northern hemisphere increased by 0.6°C. Hadley Center indicated that global warming reached its first maximum late in the 40s of XX century followed by decrease of global surface air temperature up to the middle of 60s. Later on further rise in temperature has been noted. As compared to the 60s warming on the continents made up 1.6°C. The Third National Report of the Russian Federation made by Inter-departmental Commission on climate change noted that for the last century warming in Russia made up 1°C, with rate increase over the past 50 years. Rise in temperature is distinct in winter and spring, not so evident in fall and more intensive to the east of the Urals. Mean monthly total precipitation variation over the century was not great. On the whole climate change in Russia can be characterized as "warming followed by aridization".

The most reliable assessment of climate change can be done due to observations carried out at weather stations. The stations that perform long-term observations, e.g. Barnaul weather station, the oldest one in Asia, are of particular interest. The investigations provide support for the fact that the station can be considered as a benchmark one for the territory of intercontinental regions, especially of Central Asia. Correlation coefficients of annual air temperature with Altai Krai weather stations make up 0,86-0,98, annual precipitation amount - 0,67-0,72. Annual temperature correlation with Minusinsk is 0,68, with Urumchy and Altai, Xinjiang (China) - 0,57 and 0,88.

In case of linear approximation (*Fig. 1*) annual air temperature for 166 years at Barnaul weather station demonstrates increase by 2.8°C or 1.8°C for 100 year period. Raise in temperature is rather noticeable: from 0.0147°C/per year during 1838-1958 up to 0.0336°C/per year for the last 44 years. Warming is manifested mostly in winter and spring: average temperature of the warm period (IV-X) increased by 2.3°C for 166 years or 1.4°C for 100 years while average temperature of the cold period (XI-III) has increased by 3.4°C or 2.2°C, correspondingly. The most pronounced changes are marked for January (by 4.8°C), March (4.4°C) and April (4.5°C). As for July, August and September such an increase was minimal, i.e. 1.6°C, 1.9°C and 1.5°C, correspondingly.

From 1970 till 2003 the sum of active air temperature ($\sum t>10°C$) raised by 260°C that corresponds to annual temperature increase approximately by 0.8°C. By the end of 2025 average annual air temperature may raise by 1°C at Barnaul station that equals to increase of active temperature sum $\sum t>10°C$ by 300°C.

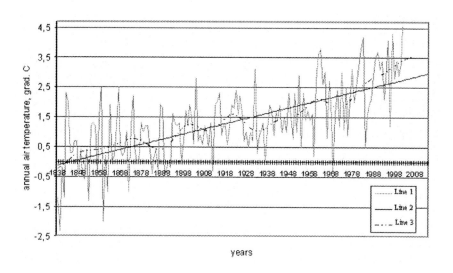

Fig. 1: Annual air temperature, Barnaul weather station, 1838-2004
   Line 1 – annual values, line 2 – linear trend, line 3 – 11- moving summer mean

Due to an increase in recurrence of extreme phenomena it is important to trace the change of absolute temperature values. Linear trend of absolute temperature maximum from 1928 to 2003 (76 years) is characterized by 0.8°C increase. Changes for absolute minimum are most remarkable and make up 3.4°C; the major raise was marked till 1965 but starting from 1989 the trend mark has changed (*Fig. 2*).

Regional Climate and Environmental Change in Central Asia 23

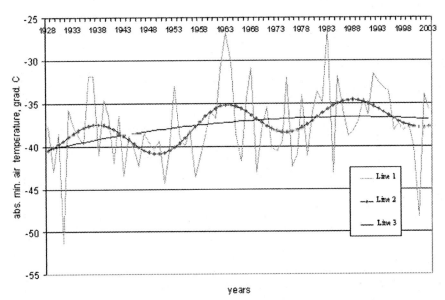

Fig. 2: Minimums of absolute air temperature, Barnaul, 1928-2004

The driest period in annual precipitation distribution (Fig. 3) was observed in 1850-1873. After secular maximum in 1907-1912 the descending moisture branch was marked with dry periods in 1923-1936, 1947-1957 and 1969-1984 that alternated with periods of increased moistening.

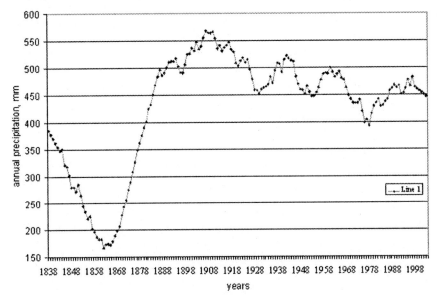

Fig. 3: 11-year faired curve of annual total precipitation value, Barnaul, 1838-2004

Natural rhythm in precipitation amount distribution caused deterioration of vegetation development conditions in the period from 1951 to 1955 when catastrophic or significant dryness of the warm period occurred. Hence virgin lands' ploughing up in Altai was performed under extremely unfavorable climatic conditions. It could bring to onset of desertification processes as evidenced by sharp raise in a number of dust storms at a later time. The soil-conserving system introduced in 1969-1972 significantly mitigated these processes. However, natural aridization of the territory leads to increase in number of dust storms again (*Fig. 4*).

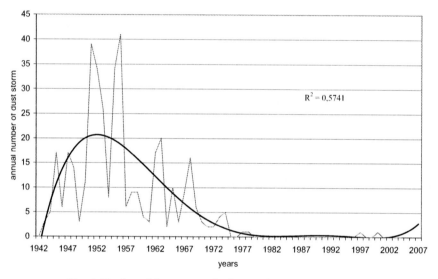

Fig. 4: Number of dust storms: the facts and polynomial trend, Barnaul

From 1977-1978 the ascending moisture branch of a new secular cycle started; here the damp period (1986-2002) was marked. Currently the intrasecular cycle of decreased moistening as well as winter with little snow take place. The data obtained at Ulangom st. situated in the Ubsunur lowland are similar to minimum of moisture in 1970s and present-day sharp raise in annual precipitation sum. The decreased moisture level may remain for some years as demonstrated by essential correlation of precipitation amount and solar activity level: in the years of minimum (the nearest is expected in 2006) probability of drought occurrence is very high (*Fig. 5*).

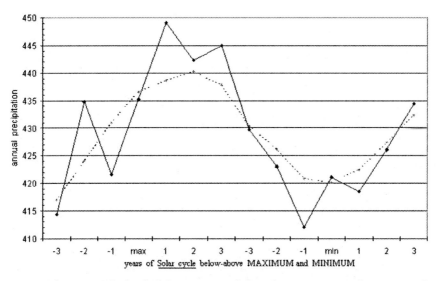

Fig. 5: Fluctuation of annual total precipitation below-above maximum and minimum within 11-year solar cycle

Thus regional peculiarities of climate dynamics in Central Asia including the Altai-Sayany mountain country are characterized by more intensive (than for Russia as a whole) rise in air temperature mostly due to winter severity. Cyclic recurrence of changes as thermal regime as the territory moistening has been revealed. All this may contribute to considerable development of desertification processes in this region as well as to ratio change of areas covered with vegetation and landscapes since temperature increase rates take the lead over precipitation amount ones. Even at present the earlier descent of snow cover reduces albedo of the underlying surface altering its heat balance that brings to fall in heat amount needed for snow melting. Increased wind velocities in March- April arid period contribute to fast reduction of moisture resources in soils. Crop sowing should be arranged in earlier and shorter time though great likelihood of late spring and early autumn light morning/night frosts is. The glaring example is when in September 26-27, 2004 huge snowfall took place. The snow cover height in Barnaul city reached 21 cm under temperature of -5°C.

The pilot analysis of areas' ratio with current certain thermic resources and in case of annual average air temperature raise by 1°C shows the following: if nowadays in Altai thermic conditions- warm (2000-2200°C - 40%), very warm (2200-2400°C - 26%) and moderately warm (1800-2000°C - 22%) are typical, in 2025 very hot (>2400°C - 49%), very warm (31%) and warm ones may be marked (13%). Substitution of forest-steppe zone by arid steppe one that in its turn may become dry steppe is probable within the Altai Krai territory. Instead of dry steppes desert steppes can appear.

At present $\sum t>10°C$ $<1000°C$ (cold belt), 28% - 1000-1600°C (moderately cool belt) are typical for 58% of Altai Republic territory. In 2025 reduction of cold belt lands by 14% and increase in areas of moderately cool belt by 8% may occur. Moreover, territory areas may be reduced by 6% where there is no $\sum t>10°C$ with appearance of new categories (moderately warm) - 8% (1800-2000°C) and 3% of warm one (2000-2200°C). Probable increase in precipitation amount may be insufficient to keep normal moisture conditions under high level of heat provision, especially after the maximum in 2028-2030 when the descending moisture branch of the secular cycle is to start. Leaving aside the problem of permafrost degradation, .let us note that major changes are expected in arid areas where aridization becomes more intensive against the background of anthropogenic impact (excessive pasturing).

Further cutting down in mountain glaciation area may also happen. The models of multiple regression of Gebler's glacier tongue retreat value (the Katunsky ridge) with such factors as average summer air temperature and annual precipitation sum (Fig. 6) allowed us to calculate the rate of maximum response of the glacier to temperature change in summer as 6 years, precipitation- 20-23 and 82-88 years. The similar models have been developed for other glaciers as well. The nearest significant reduction of the glacier tongue is expected in 2008 as the response to exceptionally warm 2002 year.

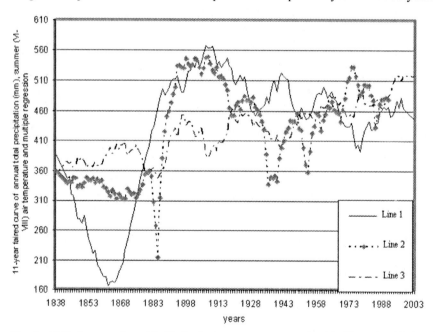

Fig. 6: *Multiple regression of Gebler's glacier tongue retreat value (the Katunsky ridge) with average summer air temperature and annual total precipitation. Line 1 – 11-mean moving summer values of annual total precipitation, line 2 – glacier tongue retreat value; line 3 – 11-moving summer temperature*

THE REASONS AND CONSEQUENCES OF CLIMATE CHANGES

V. F. LOGINOV
The State Research Institution "Institute for Problems of Natural Resources Use and Ecology NAS of Belarus", Minsk, Republic of Belarus

Modern knowledge of the reasons of climate change is not specifically complete [4]. This especially concerns estimation of the role of "small" climate-forming factors.

In literature they can be considered as external forces, such as solar activity, determining influx of electromagnetic radiation and charged particles, and also volcanic and anthropogenic activity determining aerosol pollution of the atmosphere [3].

"Small" climate-forming forces may also be regarded as a change of orbital parameters of the Earth, long-period dates, change of rotation speed of the Earth and others.

Let's consider only two factors which influences on the climate are a subject of hard debates. Instability of the electromagnetic and corpuscular change coming from the Sun and Space is established theoretically and has received experimental confirmation. High-precision measurements of an integrated astrophysical solar constant have shown its variability about 0,1 % at the development of large spots on the Sun. In the long-term change of astrophysical solar constant, the trend about 0,02 % per year is found out. Variations of a meteorological solar constant are by order higher and are also connected to the change of gas and aerosol structure of the atmosphere due to the influence of ultra-violet and corpuscular radiation of the Sun, space beams of solar and galactic origin, volcanic and anthropogenic activity.

The small power of additional electromagnetic and corpuscular radiation naturally results in instability of solar-atmospheric linkages. Sometimes this circumstance serves as a subject for full denying such linkages. Most actively this question was studied in the 1960-1980's years. It is possible to name a dozen conferences devoted to this question and, at least, dozen monographs, published in the USSR, the USA and other countries [3]. The works, executed in the 1990's years, testify that an essential promotion in the decision of this debatable question is not observed. The majority of the ideas stated in the 1960-1980's years have not reached the level of theories, the conventional laws and developed mechanisms of influence of solar agents on weather and climate. The level of researches of solar-atmospheric connections of the 1960-1980's regarding the profoundness of the analysis and scope of the used materials frequently is higher than that of the researches executed last years [1].

As one of the examples of possible influence of solar activity on the atmosphere of the Earth let's regard the schedule of the change of relative numbers of Volf and the area of ozone "holes" in Northern Hemisphere (Fig. 1). From fig. 1 it follows, that average area of regions with low ozone contents (<300 units of Dobson) in Arctic regions experiences cyclic changes of the specified characteristic on the background of ascending trend of the area of regions of low ozone content till 1995. The specified cyclic changes are in antiphase with Volf 's relative numbers, i.e. at high values of Volf 's relative numbers ozone "holes" are filled. Hardly sharp decrease in the area of ozone holes in Northern Hemisphere can be related only to the decrease in emissions of ozone-breaking substances after 1995, as lifetime of ozone-breaking substances is essentially more than several tens of years, whilst conditions of the Montreal Report on the reduction of the emissions of ozone-breaking substances have started to be carried out 10 years prior to the period of sharp filling ozone "holes" in Northern Hemisphere (1996). Note, that in Southern Hemisphere the area of ozone "holes" does not decrease.

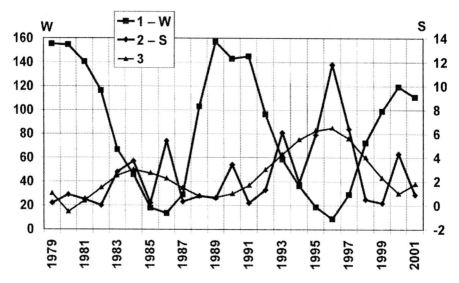

Fig. 1: Area of ozone holes (< 300 Dobson units) mln. km^2 : 1 – relative Volf numbers (W); 2 – observed values of ozone holes area (S); 3 – leveled ozone holes area values

Possible mechanisms of influence of solar activity on the climate are defined first of all by influencing on the change of properties of the atmosphere by the agent. If a solar constant changes due to ultra-violet part of electromagnetic radiation, a real mechanism then can be the mechanism connected to radiation-chemical processes and, first of all, to the change of ozone content in the atmosphere. If the agent is a solar and galactic space beam, the mechanism can be connected to the influence of such beams on parameters of atmospheric electricity and condensation of vapor in the atmosphere. Unfortunately, none of the suggested in 1950-1980's years mechanisms has received essential development.

Issues of influence of aerosols of volcanic and anthropogenic origin on climate are more advanced, physical mechanisms of their influence on climate do not cause great objections [1, 3].

Comparative estimation of the decrease of direct solar radiation after eruption of volcanoes at stations of the CIS has shown, that it varies from 2-4% (volcanoes Fouego, Soufriere, Saint Helens, Adelaide, Agung) up to 10-25% (volcanoes Montpelier, Katmai, Krakatoa, El-Chichon, Pinatubo). In high latitudes in winter after eruption of the volcano El – Chichon, the anomaly of direct radiation reached 26.5%. Last eruption of the volcano Pinatubo has led to the decrease in direct radiation more than by 25%. Decrease in values of total solar radiation after such large volcanic eruptions as El-Chichon and Pinatubo has made 4.5-5 % [3]. Anthropogenic component in the change of summarized radiation is about 0.03% per year. In cities the rate has 2-3 fold increase depending on city pollution degree.

Total radiation on the territory of Belarus in overwhelming number of months, and for the year in general, decreases.

The table provides gradients of the change of total radiation on meteo-stations Vasilevichi (countryside) and Minsk (population 1.75 mln.).

Table 1: Gradients (α, (Mj/m^2) /10 years) of change of total radiation on meteorological stations Vasilevichi and Minsk

Meteorological stations	The period of averaging					
	I	II	III	IV	V	VI
Vasilevichi	-4,52	-2,50	-4,81	-5,34	-1,10	-15,0
Minsk	-3,29	-5,15	-17,0	-2,91	-4,03	-23,0

VII	VIII	IX	X	XI	XII	Year
-13,9	-1,77	-10,2	3,22	0,48	0,91	-54,5
-15,3	-3,29	-13,0	-1,21	-0,17	-0,66	-89,0

In rural site (Vasilevichi), the amount of the decrease in mid-annual total radiation is 1.5 times less.

Efficiency of volcanic aerosol impact on climate is determined by the breadth of volcanic eruption, circulation of atmosphere, and most important, by the type and power of volcanic eruption.

Power-mean volcanoes, which throw out into the atmosphere a lot of sulfurous aerosol, provided a decrease in global temperature by 0.05 – 0.2°C, depending on the month of year in the first 1-2 years after volcanic eruption. If we take into consideration only the largest volcanic eruptions reduction, temperature will make on the average 0.2 – 0.3°C. In winter the temperature after volcanic eruptions enhances. It is one of the known "paradoxes" of the connection "volcano - climate". There are also significant spatial features of displaying volcanic "signal" in climate change. So, in subtropical latitudes an increase of pressure is observed, and in high latitudes in winter and in spring pressure after volcanic eruptions goes down. In winter, pressure above continents becomes lower, than prior to the eruption of volcanoes, and above oceans it is higher, therefore an inflow of warm air to the Western Europe and North America occurs. In summer after volcanic eruptions, above oceans there is a pressure decline, and above continents – an increase, that weakens monsoon circulation. In spring the intensity of the Atlantic minimum and the Azores maximum raises, that intensifies the western circulation and, as a result, a rise of temperature. After large volcanic eruptions, there is an increase in quantity of precipitation in winter in West Europe, which also confirms a conclusion about enforcement of zone circulation after the eruption of volcanoes.

In the first and second year after volcanic eruptions in the Northern Hemisphere, a reduction of amplitudes of annual harmonic by the value about 0.25% from the value of mean quadratic deviations is observed.

In Belarus, after large volcanic eruptions, a decrease in temperature in a warm season and raise of temperature especially in the north of Belarus in winter is also observed.

Modern estimations show, that contribution of aerosols of volcanic and anthropogenic origin to climate change concedes to the contribution of greenhouse gases to the change of global climate approximately twice. In large cities, the aerosol can result even in the greater radiating influence, than greenhouse gases or its influence is comparable to the effect of greenhouse influence, i.e. to the essential damping of greenhouse effect in a warm season.

Last years in Belarus the growth of temperature in winter and in spring, and at night even in summer and in autumn is observed.

The revealed tendencies of the change of extreme temperatures are in line with physical representations that due to the increase of content of greenhouse gases in the atmosphere there is an increase in long-wave radiation returned to a spreading surface, and, as a result, rate of increase of night minimal temperatures is higher than that of day time maximal temperatures raise at night.

The specified features of temperature change in a global scale of the last decades are in line with known ideas of the theory of climate. Increase of temperature might be most expressed in high latitudes, that is caused by the albedo feedback contribution and influence of the strong gravitational stability created by cooling near the Earth's surface which suppresses convection and transfer of long-wave radiation, resulting in concentration of stipulated by the raise of carbonic gas heating in a thin near-surface layer. In tropics, warming on a vertical is limited by influence of damp convection and as a consequence, is "smeared" on a thicker layer than in high and average latitudes. This is likely to be a principal cause of the fact, that modern warming is especially expressed in a cold season in high latitudes. Though Belarus also has the greatest extent on a meridian of only 560 km, but, nevertheless, warming appeared to be mostly expressed in northern part.

The difference of temperature " North - South" in a warm half-year (IV-X months) comprises about 1.4° C and varies in time a little, exception is made by 1907 (fig. 2a). In a cold season (XI-III) for the last 70 years the difference of temperatures "North - South" is about 1.7° C, and at the end of the 19-th - beginning of 20-th century was about 2.5° C (fig. 2 b). Such a big distinction can partially be connected to the insufficient density of the meteorological network at this time. However, presence of big and small distinctions of temperatures "North - South" corresponds to the epoch of climate cooling and warming. The minimal distinctions of values of temperature "North - South" were marked during modern climate warming (1971-2002) and during the period known as "warming of Arctic regions" (1921-1940) (fig. 2 d). In a warm season (IV-X month), differences of temperature "North - South" are also minimal during the two known climate warming of 1921-1950 and 1971-2002 (fig. 2c). Therefore, our data confirm a popular statement, that anthropogenic climate warming connected to greenhouse gases raise should be more expressed in high latitudes and cold season.

The difference of temperature "West - East" in cold and warm seasons, and also during separate periods of the year (VI-VII and II-IV months), decreases for the period of instrumental observations (fig. 3 a, b, c, d). In a warm season (IV-X month) in last decades the difference of temperatures "West - East" is close to zero whereas at the end of the 19-th beginning of the 20-th centuries in the West the temperature approximately was by 1° C higher than in the East (fig. 2a). In June and July, since the 1920's years of the last century, the temperature in the East became higher than in the West of Belarus (fig. 3 b).

In a cold season (XI-III month), the difference of average temperature "West - East" decreases almost by 1° C (fig. 3 c), and in the spring even it is higher (fig. 3 d). Thus, the spatial heterogeneity of the temperature in the direction "West –East" in the cold and warm seasons decreases.

The greatest expressiveness of anthropogenic "signal" in more continental areas relates to the circumstance, that in the centers of continents, more favorable conditions are created to take in a long-wave radiation; the warming is more intensive at night.

On oceans absorption of direct solar radiation occurs in a layer, and long-wave - in a surface film. These stimulates raise of evaporation and, as a result, reduction in temperature of ocean surface. Experimental data as stated above, testify that warming has not been revealed in some oceanic areas of Southern Hemisphere, and also North Atlantic. Even at the greatest extent of Belarus (about 650 km), the difference of temperature in western and eastern areas of Belarus appeared appreciable.

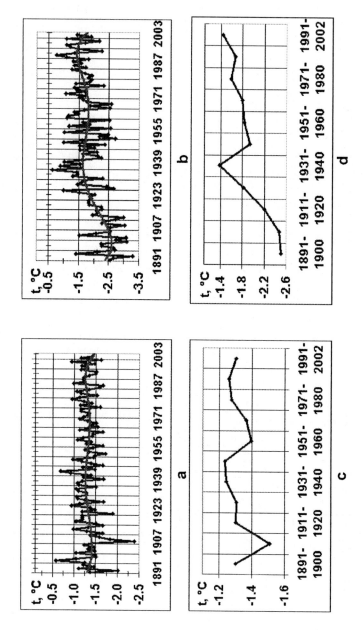

Fig. 2: Change of a difference of temperatures «North - South»; a - in the warm season (IV-X); b - in the cold season (XI-III); c - in the warm season (IV-X) on decades; d – In the cold season (XI-III) on decades

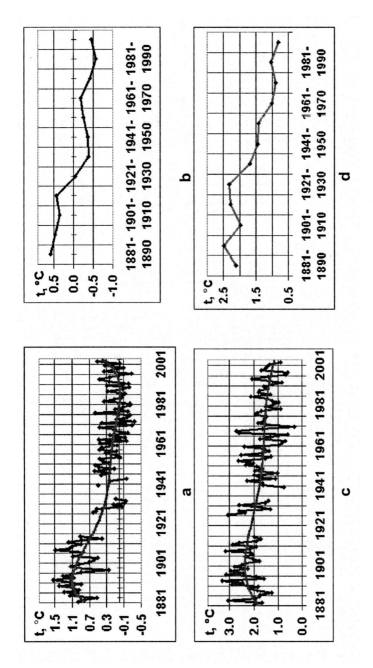

Fig. 3: Change of a difference of temperatures «West - East»:a - in the warm season (IV-X); b - in June on decades; c – In the cold season (XI-III) ; d – in February – April on decades

Let's consider one of the natural factors of climate change - a large-scale interaction of the ocean and the atmosphere. The characteristic of such interaction is the index of intensity of North Atlantic Oscillation (NAO). The intensity of this oscillation is controlled by the intensity of Icelandic minimum and the Azores maximum. Change of values of NAO index is given on fig. 4. [5]. From the fig. 4 it follows, that from the middle of the 1960-s of the 20-th century, the raise of the specified characteristics is observed and the amount of cyclonic systems raises annually with pressure in the center of 950 gPa.

Correlation factors between the air temperature in Belarus in I-IV month and NAO intensity are statistically significant. However, time dynamics of correlation factors of January – April temperature and normalized NAO index is extremely unstable: especially close correlation of air temperature and NAO index is observed at high values of NAO intensity within the last decades (Fig. 4).

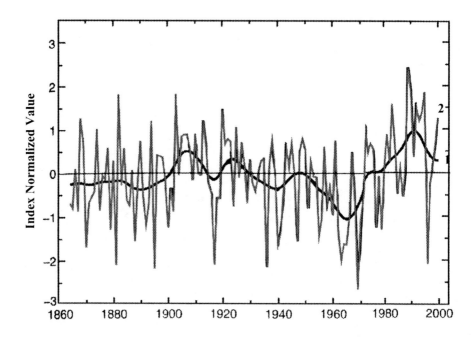

Fig. 4: Values of an index of North Atlantic fluctuations (December – March) from 1864 to 2000: 1 – the smoothed values of an index with use of 21-dot binomial filter; 2 – the values of an index normalized on average-quadratijc deviation (the norm is taken for 1900–2000)

Thus, large-scale interaction of ocean and an atmosphere defines an essential part of variability of temperature of Belarus in a cold season (I-IV month).

Special interest, as shown above, represents emphasis of anthropogenic "signal" with use of data of daily observation of temperature, the maximal and minimal daily values of temperature, and also changes of amplitude of a daily course for the last several decades. The idea of asymmetry of trends of extreme temperatures of link with influence of anthropogenic factors has been offered more than ten years ago by T. Karl [5, 6].

Our results have confirmed presence of temperature trends asymmetry, albeit they showed essentially smaller extreme temperatures trends asymmetry.

Raise rate of minimal (night) temperatures really appeared to be greater, than maximal (day time) ones [1].

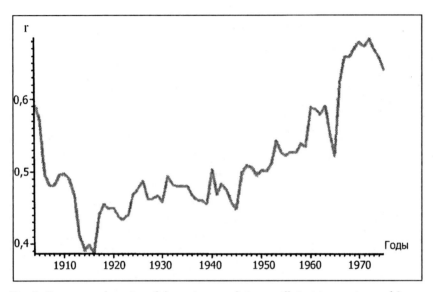

Fig. 5: Temporary dynamics of the series correlation coefficient temperatures of January - April and normalized NAO

On the average in the Republic daily minimal temperatures increase by 30% faster than maximal in winter. Rate of change of maximal and, especially, minimal temperatures is the greatest in the North of the Republic. In summertime of the year a situation is less certain, albeit in this case daily minimal temperatures increase faster than maximal ones.

This conclusion is well coordinated to famous physical notions that due to the raise of the contents of greenhouse gases in an atmosphere, there is an increase in long-wave radiation returned to a spreading surface. As a result, speed of increase of the minimal (night) temperatures is higher than the speed of increase of day time maximal temperatures. In the latter case, as show our results, aerosols are damping the scope of warming.

The marked decreases of day time and increase of night temperatures in cities entail reduction of amplitude of a daily course of temperature on urbanized territories.

The emphasis of anthropogenic "signal" is especially perspective in cities' climate change.

The carried out analysis of maximal and minimal daily temperatures showed, that maximal city temperature decreases in comparison with temperature in a countryside, and the minimal temperature in cities, on the contrary, rises. Especially precise distinctions of the maximal and minimal temperature in cities and countrysides are marked in summer. Differences of amplitudes of a daily course of temperature at stations "Maryina Gorka - Minsk", "Vasilevichi-Gomel" are also the greatest in summer [1,4].

Let's consider possible reasons of the received distinctions of temperature in large cities and countryside.

Principal cause of the received distinctions of the maximal temperature in large cities and countryside is the raise of aerosol pollution of cities.

At strong aerosol pollution in cities there is an easing of total solar radiation and, as a result, especially essential reduction in temperature in a warm season in the afternoon.

Thermodynamic conditions and increase of aerosols contents which serve as nucleus of condensation, promote formation of convective clouds above a city in the afternoon.

As K. Y. Kondratjev's researches have shown, in urbanized areas there can be polluted clouds which provide excessive absorption of radiation. Absorption of radiation in such clouds can reach 25 Wtm^{-2} [1]. A proper direct and indirect influence of anthropogenic aerosol through the formation of additional cloudiness results in the reduction of maximal temperatures and, as a result, to the reduction of amplitudes of temperature daily course above the city. The differences of amplitude of a daily course of temperature "Maryina Gorka - Minsk" and "Vasilevichi -Gomel" will also increase.

Fig. 6 shows that differences of maximal temperatures "Maryina Gorka-Minsk" within one year exceed by absolute value those of minimal temperatures on the average for a year, and differences of maximal temperatures "Maryina Gorka - Minsk" during all seasons of the year - are positive, i.e. in the afternoon a suburb is warmer than a city. Differences of maximal temperatures "Maryina Gorka - Minsk" have the greatest values in April - May and June - August. In winter season these differences are close to zero.

Peculiarities of minimal temperatures differences change "Maryina Gorka - Minsk" in annual course are a little bit different. Differences of minimal temperatures "Maryina Gorka - Minsk" are negative during all seasons except for spring. They reach the greatest values, as well as in case of an estimation of differences of maximal temperatures, in summer.

Thus, in the most part of the year, a night city is warmer than a suburb. In spring the differences of values of minimal temperatures " Maryina Gorka - Minsk " are close to zero.

Differences of amplitudes of daily course of temperature "Maryina Gorka - Minsk" are maximal in summer and minimal in winter.

The least value of differences of maximal daily temperatures "Maryina Gorka - Minsk" occurs in winter and autumn, and the greatest - in summer and spring, when repeatability of the cloudy sky is the least.

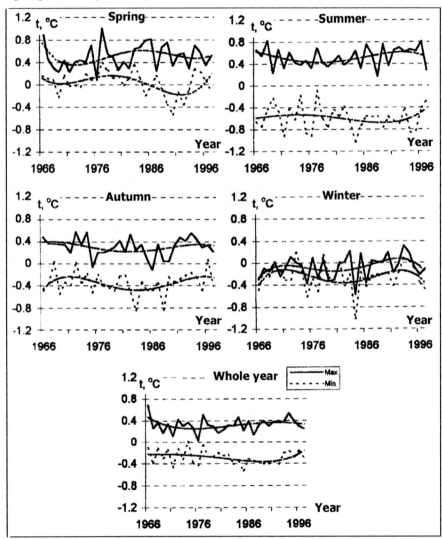

Fig. 6: Differences of maximum (1) and minimum (2) temperatures on the stations Maryina Gorka - Minsk

It is known, that a snow cover with an albedo of higher than 0,6 and being kept in the suburb in March, can result in the reduction of differences of maximal temperatures "Maryina Gorka - Minsk". For this period, a reduction of differences of maximal temperatures is observed, and for minimal temperatures such distinctions are not traced.

Thus, an aerosol effect in the change of maximal temperature and amplitude of a daily course of temperature of the specified pairs of stations is the greatest in the polluted city in summer and in spring during small cloudiness. During other seasons (winter and autumn), strong cloudiness shades aerosol effect in the temperature change. The character of the change of differences of minimal daily temperatures "Maryin Gorka - Minsk" during various seasons as it is seen on fig. 6, differs from the character of change of differences of maximal temperatures on the mentioned above two pairs of stations.

First, minimal daily temperatures in the majority of seasons are higher in Minsk, in comparison to minimal temperatures in Maryina Gorka.

Second, the greatest by value differences of minimal daily temperatures "Maryina Gorka - Minsk" are observed in summer, decrease in autumn and become close to zero in spring. To explain the revealed character of change of differences of minimal daily temperatures "Maryina Gorka – Minsk", it is possible probably to proceed from seasonal change of the intensity of island of heat in the city. A greater discharge of anthropogenic heat in winter results in the increase of maximal intensity of island of heat in a city in winter; it can reach summer values or surpass them.

In spring, a countryside starts to get warm more intensively than a city, having a greater thermal capacity, which is heated up more slowly and monotonously. The difference of minimal daily temperatures "Maryina Gorka - Minsk" becomes close to zero.

The character of change of differences of amplitudes of a daily course of temperature "Maryina Gorka - Minsk" can also depend on a sign of relation of minimal daily temperature with overcast (cloudiness) during various seasons. As it is known, the relation between the minimal daily temperatures with overcast is negative in summer and in spring [5]. The temperature with aerosols content has the same sign of relation. In autumn and in winter, the relation of minimal daily temperature with overcast is positive and overcast as a more powerful factor of temperature change, during these seasons can shade substantially relation of temperature with the change of aerosols content in the atmosphere.

Besides, variability of differences of maximal and minimal daily temperatures depends on wind velocity. Greater by value reduction of wind velocity from the middle of 1970's years in cities in comparison with a countryside should also result in the increase in differences of daily temperatures "city - countryside".

Thus, the decrease of the amplitude of daily course of temperature in big cities in comparison with countryside, the fall of maximal (day time) temperatures and raise of minimal (night) temperatures can be mainly explained by the impact of the increasing aerosols content in the atmosphere of cities.

However, such factors, as:
- Character of change of intensity of island of heat in a city within a year;
- A greater, in comparison with countryside, decrease of wind rate in cities;
- Change of the character and narrowness of relations of cloudiness and temperature during various seasons of the year can complicate and shade a relation of temperature with the change of aerosols contents in the atmosphere. The optimal conditions to display the influence of aerosols on temperature are developed in summer in the afternoon.

The difference of amplitudes of daily course of temperature "Maryina Gorka - Minsk" at annual averaging increases by 0.5° C in the 1960-70's years up to 0.7° C in 1990's years (Fig. 7). Accordingly, the value 0.5 – 0.7° C can be regarded as aerosol "amendment" to modern climate change. In summer, this "amendment" is about 1.0° C, and is close to zero in winter. It means that on the urbanized territories a modern climate warming in a warm season due to the increase of greenhouse gases is close to zero, as it is damped by aerosol pollution of the atmosphere.

In winter, when repeatability of cloudy sky averages 70%, the contribution of aerosol pollution to the change of temperature is minimal. The aerosol damping of warming at this time of the year is also minimal and the increase of temperature, related to the raise of greenhouse content gases in the atmosphere, becomes most notable. Aerosols also enforce warming at night (in winter) due to the reduction of effective long-wave radiation from more polluted atmosphere of the urbanized areas. Besides, winter thermodynamic conditions in the atmosphere promote display of a greenhouse effect in this time of the year.

Executed by us, researches of amplitudes of annual temperature course testify to the change of amplitude of annual temperature course for the last several decades. This amplitude reduction occurs due to the rise of winter temperatures and as a result, leads to the reduction of Belarus climate as continental [1].

One of the threats to Belarus ecological security relates to climate change. Changes of climate render great influence on agricultural, forest and water economy of the country, construction, transport, breeding, depredators of agricultural and forest economy, health of the population etc.

The losses of productivity from weather and climatic conditions in separate years in relation to average productivity regarding regions can achieve

45-50% for summer barley and 35-40% for winter rye. The change of productivity of grain cultures caused by deterioration of climatic conditions occurred at the end of 70's, despite of farming practices of those years.

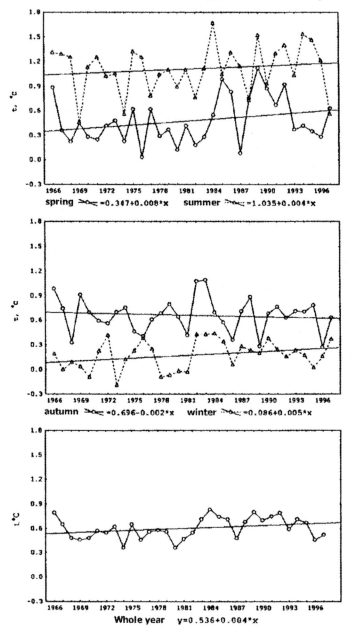

Fig. 7: *Differences of amplitudes of daily temperature course Maryina Gorka – Minsk*

Climatic conditions from 1984 to 1992 on the background of high farming practices favoured to productivity rise. The productivity of winter rye and summer barley in the majority of districts exceeded 30 centners per hectare. After 1992, the secondary fall of productivity caused not so much by deterioration of climatic conditions, but economic situation and related to it a decrease of the level farming practices are marked. The researches have shown, that the higher amounting variabilities of productivity of cultures caused by the level of farming crop, the lesser influence of weather and climatic conditions on productivity of a cultures is.

Our economic losses from unfavourable weather and climatic phenomena in absolute expression grow from year to year. In series of last years the country left on a level of the minimal food safety (5500-6500 thousand tons), and in other years production of grain reached about 60 % of this level. Cost of climatic component in the change of grain collection makes not less than 120 billion US- dollars. If to proceed from the modern scenarios of climate change, connected to "greenhouse" gases and aerosols increase in the atmosphere, then agro-climatic analogues on heat provision ability in modern forest-steppe of Ukraine should be regarded.

The productivity of modern extensive agriculture will tend to decrease (at cultivation of usual modern set of farm crops). The productivity of high-intensive agriculture will tend to increase.

By the middle of XXI century, the temperature can increase by $1 - 2°$ C, and by the end of the century - by $3 - 4°$ C. Some ecosystems have no time to adapt to quickly changing climatic weather conditions. As a result, some kinds can disappear absolutely, that, naturally, will result in the reduction of biological diversification; the large-scale change of the climate at the end can destroy the system of international ecological safety.

For the last 11 years (1992 - 2003), only in 1998 and 2000 in any of Belarus areas no droughts has been marked. The droughts in 1992 and 2002 were especially large-scale. However in the first half of warm season (May - June) on the territory of Belarus no essential changes of repeatability of the drought phenomena occurred [1].

During the term of active amelioration and following years (1965-1995) there was an essential augmentation of frosts number per various months of the year.

If in the period 1946 to 1964 the quantity of frosts in the North was 2.2 times higher than in the South, then in the period 1985 to 1995 the quantity of frosts in the North and South of the country became commensurable; depending on the area, this interrelation varies from 1.0 up to 1.3. Quantity of frosts in the season of active amelioration (1965-1984) was higher, than those in pre-ameliorative one. This distinction was greatest in the South of the country.

Even in June peat surface frosts might be observed every 2-3 years, while on mineral soils they are registered in the South on the average 20-50 years, and in the North once per 10 years. Frosts are probable on peat deposits and in July - once 10 years. On mineral soils in the last 50 years July frosts are not registered.

Average duration of a frostless season above drained peat lands is by 15-20 days less than above mineral soils of the South of the Republic, and on the average by 5 days less than in the North of the Republic.

In some years, soil frosts in southern areas are even more intensive, than those in boreal ones. This increasing frost danger of austral areas of the country is due to its origin of incorporation into farming rotation of drained peat-mire soils.

In fluctuations of precipitation in different areas of the republic an essential diversity is found out. Mean annual sums of deposits decrease in the central and austral parts of the republic especially in the second half of a warm season, whereas in its boreal part the insignificant increase of precipitation within the last 20 years is marked. As a whole in the republic for the post-war years the fall of precipitation in the amount of about 60 - 80 mm is marked.

The enhancement of precipitation in boreal part of Belarus is marked in winter and in the beginning of spring, and also in the most part of summer season (June, July).

The increase of mean annual temperature by 1°C (within the whole term) results in augmentation of growing season by 10 days and sums of temperatures by $200°$ C, that corresponds to the shift in latitude (to the north) of more austral climatic conditions by 150-200 km. Thus, the climatic zoning of the country will change, there will be an essential alteration of dates of phenophases (especially in spring), and the growing season will be extended. Threat of late-spring (May) and early-autumn frost however is kept, that will require appropriate selection work to produce frost-resistant and drought-resistant kinds of traditional cultures.

On the background of likely climate changes, there is a necessity and importance to consider the influence of varying agro-climatic conditions. Thus it is necessary to take into account the following circumstances:
- As a result mean annual rise of air temperature the repeatability of extreme levels of heat and humidity will increase, that negatively will have an effect on the development of agricultural crops. The decrease of productivity of basic agricultural cultures due to unfavourable weather conditions can reach 50-60%, and in separate years even more. Basic fall of productivity (especially summer grain cultures) will be caused by arid conditions;
- The augmentation of repeatability of warm winters for the last years has resulted in essential changes of conditions of winter cultures wintering.

The augmentation of duration and heat-provisioning of growing season opens up the following opportunities:
- Application of more fruitful late-ripe kinds and grain cultures and vegetables;
- Terms shift of sowing of summer cultures to earlier time. It allows more effectively to use moisture reserves in a soil after spring snow - thawing, will result in more early maturation of a yield, that will increase opportunities of stubble-by-stubble agriculture. But it is necessary to take into account risk of May frosts, therefore cultures should be frost-resistant;
  - Progression to the North of cultivation region of warm-loving vegetable cultures - cucumbers, tomatoes;
  - Expansion of cultivation areas of summer rape.

In connection with drought augmentation it is necessary to:
- Intensify works on building new sorts and expand the use of drought-resistant cultures;
- Expand the regions of shower and irrigation agriculture;
- Further develop insurance system from consequences of droughts;
- Expand sowing areas of corn and millet.

Warm winters will promote the best wintering of depredators, originators of illnesses of plants, weed green enhancement. All these will require development of new measures of a pest control, protection of plants. Probable climate warming and related to it, changes of territorial distribution of atmospheric precipitation will cause reconsideration of cadastre system of the assessment of grounds. As a whole, a notable structural rearrangement of agricultural lands and tillage grounds will be required in the state. Probably, in an austral part of the country, the area of tillage land will be reduced both on predominant mineral soils of light structure, and due to the expense of impossibility to use drained peat soils as a tillage one. As a whole, the acclimatization of the agricultural sector to changes of climate is required.

Climate warming will have an effect on conditions of water consumption in agriculture. It will result in deterioration of conditions of humidification of soils and augmentation of evaporative power. On ameliorated soils, it will cause a decrease of water-control effect of irrigative amelioration. Water-supply of irrigation and drain- humidity systems will require measures on regulation of a drain, outside water supply, drainage waters reuse.

The problems of power resources especially renewed power sources are closely linked to climate change.

The decrease of speed of wind by 15-20% for the last 20-25 years is registered, that reduces opportunities of wind power use.

The assessments of influence of climate change on water economy are reduced to the following: the decrease of atmospheric precipitation by 5 % can result in the decrease of average charge for the hydrological year by 4.5-8%, and decrease of precipitation by 10% - to the decrease of drain by 7-16 of %. The increase of air temperature at constant precipitation results in insignificant decrease of drain (3%). The simultaneous count of temperature rise by $2°$ C and precipitation decrease by 10% results in the decrease of a river drain by 13-14% [1].

The account of the change of river drain and evaporations for various scenarios of temperature and precipitation has shown, that river drain can decrease from 10 up to 45% (July).

Forecasted climate warming will cause the next negative reaction of both water ecosystems as a whole and their separate parts, which especially will have an effect on flood-plains of rivers - most sensitive landscapes.

When "thermal load" on the rivers and reservoirs is observed, one can expect acceleration of eutrophication processes. The warming will differently have an effect on fish stores, depending on what depths the fishes live on. The greatest changes can be expected in shallow lakes.

At the decrease of water levels in rivers and lakes there will be an augmentation of concentrations of Cesium -137 and Strontium -90 in surface water sources of Dnieper, Pripyat' basins.

The analysis of the influence of climate change on power has shown that the heating season was already reduced by 6-9 days basically in connection with its earlier terminal. Average temperature of the heating period (more in the North) has increased by 1- $1.5°$ C. This has resulted in the decrease of the sum of grade-days by 9-11%. It was in appropriate way reflected in the charges for heating fuel.

At the increase of mean annual air temperature from 0.5 up to $3°$ C, the heating season decreases by 6 days at the increase of temperature by $0.5°$ C and by 36 days at the increase by $3°$ C.

Decrease of heat losses, and consequently, the economy of fuel at the increase of temperature by $0.5°$ C will be 3.5%, and at the increase of temperature by $3°$ C -15.3%.

REFERENCES

1. Belarus Climate Changes and their Consequences. Under general edition of V.F. Loginov. Minsk, "Tonpik", 2003, 330 p.
2. Belarus Climate / Under edit. of V.F. Loginov. Mn., 1996.
3. Loginov V.F. Reasons and Consequences of Climatic Changes, Mn., 1992, 320 p.
4. Loginov V.F., Mikutski V.S. Variation of Extreme Daily Temperatures of Belarus in the Conditions of Various Anthropogenic Loads. Natural Resources, # 1, 1997, p. 129-134.
5. Climate Change 2001, IPCC, UNEP, Cambridge. Univ. press, 2001.
6. Brazdil R., Machu R., Budikova M. Temporal and Spatial change in maxima and minima of air temperature in the Czech Republic in the period of 1951–1993. Contemporary Climatology. Ed. Brazdil R. and Kolar M., Brno, 1994. P. 93–102.
7. Karl T.R., Kukla G. et al., Global Warming: evidence for asymmetric diurnal temperature change. Geoph. Res. Lett., 1991, v. 18. P. 2253–2256.
8. Kukla G., Karl T.K. Recent rise of the night time temperatures in the Northern Hemisphere. DOE Research Summary, 1992, 14. P. 1–14.
9. Karl T.K., R.G. Baldwin and M.G. Burgino. Time series of regional seasons averages of maximum, minimum and average temperature, and diurnal temperature range across the United States: 1901 –84. Historical Climatology Series 4–5, 1988. NCDC, NOAA, National Environment Satellite, Data and Information Service, Ashville, North Carolina.

CLIMATE CHANGE CONSEQUENCES IN STEPPE - FOREST TRANSITION ZONE IN MORAVIA

M. KLIMANEK
Mendel University of Agriculture and Forestry, Faculty of Forestry and Wood Technology, Department of Geoinformation Technologies, Brno, Czech Republic

Abstract: There are climatic changes registered in the area of South Moravia. They can be documented by both climatic values and evaluation of phenological data. The elongation of the big vegetation period is especially important. It is particularly caused by its earlier beginning over the last 50 years period. At the same time vegetation summer would begin earlier and earlier. A growth in number of days (frequency) with effective temperatures over +5°C and over +15°C was revealed. A gradual increase in yearly totals of effective temperatures over +5°C is remarkable. The above facts represent a change in growth conditions of plants, especially forest trees. Direct relationship has been proved between the onset of phenophases and mean air temperatures. The increase in need of humidity for plants in steppe area of South Moravia in respect to elongation of assimilation period can be documented as well. The water regime of ecosystems has considerably changed after the hydraulic engineering adjustments and transformation from nival regime to evaporative regime. The future utility of the ecosystems is unthinkable without effective revitalization measures. The above stated facts indicate a change in growth conditions of plants, particularly long-term adapted forest trees.

Keywords: climate change, phenology, effective temperature

1. INTRODUCTION

The assimilation period of most plants is determined by the beginning and the cessation of air temperature +5°C (big vegetation period) or by the temperature +10°C for more thermophilic kinds (main vegetation period) in our conditions. The beginning and the cessation of an average daily air temperature of +15°C define the vegetation summer. The main vegetation period and vegetation summer are seasons of intensive growth of all cultures. It is the season of reproduction, ripening and harvest of most cultures.

Hitherto results indicate a direct dependence of the plants' need of humidity on the length of their vegetation period as well as heat requirement. Longer vegetation period means more humidity needed by plants or the more water is needed for irrigation.

Table 1: Assimilation period in altitudal zones (Kurpelova, 1979)

Season with average air temperature	Altitude [m above s.l.]		
	100	200	300
Higher or equal +5°C	243	233	225
Higher or equal +10°C	188	177	165

The need of humidity in altitudes over 100 m is on average 600 – 670 mm in the big vegetation period and 500 – 580 mm in the main vegetation period. The fact is that in altitudes up to approximately 350 m the need of humidity exceeds the amount of precipitation. The difference is approximately 100 – 200 mm over a vegetation period. It is especially hardwood species with the overall foliage phenophase in early spring and with long vegetation period that need most humidity (Bagar, 1996).

The actual evaporation in South Moravia makes 66% of the annual potential evaporation. The difference between the potential and actual evaporation is 246.5 mm and 211.0 mm in the vegetation period (Cerveny, 1984).

Table 2: Seasonal precipitation totals, potential evapotranspiration and humidity deficit (May – October) in "Lednice na Moravě" (Cermak, 1995)

Year	1972	1973	1974	1978	1979	1982
Precipitation [mm]	230	184	319	207	275	313
Potential evapo-transpiration [mm]	452	566	439	579	552	616
Humidity deficit [mm]	222	383	120	372	276	303

The difference between the evapotranspiration and precipitation indicates the shortage or the surplus of water expressed by the climatic index of irrigation. In our case it is the question of lack of humidity in the vegetation period, i.e. 259 mm for the period of month April – October and 383 mm a year.

The run of potential evapotranspiration and precipitation in the vegetation period is documented by the precipitation deficit of 314 mm. The humidity deficit value is in fact even higher by the values of surface and sub-surface runoff (Cerveny, 1984).

Also evaluations of phenological observations of The "Hrušovany u Brna" site prove an earlier beginning of the first phenophases of the monitored plants as well as a higher need of irrigation, especially in summer months (Bagar-Klimanek, 1999, 2000).

2. METHODS

The report is based on processing average air temperature data of the monitoring stations of CHMI (Czech Hydrometeorological Institute) in southern Moravia. Each calendar date of a year was numbered to create a continuous figure succession of days of the year.

In the second phase, effective temperatures over +5°C as well as cumulative effective temperatures for each year were calculated out of daily average air temperatures for each year of the monitored period. The data were ranged in tables for computer processing. Regression lines were carried out in graphs based on linear or polynomial regression computer data processing and trend analysis.

3. MAIN CONCLUSIONS

- The big vegetation period (temperatures over +5°C) over the monitored period would gradually start earlier. In 1961 it was approximately the 78th day of the year (May 19), in 1998 it was already the 55th day of the year (February 24), which is 23 days earlier (Graph 1).

- The length of vegetation period (temperature over +5°C) was getting longer, which probably relates to global warming or to a climatic change. If it was approximately 245 days at the beginning of the 1960s, it was already approximately 270 days in 1998, which is an increase by about 25 days (Graph 2).

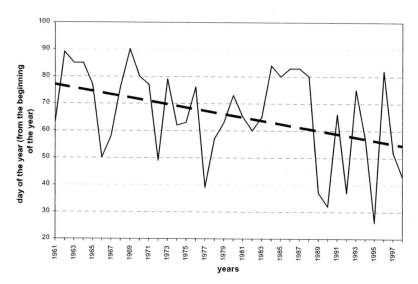

Graph 1: The beginning of effective temperatures over +5°C, "Lednice na Moravě"

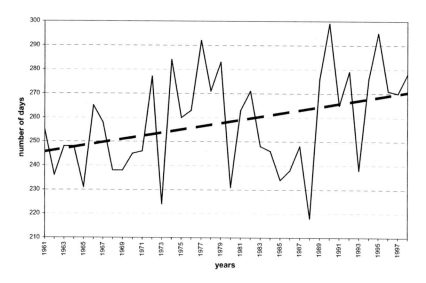

Graph 2: The length of duration of effective temperatures over +5°C, "Lednice na Moravě"

- The cessation of the big vegetation period is accompanied with considerable fluctuations in individual years, but on average it is the 323rd day of the year (November 19).

- Between 1961 – 1998, the length of the main vegetation period (temperatures of over +10°C) was gradually getting slightly shorter from 185 days a year in 1961 to approximately 180 days a year in 1998, which is 5 days.

- The beginning of the main vegetation period (temperatures over +10°C) is accompanied with considerable fluctuations. On average, the beginning of the main vegetation period was prolonged by about 8 days a year in 1998 in comparison with 1961.

- In the mid 1960s, at the beginning of the 1970s and at the turn of the 1980s, the cessation of the main vegetation period (temperatures over +10°C), was accompanied with considerable fluctuations. On average, the period elongated by about 4 days in 1998 in comparison with 1961.

- The beginning of vegetation summer (temperatures over +15°C) would come earlier over the monitored period. If it was on average the 143rd day of the year in 1961 (May 23), it was already the 120th day of the year (April 30) in 1998, which means a shift by about 23 days (Graph 3).

- The length of vegetation summer (temperatures over +15°C) was to a great extent getting significantly longer. If it was 117 days in 1961, in 1998 it was already about 140 days, which represents an increase by about 23 days (Graph 4).

- In individual years 1961 – 1998, the cessation of the vegetation period is accompanied by fluctuations, even though on average it is the 261st day of the year (18 September).

- Following the frequency rate (number of days) with temperatures over +5°C in 1961 – 1998, we can see a gradual increase in the number by about 25 days in 1998 in comparison with 1961.

- The number of days (frequency) with a temperature over +10°C increases very slightly in the years 1961- 1998.

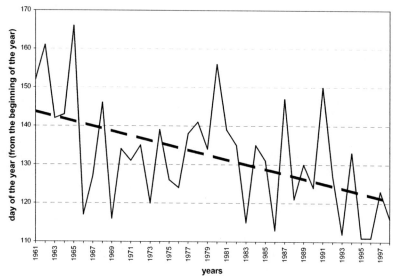

Graph 3: The beginning of effective temperatures over +15°C, "Lednice na Moravě"

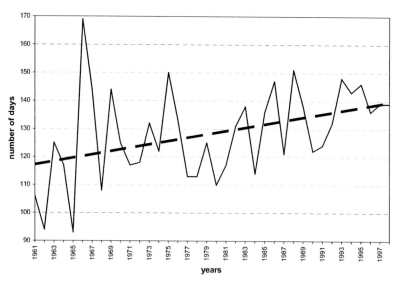

Graph 4: The length of duration of effective temperatures over +15°C, "Lednice na Moravě"

Climate Change Consequences in Steppe-Forest Transition Zone

- The number of days (frequency) with a temperature over +15°C in the monitored period gradually considerably increased.

- Annual totals of effective temperatures over +5°C in 1961 – 1998 were gradually increasing. Nevertheless the increase in temperatures was not balanced (linear). In 1980, there was certain stagnation with a slight drop in comparison with 1970. The increase in the temperatures is obvious from the 1980s on, even though there were fluctuations in some years. However, the increase did not finish with the end of the monitored period, i.e. 1998. The run of effective temperatures between April – October is similar. It is interesting that the difference in effective temperatures over +5°C during the whole year in comparison with a classical vegetation period (April – October) is almost linear over the whole period of 1961 – 1998 with fluctuations in some years. It is essential that in 1998 the effective temperature values over +15°C, increased by about 180°C in comparison with 1961. To have a comparison, this value represents a cumulative sum of effective temperatures over +5°C over the period of the warmest years 1995 and 1996 for the month of April (Graph 5).

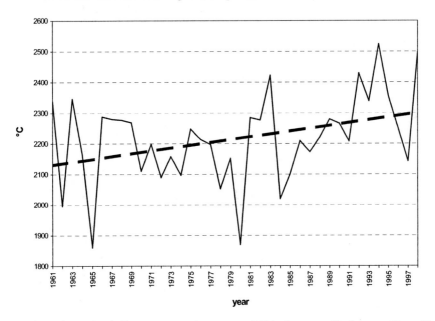

Graph 5: The totals of effective temperatures over +5°C in the years, "Lednice na Moravě"

- Monthly averages of daily relative air humidity values (1971 – 1999) have a degressive character. Relative air humidity (annual values also in the course of the vegetation period) decreased considerably in the last years in particular in comparison with 1971. In 1999 the relative air humidity decreased by about 6% in comparison with 1971. Should we compare the run of the relative air humidity of the months of 1901 – 1950 period (a long-term 50-year average) and monthly values of 1998 and 1999, we can see a considerable drop in the values. The biggest difference (drop) was observed in May and August, representing a drop by about 10% (in 1998 and 1999) in comparison with values of 1901 – 1950 long-term 50-year average.

- In 1984-1999 the sum of global radiation gradually increased, especially in May and August (Graph 6).

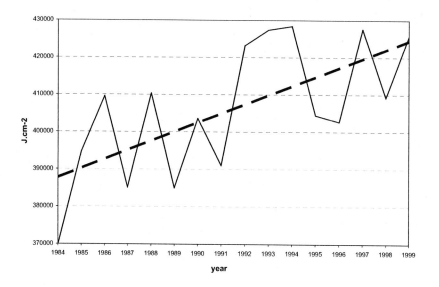

Graph 6: The sum of global radiation in the years, "Znojmo-Kuchařovice"

- We also consider the gradual drop in relative air humidity over the last couple of years as significant. In May and August 1998 and 1999 the relative air humidity was 10% lower in comparison with the values of a 1901 – 1950 long-term average. The increase in length of the big vegetation period along with the increase in annual totals of effective temperatures over +5°C is responsible for a higher need of humidity in addition.

- The annual precipitation totals were lower in recent years (1998, 1999) than the values of the 50-year long-term average (1901 – 1950). In May, August and winter of the monitored period, the drop in precipitation volume was particularly remarkable. Rainy weather in July 1997, the inrush of water and consequent floods had a huge impact on the course of the regression line of the monitored year series.

- The absolute frequency values of the wind direction are outlined clearly in the vegetation period (April – October). North-West airflow is twice as frequent in comparison with the South-East wind direction. The frequency of average wind speed SE, ESE, and SSE directions exceed the NW, WNW and west direction of wind approximately threefold. The maximal frequency of the Southeast wind direction exceeds the Northwest direction twice. The influence of drying winds is significant.

- The decrease in groundwater level in The Lednice floodplain is caused not only by the abstractions but also by the climatic influences negatively affecting growth conditions of forest trees in certain months of the year. It is especially the gradual growth in average air temperatures over the vegetation period. The average air temperature in the vegetation period increased by about 1.5°C over the 1971 – 1999 time series. The highest increase in average air temperatures was registered in May and from July to October. The amount of available groundwater limits the life of the floodplain forest ecosystem accordingly. The natural conditions can not be modified satisfactorily enough to increase the precipitation volume to balance the floodplain forest requirements of water. The deficit is caused by higher average air temperatures, global radiation etc.

- Phenological observations were conducted of the onset of the "first flowers", "first leaves" and "general yellowing of leaves" phenophases. Direct relationship has been proved between the onset of phenophases ("first flowers" and "first leaves") and mean air temperatures. It has been found the gradual earlier onset of the phenophase "general yellowing of leaves" occurred. The sum of effective air temperatures higher than 5°C after the onset of the "general yellowing of leaves" phenophase has an increasing trend in the majority of species.

4. SUMMARY

There are climatic changes registered in the area of South Moravia. They can be documented by both climatic values and evaluation of phenological data.

The elongation of the big vegetation period is especially important. It is particularly caused by its earlier beginning over the 1961 – 1998 period. At the same time vegetation summer would begin earlier and earlier. A growth in number of days (frequency) with effective temperatures over +5°C and over +15°C was revealed. A gradual increase in yearly totals of effective temperatures over +5°C is remarkable. The above facts represent a change in growth conditions of plants, especially forest trees. The increase in need of humidity for plants in floodplain area of South Moravia in respect to elongation of assimilation period can be documented as well.

Acknowledgements

Author thanks the support of the Faculty of Forestry and Wood Technology research program No. 434100005 (Grant of the Ministry of Education, Youth and Sports) and grant No. 526/03/H036 from Czech Science Foundation (GA CR).

REFERENCES

1. Bagar, R. Vlivy klimatických a srázkových zmen na lesní ekosystém, porovnání se základními růstovými faktory rostlin a potenciální škody suchem podle souborů lesních typů v období 1971-1995 pro Lesy České republiky lesní závod Zidlochovice. Brno (Czech Republic): Forest Management Institut Brno, 1996.
2. Bagar, R.–KLIMANEK, M. Vyhodnocení fenologického pozorování z lokality Hrušovany u Brna. Acta universitatis agriculturae et silviculturae Mendelianae Brunensis. Brno (Czech Republic): Mendel university of agriculture and forestry Brno, 1999, vol. XLVII, no. 3, pp. 45-56. ISSN 1211-8516.
3. Bagar, R. – KLIMANEK, M. Efektivní teploty pro dosazení nekterýchfenofází vybraných bylin, keřů, lesních a ovocných dřevin pro LČR. Brno (Czech Republic): Lesní závod Zidlochovice, 2000.
4. Czech Hydromeorological Institute. Klimatické a srazkové hodnoty monitorovacích stanic ČHMÚ. Podnebí ČSSR, tabulky HMÚ, Praha (Czech Republic): ČHMÚ, 1961, 2000, 2003.
5. Cermak, J. Spotreba vody porostem luzního lesa na jzní Morave, její zmeny vlivem poklesu hladiny podzemní vody a moznosti systematické kontroly (refarát na mezinárodním vedeckém semináři „Ochrana luzních lesu jizní Moravy se specifikou lesního hospodárstvi"). Zidlochovice (Czech Republic), 1995.
6. Cerveny, J. Podnebí a vodní rezim CSSR. Praha (Czech Republic): SZN Praha, 1984.
7. Kurpelova, M. Vzťahy fenologických javov a meteorologických charakteristík v matematicko-štatistickom spracovaní. Meteorologické zprávy 32, č. 5, s. 143–148, 1979.

CHAPTER 2
SUSTAINABLE LAND USE AND REGIONAL DEVELOPMENT

ASSESSMENT OF SUSTAINABILITY OF ECOLOGICAL - ECONOMIC SYSTEMS BY INDICATORS OF SUSTAINABLE DEVELOPMENT

F.M. RODINA
"Sustainable Development of Environment" Department, Kyrgyz Russia Slavik University, Kyrgyzstan

Assessment of correlation between environment and human activity makes it incontestable that degradation of environment renders a growing impact on life quality: people's health, poverty level, economic development and even national safety. So, for example, pollution of river reservoirs lead to productivity reduction of irrigated territories and, as a consequence, to poverty increase of countrymen.

In 2001, Kyrgyzstan received a note from the Ministry of Foreign Affairs of Kazakhstan about the inadmissibility of pollution of the transboundary rivers Chu, Talas and Assa. If cardinal measures are not undertaken, this conflict may grow further and become an issue subject to national safety.

However, so far the seriousness of ecological degradation impact on the above listed factors of life quality are studied very scarcely, there are no authentic and clear methods for identifying its quantitative parameters, methods for identifying "critical load" on environment and "environment potential capacity" (1). All of these lead to an incorrect estimation of possible development of the serious irreversible phenomena for humans and environment.

Thus, in Kyrgyzstan where the achievement of food independence is declared as a basic principle of country development, and where in 2003 the contribution of agriculture in the whole with services and processing industry in gross national product amounted 70%, there is a tendency of reducing ground area for the production of grains. Calculations of experts on sustainable development show, that now 0.22 hectares of arable lands (at the availability of 0.26 hectares per capita) provide minimum of food calories per capita (according to the data of WHO – 2100 kcal/day). And at the development of this tendency already by 2015, only 0.3 hectares per capita might provide this minimum, because the caloric content of products from arable lands depends on grain crops almost by 70%.

We see here an underestimation of food potential of the country's arable lands, and this might lead to rather complicated consequences.

There are no simple, clear, but rather exact methods of defining quantitative characteristics of natural resources' potential. And this puts serious obstacles in the way of realistic assessment of countries' development prognosis' state, based on exploitation of these resources. It is not always possible, using different development models, to convincingly and reasonably answer the questions: Is there enough water for us? Is there enough food for us?

We undertook an attempt to develop a method, which allows giving a reasonable and simple answer to these questions with the help of indicators of sustainable development.

In Central Asia experts of a regional network of sustainable development have implemented a UNDP pilot project «Adaptation and Examining of Sustainable Development Indicators in the Basin of Aral Sea».

It is known that the last list of indicators suggested by the United Nations Commission of Sustainable Development (UN CSD) included 59 parameters. Mr. J. Spangenberg, the Vice-President of the European Institute of Sustainable Development, has selected indicators out of this list describing sustainability of ecosystems in the whole, as well as sustainability of water- and land use. Following his recommendation, we have tested 30 indicators by the coordinated criteria. It was found out, that in the region only 13 out of 30 indicators exist in the national statistics. According to the developed rating scale 5 indicators out of 13 have a leading position:

- Population growth rate
- Internal consumption per capita
- Percent of arable lands irrigation
- An annual water-intake of surface and underground waters in % from available stocks
- Arable lands per capita

All countries of the region following the method accepted in the UN Blue Book and UN CSD recommendations on each of these indicators have developed methodological sheets accepted by the national statistics committees.

The «Population Growth Rate» indicator, certainly, is the most significant parameter describing long-term stability. It is connected with people distribution and usage of natural resources, including reduction of potentials.

Population growth rate in Central Asia varies from 1.3% in Kyrgyzstan up to 2.2% in Turkmenistan; it is rather high and proves the instability of development in the region.

The Global Agenda for XXI determined critical value of the «Internal Consumption per Capita» indicator equal to 40 litres per capita per day (according to the materials of the World Conference on Sustainable Development in Johannesburg - 50 l/capita/day). In all countries of Central Asia there are settlements where accessibility to safe water is lower than these indicators. And this proves the presence of centers of unstable development within this territory.

The «Percent of Arable Lands Irrigation» indicator is extremely important for the Central Asia region, because if it is calculated properly and exactly, it serves as a basis for objective political decision - making in the sphere of interstate water division. For example for the transboundary Talas River (Kyrgyzstan, Kazakhstan), each country's water-division equals to 50%. The irrigation percent of arable land area for Kazakhstan's territory of the Talas pool was defined to 30% and Kazakhstan in its programmes has foreseen significant increase of arable land area in this basin. But the request of Kazakhstan for revision of water-division was not supported, as in the Kyrgyz part of the basin it was foreseen to take 100% of its share for the perspective development of irrigated agriculture.

Is there enough water in Central Asia for us? The certain answer is such: it is enough until the value of the «Percent of Water-intake from Available Stocks» indicator will not exceed 50%. It is known, that in Kyrgyzstan the percent of water-intake from available stocks does not exceed 10-15% and fortunately it is far till the destruction of the country's water ecosystems. We use even the set interstate quota only for 45-50%. Even if it is used fully, water-intake in the country also will not exceed 25% from available stocks of water resources.

In such Central - Asian countries as Uzbekistan and Turkmenistan where water-intakes are close to 100%, the situation already is catastrophic. And taking into account that the population growth rate in these very countries of Central Asia is the highest, the situation will become even more complicated.

Why is the limiting volume of 50% of water-intake accepted? Because excess of this very limit in the 60-s in the Aral basin led to noticeable changes in the ecosystem of this basin, and then to the greatest ecological catastrophe of our time – the Aral Sea catastrophe.

Approaching the definition of water-intake quota with a real degree of precaution, this limit should be even lower. In the Law of the Kyrgyz Republic «On Sustainable Development of Issyk-Kul Ecological-economic System», signed by President Akaev in this August, and developed with our participation, the water-intake quota is equal to 30%.

It is not that easy to answer the question: Is there enough food for us? The method to estimate land resources quality by the growth class existing in the country gives an idea about a degree of lands degradation and desertification, but does not clearly answer the stated question.

Calculations of "consumer's basket", other financial-economic methods, due to the instability of transitional financial system, do not allow making an adequate and highly accurate forecast on food independence.

The method based on standard, not doubtful, quantitative parameters was required. The WHO recommendation, that the minimum level of food - calories consumed by a person in average in the country should be not lower than 2100 calories per capita per day, has become such initial indicator. The main principle of the method is the basic condition of natural ecosystems' stability: a biomass at the lowermost level of a food chain is the greatest. Available stocks of forages on pastures and grown by humans are enough to feed only a certain number of domestic animals, their biomass in the form of meat, milk, eggs is naturally smaller and, in turn, is enough to feed also only a certain number of people. Energy transformation in the form of calories to each subsequent food chain was accounted under a known in the biological ecology "rule of 10 percent": only 10% of calories of all received by animals at eating of forages is saved in them in meat, milk, eggs and then get into a person.

At first the electronic model of social-economic development in the Aral Sea basin, developed in SRC ICWC in Tashkent by the group of V.A. Dukhovny, was analyzed. This model was based on the WHO recommendation on 2100 kcal per capita per day and built interrelations between natural resources usage and social-economic development. The model rather evidently demonstrated various variants of development. But it demanded preparation of the qualified experts in computer technologies and technologies of sustainable development and the fact that the model did not take into account distinction of political realities in the countries of Central Asia, has made it unclaimed by the decision-makers. Moreover, it became clear that our authorities are not going to rely only on the opinion of experts; they want to be sure in expert conclusions.

A task was set: to develop a method in the format "for children and ministers", which nevertheless would provide a rather high accuracy of assessment.

Eventually a calculating matrix was prepared, which is given below, with the design data for Kyrgyzstan for 2000.

Table 1: Caloric content of arable lands

"Caloric Content" of Arable Lands

Agricultural crop	Caloric content, Kcal/kg	Net product	2000 year			
			Share of crops area	Crop capacity, kg/ hectare	Harvest Kg/ hectare	Kcal/ day/ hectare
Wheat	3400	0.82	0.401	2340	938.3	7167.4
Rice	3880	1.0	0.005	2630	13.15	140.0
Corn	3400	0.55	0.052	5580	290.2	1486.6
Barley	3400	0.1	0.068	2140	145.5	135.6
Oats	3400	0.1	0.001	2190	2.19	2.04
Total grains			Σ 0.527			Σ 8931.6
Potato	800	0.7	0.054	15100	815.4	1251.0
Melons and gourds	400	0.7	0.037	15700	580.9	445.6
Fruits, barriers	450	0.9	0.004	3780	15.12	16.8
Sugar beat (through granulated sugar)	3700	1	0.026	1699.4	44.2	333.0
Forage crops	680	0.1	0.21	5680	1192.8	222.2
Industrial crops(oil-yielding crops)	8990	1	0.086	78.04	6.71	165.3
Net vapor	-	-	0.056	-	-	-
Total without grains	-	-				2433.9
TOTAL						Σ 11365.5
Required level of arable lands per capita in hectares for provision 3000 Kcal						0.26
The same for 2100 Kcal						0.18

The matrix allows, with a sufficient degree of accuracy, at existing land tenure, structure of crops and productivity of agricultural crops to determine: how many food calories is produced for a person by 1 hectare of arable lands and pastures, how large ground area is required per capita to provide each person with a recommended minimum of food calories (2100 kcal /day/capita), or a normal high-grade level (3500 kcal /day/capita).

Comparison of these design data with actual ones allows drawing a conclusion on stability or instability of agricultural ecosystem, on capacity of food independence and safety. Using the matrix, we can calculate: how many agricultural animals can afford the ecosystem avoiding degradation, i.e. it is possible to determine the capacity of ecosystem for the allowable number of agricultural animals.

At various cost of agricultural crops products using the suggested matrix, it is possible to calculate rather precisely, will there be enough food for us if to produce non-food crops in large volumes, and after sallying them, whether it would be possible to buy both food and meet other necessities.

Work on perfection of this method continues. Work on definition of critical values of other indicators of sustainable development goes on as well. It allows keeping hope for progress in the decision-making process on development of the countries in view of the ecological component.

OVERVIEW OF NATO CCMS PILOT STUDY ON ENVIRONMENTAL DECISION-MAKING FOR SUSTAINABLE DEVELOPMENT IN CENTRAL ASIA

M. KHANKHASAYEV, R. HERNDON, J. MOERLINS AND C. TEAF
Institute for International Cooperative Environmental Research
Florida State University, Florida, USA

Abstract: An overview of the CCMS Pilot Study on "Environmental Decision-Making for Sustainable Development in Central Asia" is presented. This CCMS pilot study is focused on the following countries of the Central Asia (CA) region: Kazakhstan, Kyrgyzstan, Tajikistan, Turkmenistan and Uzbekistan. The presentation discusses the pilot study objectives, historical information concerning establishment of this pilot study, technical issues addressed, and overall approach focused on capacity building, as based on declared Central Asian needs for education; training; and technical assistance (involving decision-making tools such as risk assessment, cost-benefit analysis, landscape science tools, decision support software, structural/functional analysis, etc.).

As a key accomplishment of this pilot study, the role of risk assessment in environmental decision-making process was addressed. This component of the pilot study was focused on risk assessment as a tool for water resources management decision-making in Central Asia. Water resources, both in terms of water quality and water quantity, are of critical importance for sustainable development in Central Asia, as well as in other parts of the world. Risk-based approaches can play an important role as part of the overall decision-making process related to sustainable resources use in the Central Asia region.

Keywords: environment; sustainable development; decision-making process; Central Asia; risk assessment; risk management; risk communication; landscape sciences; water and health problems

1. INTRODUCTION

The main topic of the present workshop, ecologically acceptable land use as the basis for sustainable development of the mountain territories of Altai and Mongolia is a complicated and multi-dimensional socio-economic and political problem. The collapse of the Soviet Union in 1991 has strongly effected all of the countries of the former Soviet Union and in this region as well. These countries have entered into a long transition period which has been characterized by often severe economic problems and hardships.

The purpose of this presentation is to provide an overview and summarize the primary findings of the NATO CCMS Pilot Study on "Environmental Decision-Making for Sustainable Development in Central Asia", which was initiated in March of 2001. We believe that many of the results and recommendations of that Pilot Study are relevant to the issues that were discussed at the present NATO ARW on sustainable development of the Altai Region.

In March 2000, in Chimbulak, Kazakhstan, regional experts from the five CA countries concurred on a number of regional environmental problems and priorities for the Central Asia subregion. Among these problems and priorities are the following[1]:

- General Shortages of Water Resources
- Transboundary Pollution Via Shared Water Bodies
- Catastrophic Changes to the Hydrologic Regimes of Rivers
- Risks from Dam Degradation and/or Dam Failure
- Degradation of Arable and Pasture Lands
- Transboundary Transfers of Wastes Resulting in Subsequent Mismanagement (i.e., subsequent contamination of water and soils)
- Pollution from Oil & Gas Extraction Activities
- High Incidence of Human Health Problems from Poor Quality Drinking Water

Most of these problems have components that are transboundary and perhaps even global in nature, aspects of which could be more effectively addressed by sharing information and experience. Currently, there is a general lack of capacity concerning the collection and assessment of environmental data, the transformation of these data into useful information, and the effective dissemination of this information.

The World Development Report[2] from 1999 indicated that approximately 40% of the population of the Central Asia subregion lives below the poverty line. Selected comparative statistics on economics and health in the Central Asian region are given in Table 1. These data show not only wide variations in the GDP (Gross Domestic Product) and Total Health Expenditure per capita between these countries and the western countries such as USA and Germany, but also a substantial difference in those parameters among these countries. These data pointed out a striking correlation between infant mortality and the total health expenditure per capita. The transition in Central Asia from the former administrative "command economies" to present market-based economies continues to be a very painful and difficult process. In this situation, the issue of sustainable development of the countries under such a transition period has become very important.

Ecological conditions in the Central Asia subregion have become of critical importance. At the present stage of economic decline over the transition period, many of these problems require attention and economic support from the international community[3].

2. BACKGROUND

What is the NATO Committee on the Challenges of Modern Society (CCMS) and its pilot studies? The CCMS was established in 1969 in order to work on the solutions of social and environmental problems. The CCMS represents a unique forum for the exchange of information on both civilian and military environmental matters. The Committee meets twice a year in plenary sessions and annually with the Euro-Atlantic Partnership Council (EAPC) countries. The Committee does not engage in research activities; rather its work is carried out on a decentralized basis with other organizations, through its pilot studies. Subjects for pilot studies cover a large spectrum dealing with many aspects of environmental protection and the quality of life, including defense-related environmental problems. The participation of NATO members and EAPC countries in the pilot studies is on a voluntary basis. As a part of the activities of the pilot study, workshops, seminars or international conferences may be held. Reports on the progress of studies are submitted to the CCMS by pilot countries at regular intervals. Upon completion of a study (which normally takes three to five years) a summary report is submitted to the Committee members and then forwarded to the North Atlantic Council. A technical report usually is published by the pilot group and made available on a worldwide basis to anyone expressing interest. The list of publications available can be obtained from the CCMS Secretariat. Additional information on the NATO/CCMS Program can be found at http://www.nato.int/ccms/.

Since March of 2001, the Central Asian NATO/CCMS Pilot Study has conducted the following five meetings [4-8]

1. "Planning Meeting", 26 February - 1 March 2001, Silivri - Istanbul, Turkey;
2. Working Group Meeting, 16-20 March 2002, The Belgian Nuclear Research Centre Headquarters (SCK-CEN), Brussels, Belgium;
3. Working Group Meeting on "Landscape Sciences", 22-24 Sept. 2002, Almaty, Kazakhstan;
4. Working Group Meeting "Landscape Science and Public Health Issues for Environmental Decision-Making in Central Asia", 17-18 March 2003, Brussels, Belgium;
5. Working Group Meeting on "Water and Health Issues in Rural Areas of Central Asia", 4-5 November, 2003, Almaty, Kazakhstan.

In conjunction with this Pilot Study, a NATO Advanced Research Workshop on Risk Assessment as a Tool for Water Resources Decision-Making in Central Asia was conducted on September 23 - 25, 2002 in Almaty, Kazakhstan.

Our main partner in the Central Asian region for this Pilot Study is the Regional Environmental Centre for Central Asia (CAREC) located in Almaty, Kazakhstan. The CAREC is one of a number of Regional Environmental Centres established in Central and Eastern Europe and some countries of the Commonwealth of Independent States (Russia, Georgia, Ukraine and Moldova). The CAREC was established under the decision of the Fourth Pan-European Conference (1998) in Aarhus, Denmark on the initiative of the Central Asian states (Republic of Kazakhstan, Kyrgyz Republic, Republic of Tajikistan, Turkmenistan and the Republic of Uzbekistan). In 1999 the governments of the CA countries decided to locate the headquarters of the CAREC in Almaty, Kazakhstan, and set up its branches in the Republics of the region. The CAREC encourages public participation in environmental decision-making processes and facilitate co-operation between different sectors of society – non-governmental organizations (NGOs), governments and the business community. More information about the CAREC can be found on the website: www.carec.kz.

Table 1: Selected comparative statistics on Central Asia

Country	Population	Area	GDP	GDP Growth	Inflation	Total Health Expenditure	Infant Mortality	Life Expectancy
			2001 Int. Dollars	%	%	2001 Int. Dollars	per/ 1,000	years (M+F)
	2002	Sq. miles						
Kazakhstan	14.8m	2,700,000	6,855	9.5	6.4	211	19.6	65.6
Kyrgyzstan	5.0m	77,161	2,435	5.3	2.1	145	21.8	68.7
Tajikistan	6.25m	143,100	1,155	9.2	33.0	29	82.4	65.2
Turkmenistan	4.7m	188,417	4,700	10.0	10.0	286	55.0	61.0
Uzbekistan	24.9m	117,868	2,300	2.0	50.0	86	57.5	60.1
Germany	82.0m	137,820	25,950	0.6	2.5	2,754	4.5	77.0
United States	285.9m	9,600,000	36,650	0.3	2.8	4,499	8.0	77.0

Sources: U.S. Dept. State; Bureau of European and Eurasian Affairs (May 2003); World Health Organization (WHO Selected Health Indicators)

Dollar Values are in International Dollars (2001)

3. MAIN OBJECTIVES OF THE PILOT STUDY

Environmental problems can impact significantly on human health, economics and political stability of countries and regions around the world. Shared watersheds and river systems and atmospheric systems can place the environmental goals of an individual country in conflict with regional goals. Environmental degradation also can generate conflict and instability. Scarcities of croplands and fresh water could induce migration of the population, aggravate tensions along ethnic, racial, and religious lines, and increase wealth and power differentials among groups (see, e.g.[19]). The Central Asian pilot study is focused on the environmental decision-making process, i.e., on the decision-making process that could protect and/or preserve the environment, and this way, provide the basis for sustainable social and economic development of the Central Asian region.

The countries of Central Asia declared their independence in the 1990's, and many of them are still in transition from the former Soviet Union "command" style economic systems to more free-market economic systems. However, these countries inherited from the Soviet system the administrative/regulatory approach to environmental decision-making. This decision-making method is not effective in the rapidly changing socio-economic and political environment. The CA countries, facing strong economic declines, are not able to provide sufficient funding to support the administrative/regulatory decision making system. In addition, the "command" style decision making process excludes the public and stakeholders from the decision making process which creates significant problems and barriers for future effective resource allocation.

The first planning meeting for this Pilot Study was conducted on 26 February - 1 March 2001 in Istanbul, Turkey[4]. The purpose of the meeting was to determine the most effective manner in which this Pilot Study could address the key issues related to environmental decision-making from the perspective of Central Asia. Based on the Pilot Study presentations and discussions, including specific requests of the participants from Central Asia, it was concluded that the Pilot Study could be most effective in working with the Central Asian countries in the areas of education, training, and technical assistance. It was, therefore, established that the main objectives of this Pilot Study would be:

- To learn and analyze the approaches and processes used in the Central Asian Countries for environmental decision-making;
- To provide information to the Central Asian participants on the NATO/CCMS Program through presentations by the CCMS Program Director, CCMS Pilot Study Directors and CCMS Experts in selected areas, e.g., water resources management, and use of environmental impact analysis.

4. THE ENVIRONMENTAL DECISION–MAKING PROCESSES IN CENTRAL ASIAN REGION: CURRENT STATUS

The second Pilot Study Meeting, conducted on 16-20 March 2002, in Brussels, Belgium, was devoted to an evaluation of the current processes and mechanisms used for environmental decision making in the Central Asian Countries[5]. The presentations made by the invited CA experts provided a detailed analysis which covered the following topics:

- Procedures and regulatory processes in these countries for making environmental decisions;
- Quantity and quality of data available in these countries for making environmental decisions;
- Scholarly research activities in these countries involving environmental issues.

The materials provided by the CA experts[9,10] served as the basis for in depth discussions concerning how this pilot study could assist the countries of Central Asia in their efforts to attain sustainability through improved environmental decision making. The current systems of decision-making for environmental protection in Central Asia reflect the contemporary levels of social and economic development in this region. The following main components of decision-making in the sphere of environmental protection were discussed:

1. Information for decision-making;
2. The scientific foundation for decision-making, methods of analysis, and assessment;
3. Expert procedures and associated technical documents used for environmental decisions;

4. Inter-agency coordination and public participation;
5. Training needs of environmental personnel.

The main recommendations from this meeting are summarized in Sections 4.1 to 4.5.

4.1 Information for decision-making

In order to improve the quantity and quality of environmental data and information, it is recommended that:

- National and sub-regional integrated information databases be established on environmental protection and sustainable development, on the basis of a unified ideology;
- The status of existing information systems and databases in the sphere of environmental protection, and their practicality for decision-makers be improved; it would be necessary to start with the inventory of the existing ecological information. Systematic publication of the inventory of accessible information will add better access to ecological information;
- Strategies for coordinated efforts and information exchange among state and local organizations, NGOs, business entities, and scholars be implemented;
- The mechanisms of distributing information need to be improved, taking into account existing communication technologies, as well as level of technical information of the public;
- A network of libraries with free access to ecological information be developed.

4.2 Scientific basis for decision-making process

The creative potential of scholars and engineers in the region must be combined in such a way that this potential is utilized in full. Scholarly efforts must be united and directed at conducting a complex inventory of the status of the environment, and setting up ecological frameworks for economic and other types of activity in the region, as well as for new economic mechanisms for ecosystem preservation. The following are specific recommendations concerning the scientific and technical needs of the region:

- A special public council consisting of environmental experts and scholars in the region should be convened in support of efforts by Central Asian countries to prepare and realize a regional strategy for sustainable development;
- A system of educating and raising the qualifications of environmental specialists should be re-established; a system of ecological education should be developed;
- Planning for nature utilization through mechanisms of economic assessment;
- Compulsory auditing and ecological insurance of ecologically risky economic activities should be introduced, with the utilization of risk assessment methodologies for the environment and public health as a tool for decision-making;
- A network of learning resource centers should be established in order to educate the public and provide information on "cutting-edge" environmental technologies.

4.3 Technical materials and scientific expertise for environmental decisions

Technical materials and scientific expertise used for the assessment of environmental problems, environmental auditing, and for other related needs are the foundations for effective decision-making in conjunction with inspections of planned commercial or industrial activity. In all countries of the region, there is a legal procedure referred to as "ecological expertise". The following are recommendation to improve environmental decision making in Central Asia:

- Utilize principles of strategic ecological assessment when a decision is being considered; and utilize ecological auditing;
- Integrate economic assessment of natural resources and the environment into the decision-making process;
- Conduct the analysis of existing assessment methodologies and establish the most efficient ones as a necessary legal requirement to utilize them as instruments for decision-making, especially in solving trans-border problems;
- Integrate public health risk assessment into the legal procedures for decision-making;

- Strengthen control, inspection and ecological expertise services in the region: their legal and normative base must correspond to the necessities of ecosystems and must be supported by technical measures and parameters;
- Widen and strengthen the practice of including public ecological expertise into decision making in order to provide better utilization of the "Aarhus Convention" [21] mechanisms.

4.4 Inter-agency coordination and public participation

The issues of inter-agency coordination and the involvement of the public and other interested parties (stakeholders) are very difficult problems for the countries in the region, taking into account traditional (for the previous system of management) mechanisms for decision-making.

The following are recommendations related to public participation in the environmental decision making process in Central Asia:

- Develop a legal mechanism for public participation at all of the most important stages of decision-making in the sphere of nature utilization and projects having potentially hazardous influences on the environment;
- Develop and implement a system for informing and integrating the general public in environmental decisions;
- Develop a system of management for environmental protection on the foundation of political priority goals in the field of sustainable development;
- Conduct an analysis of existing environmental protection in order to improve the role of all affected parties in the decision-making process;
- Develop and fully implement the intended mechanisms of the Aarhus Convention (introduce them at the level of national statutes and acts);
- Convene and place into practice "Overseeing Councils" to manage programs and projects with NGO participation.

4.5 Training of environmental personnel

Many legal instruments and political strategies in the region have established the need for improved ecological education and training. This includes the levels of pre-school, middle school, and high school, as well as professional, and higher education. At the same time, there is not always a connection between these educational institutions and the Ministries of Environmental Protection. As a rule, such coordination only exists at the level of separate seminars and projects supported by donors. In most of the Central Asian countries, economists, environmental auditors, information technology managers, and environmental lawyers lack correct knowledge of ecological management. In terms of the training needs of the region, the following recommendations are offered:

- Develop national and local branch systems for educating and improving the qualifications of environmental specialists;
- Unify the efforts of the existing centers, programs, and projects for educating specialists through development of a network and a system of information and experience exchange;
- Re-vitalize the system of education for increasing the qualifications of environmental specialists;
- Introduce a system of continuous environmental education from kindergarten through higher education.

From these recommendations (Sections 4.1 through 4.5), it is clearly seen that the countries with transitional economies represent unique opportunities improve significantly traditional systems of education and economic planning and project assessment.

5. RISK ASSESSMENT & RISK MANAGEMENT

The process of risk assessment[11,12], when incorporated into the decision-making process, permits the prioritization of sites that represent the greatest potential risks, and allows the most effective use of financial/technical resources allocated for remediation or conservation of those sites.

The evaluation and resolution of environmental contamination problems is most effective when it involves application of three complementary elements: risk assessment, risk management and risk communication. The relationships among these three important elements of environmental risk and decision-making are illustrated in Figure 1.

By optimizing the contribution from each element, it is possible to arrive at a solution for public evaluation which is technically sufficient and cost-effective. These three components of the evaluation and resolution process address the following:

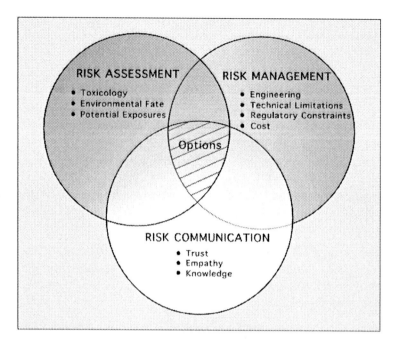

Fig. 1: Elements of environmental decision-making

- Understanding the extent and magnitude of possible negative consequences from realization of proposed projects/or hazards (e.g., chemical contamination);
- Development of practical and technically effective qualitative or quantitative goals and decisions regarding environmentally safe implementation of projects or exposure levels, coupled with methods to minimize or control potential exposures (e.g., remediation, engineered structures, restrictions on resource use);
- Dissemination and discussion of information among scientists, regulators, politicians, and the public, regarding potential risks and available management strategies.

Experience around the world has shown that while, in theory, it may be possible to address a contaminated site by using only one or only two of these three elements, in practice the best results are achieved by implementing coordinated aspects of all three elements in appropriate balance.

5.1 The risk assessment process

Risk assessment is a methodological approach that provides a framework to acquire, organize and interpret environmental data in order to provide the foundation for decisions regarding protection of human health and environmental quality. Stated differently, risk assessment is the "characterization of the potential adverse effects of human exposure to environmental hazards" [13]. This characterization of potential risks can be qualitative (e.g., exposure assessment) or quantitative (e.g., numerical risk estimates). Risk assessment approaches have been practiced widely from about 1970 about three decades in the context of contaminated sites, including the establishment of acceptable criteria for application to protection and rehabilitation of water resources, food supplies, air quality and the environment. Paradigms for organizing the elements of a risk-based approach have been articulated by such agencies as the World Health Organization[14], the U.S. Environmental Protection Agency[15,16], the European Commission [17] and the U.S. National Research Council [13, 18].

Legislative and regulatory requirements that influence remedial actions at contaminated sites or for natural resource management differ among countries. Existing statutes and the rules that control the implementation of these requirements in many cases recognize the value and applicability of risk-based approaches. An advantage of developing health-based target concentrations is that the process permits prioritization of sites that represent the greatest potential risks, and optimizes the likelihood that financial/technical resources are used most effectively to address problems or needs related to these sites.

5.2 Risk management

Risk management is a decision-making process that applies the information from qualitative or quantitative risk assessments to establish a program that protects human health and the environment in the context of available technical and economic resources. Any action or approach which results in the limitation or control of health risk can be included in the category of "risk management" options. This may include engineering approaches that preclude or limit exposure, as well as administrative/regulatory measures that influence land use or resource use (e.g., water use restrictions). It may include remedial measures such as removal of contaminated soils that are contaminating water supplies, or containment, withdrawal and treatment of the water supply before use.

In circumstances where complete elimination of the risk is not feasible due to technical or financial constraints, it is likely that a risk management approach of some type will be implemented. Risk management approaches are best applied in conjunction with a clear statement of the potential risks and how they are being addressed ("risk assessment"). There also should be a component of public involvement which not only explains the options available and the criteria that will be used to select the response, but which provides for public comment and dialogue with individuals responsible for the decision-making ("risk communication").

5.3 Risk communication

The effective transfer of technical information regarding possible chemical hazards in the environment to non-technical audiences is termed "risk communication". In most circumstances, and increasingly so in countries such as the Central Asian republics where citizens seek a more active role in environmental decision-making, understanding the results of risk assessment approaches is essential to active participation in or at least general acceptance of risk-based environmental decisions. Communication of risk information may be troublesome for several reasons, including such factors as difficulty in understanding the technical message, lack of credibility in the messenger, unrealistic expectations on the part of the public, and difficulty in effectively explaining the uncertainty which is inherent in most risk estimates.

Effective risk communication can be defined as a process that clearly but simply describes the potential risk, explains the factors which influence the risk (e.g., frequency of exposure, duration of contact), explains what uncertainty accompanies the risk estimates (e.g., limits to knowledge about the chemical, variability in the type or degree of contamination), and provides an opportunity for the audience to ask questions or deliver comments.

6. LANDSCAPE SCIENCES AND DECISION -MAKING PROCESS

One of the more important tools available to risk assessment is landscape science[20]. The availability of high quality digitized images of the Earth's surfaces along with advanced level of computer science and technologies makes landscape science one of the most important tools of the Decision-Making process.

This Pilot Study conducted two Working Group Meetings (WGM) on the role of landscape sciences in the decision-making process. The first WGM was held on 22-24 September 2002 in Almaty, Kazakhstan[6], and the second, Landscape Science and Public Health Issues for Environmental Decision-Making in Central Asia, on 17-18 March 2003 in Brussels, Belgium[7]. One of the main goals of these meetings was to combine the efforts of the present Central Asian Pilot Study and the NATO/CCMS Pilot Study on Landscape Sciences for Environmental Assessment (chaired by William Kepner (U.S. Environmental Protection Agency), Felix Mueller (University of Kiel, Germany) and Frederick Kutz (The U.S. Environmental Protection Agency).

The discussions of these WGM have focused on the following issues:

- Basic needs of CA environmental experts for use of LS tools in environmental assessments;
- CA and other countries' experiences of using Remote Sensing Data, Geographic Information Systems (GIS) and other spatial data-based tools and analyses;
- Appropriate landscape science tools for environmental assessment and/or decision support (visualization);
- Training needs and basic computer hardware and software requirements;
- Availability of data for LS assessments (i.e., satellite and other ground over data);
- Demonstration of two recently released LS software packages (Analytical Tools Interface for Landscape Assessment: ATILA and the Automated Geospatial Watershed Assessment: AGWA).

The major findings of these two WGM were the following:

- Environmental scientists in CA have the fundamental skills and experience to embark on landscape science training and application;
- Environmental scientists in CA have demonstrated a "political willingness" to engage in a landscape science capability development initiative;
- Based on expert opinion, a likely timeframe needed to successfully implement a landscape science initiative in CA is in the range of 2-4 years.

The following strategy of the path forward has been elaborated:

- Develop a strategic and scientific plan of action related to a fully realized LS initiative in CA;
- Identify appropriate funding sources;
- Building technical capacity;
- Establish a Working Group consisting of CA experts and LS experts from an international pool of scientists vis-à-vis the CCMS Program;
- Enhance the collaboration between the CA Pilot Study and the umbrella NATO-CCMS Landscape Science Pilot Study initiative.

7. HIGH HEALTH RISK AREAS OF CENTRAL ASIA

The NATO/CCMS Pilot Study meeting on Environmental Decision-Making for Sustainable Development in Central Asia conducted on 4-5 November 2003 in Almaty, Kazakhstan, was devoted to the impacts of water, sanitation, and public health in rural areas of Central Asia[8]. The presentations were made by the CA experts on the following topics:

- Water Supply/Quality Problems in Central Asia
- Health Problems from Exposures to Contaminated Water in CA
- Scholarly Research Activities Involving These and Related Issues

These CA presentations served as the basis for in-depth discussions concerning how this pilot study could assist the countries of Central Asia in dealing with the challenges of establishing and maintaining accessibility, distribution, and quality of water supplies within the region. As with prior pilot study meetings, the goal of this meeting was to establish a network of experts with backgrounds that are applicable to addressing and overcoming problems related to environmental decision-making involving water and health issues in rural communities of CA.

The main problems and needs related to safe drinking water availability are the following:

- There is poor access, especially in rural areas, to safe drinking water supplies;
- A substantial portion of the population in rural areas takes water from open sources;
- There is ongoing degradation of the physical aspects of the water supply systems (storage and distribution components) throughout CA;
- There is a lack of, and ongoing degradation of, water purifying and sanitation systems;
- There is contamination of water sources and water supply/distribution systems (chemical and biological) throughout CA;
- There are often difficulties in distinguishing disease causation between water origins and food origins;
- There is poor coordination of land use controls with knowledge of water supply data and related information;
- There is natural occurrence and enhancement of salinity/ mineralization of water sources;
- There is a need to introduce economic tools (e.g., pricing schema) for the efficient regulation of water use;

- There is a lack of effective legal frameworks for improving and maintaining water quality.

Based on these needs and problems, the following recommendations were formulated:

- There is a need to develop/provide cost-effective methods and systems for water management and purification;
- There is a need to develop/implement cost-effective technologies for utilization and processing of agricultural and industrial waste;
- There is a need to provide cost-effective water quality control and monitoring systems, including training and development of laboratory capabilities;
- There is a need to implement planning strategies and related restrictions/controls on drainage basin development;
- There is a need to educate/train the general population on effective methods of water sanitation, personal hygiene, and conservation methods;
- There is a need to involve the public and NGOs in discussions, education, and assistance for the population regarding the protection and maintenance of good quality water supplies;
- There is a need to provide free access to data on drinking water quality and health of the population in rural areas of Central Asia; and to create a database as well as a special internet site;
- There is a need to collect all cost-effective project proposals for improving access to good quality drinking water;
- There is a need to identify funding opportunities for supporting projects on water and health issues, including the 2004 Ministerial Conference in Budapest; and
- There is a need to develop training for specialists on new effective methods and tools for water management in rural areas.

8. CONCLUSIONS

This presentation summarizes the most important up-to-date findings of the NATO CCMS Pilot Study on "Environmental Decision-Making for Sustainable Development in Central Asia", which was initiated in March of 2001. From the conclusions and recommendations formulated during these meetings on various aspects of the environmental decision-making process in the CA region, it is clearly seen that these countries represent unique and challenging opportunities to radically improve traditional systems of education, economic planning, and project assessment.

All CA countries are struggling now with severe economic problems (i.e., income, employment). There is systematic insufficiency of funds to support existing systems of environmental protection. Therefore, it is important in these circumstances to preserve the creative potential of scholars and engineers in these countries. A positive impact of Western developed countries could be in helping the CA scientists and experts to be integrated in the world's community of environmental scientists. The Central Asian Pilot Study presents only one example of such efforts. During the several years of this Pilot Study, an active international team of scientists has been informally convened which provides very active exchange of information on needs, positive experience, and possible solutions.

One of the promising tools that may improve the decision-making process is risk assessment. The methodology of environmental risk assessment is a new, rapidly developing interdisciplinary scientific approach not only in the Central Asian countries but also in other countries of the Commonwealth of Independent States. The risk assessment approach in the decision-making process permits prioritization of sites that represent the greatest potential risks, and the most effective use of financial/technical resources allocated for remediation or conservation purposes. It makes especially important the introduction of risk assessment methods in the environmental decision–making process in this part of the world.

It is important to keep in mind that the evaluation and resolution of environmental contamination problems is most effective if it involves appropriate application of three complementary elements: risk assessment, risk management and risk communication. All these three parts are important, and, by optimizing the contribution from each element, it is possible to arrive at solutions which are technically sufficient, cost-effective and transparent to public evaluation.

REFERENCES

1. Central Asia Sub-Regional Report for the World Summit on Sustainable Development, UNEP, September, 2001.
2. (www.rrcap.unep.org/wssd/documents/ 01%20CA%20Report.pdf).
3. World Development Report 1999/2000: Entering the 21st Century, The World Bank Group, September, 1999. (www.worldbank.org/wdr/2000/).
4. 2002 World Summit on Sustainable Development Regional Round Table For Central & South Asia, Report, Bishkek, Kyrgyzstan. (www.johannesburgsummit.org/html/prep_process/asiapacific_prep1/bishkek_report..pf).
5. Summary Report of the NATO/CCMS Pilot Study Meeting "Environmental Decision-Making for Sustainable Development in Central Asia" - Planning Meeting, 26 February - 1 March 2001, Silivri-Istanbul, Turkey. NATO/CCMS. (www.nato.int/ccms/pilot-studies/pilot005/).

6. Summary Report of the NATO/CCMS Pilot Study Meeting "Environmental Decision-Making for Sustainable Development in Central Asia" - Working Group Meeting, 16-20 March 2002, The Belgian Nuclear Research Centre Headquarters (SCK-CEN), Brussels, Belgium. (www.nato.int/ccms/pilot-studies/pilot005/).
7. Summary Report of the NATO/CCMS Pilot Study Meeting "Environmental Decision-Making for Sustainable Development in Central Asia" - Working Group Meeting on Landscape Sciences, 22-24 Sept. 2002, Almaty, Kazakhstan. (www.nato.int/ccms/pilot-studies/pilot005/).
8. Summary Report of the NATO/CCMS Pilot Study Meeting "Environmental Decision-Making for Sustainable Development in Central Asia" - Working Group Meeting "Landscape Science and Public Health Issues for Environmental Decision-Making in Central Asia", 17-18 March 2003, Brussels, Belgium. (www.nato.int/ccms/pilot-studies/pilot005/).
9. Summary Report of the NATO/CCMS Pilot Study Meeting "Environmental Decision-Making for Sustainable Development in Central Asia" - Working Group Meeting on Water and Health in Central Asia, 4-5 November, 2003, Almaty, Kazakhstan. (www.nato.int/ccms/pilot-studies/pilot005/).
10. R. Herndon, B. Yessekin, V. Bogachev, M. Khankhasayev, J.M. Kuperberg, J. Moerlins and I. Petrisor (editors), Environmental Decision- Making for Sustainable Development: A Central Asian Perspectives, *NATO/CCMS Pilot Study Monograph (Pre-publication)*, IICER Florida State University, March 2002.
11. V. Bogachev, K. Bolatbayeva, and B. Yessekin. Environmental Decision-Making in the Context of Environmental Protection in Central Asia. *In Ref. 6, pp. 3-19*.
12. C. Teaf, M. Khankhasayev, and B. Yessekin (editors), Risk Assessment as a Tool for Water Resources Decision-Making in Central Asia, *Kluwer Academic Publishers, NATO Science Series, Dordrecht, The Netherlands, 2004*.
13. C. Teaf and J.M. Kuperberg, Risk Assessment, Risk Management and Risk Communication: Principles and Applications. *In Ref. [8], pp. 1-16*.
14. NRC. 1983. Risk Assessment in the Federal Government: Managing the Process. National Research Council Commission on Life Sciences, *National Academy Press, Washington, DC*.
15. World Health Organization (WHO) IPCS Risk Assessment website http://www.who.int/pcs/ra_main.html. September, 2002.
16. U.S. EPA. 1991. Risk Assessment Guidelines for Superfund (RAGS). U.S. Environmental Protection Agency.
17. U.S. EPA. 2002. U.S. Environmental Protection Agency Superfund Risk Assessment website http://www.epa.gov/superfund/programs/risk/index.htm. September, 2002.
18. EC. 2002. European Commission Chemical Risks website http://ihcp.jrc.cec.eu.int/Activities/ACTChem/ACTChem.html. September, 2002.
19. NRC. 1994. Science & Judgment in Risk Assessment. National Research Council, National Academy Press, Washington, DC.
20. Th. Homer-Dixon (1994) Environmental Scarcities and Violent Conflict: Evidence from Cases. *International Security*, vol. 19, no. 1, 1994, pp. 5 - 40.
21. Sc. Miller and M. Khankhasayev M., Review of Recommendations from the Almaty Working Group Meeting on Landscape Sciences, Almaty, Kazakhstan, 22-24 September 2002. *Presented at the NATO/CCMS Pilot Study on Environmental Decision-Making for Sustainable Development in Central Asia*, Brussels, Belgium, 17-18 March 2003.

22. The UN Economic Commission for Europe (UNECE) Convention on Access to Information, Public Participation in Decision-Making and Access to Justice in Environmental Matters was adopted on 25 June 1998 in the Danish city of Aarhus at the Fourth Ministerial Conference in the "Environment for Europe" process. The Aarhus Convention establishes a number of rights of the public (citizens and their associations) with regard to the environment. *(http://europa.eu.int/comm/environment/aarhus/).*

EXPERIENCES IN THE STUDY OF LAND COVER TRANSFORMATION ON MEDITERRANEAN ISLANDS CAUSED BY CHANGE IN LAND TENURE

B. CYFFKA
Institute of Geography. Faculty of Geosciences and Geography, University of Goettingen, Germany.

Abstract: The land cover of today is transforming at an accelerating rate. There are several reasons for it in the general Mediterranean region. The principle reason on the Mediterranean islands is a change in land use. The dominant directions of land use change are urban sprawl and land abandonment. The latter often finds its reason in the first one. Tourism is the driving force. The more tourism, the more urban sprawl, which also often leads to a loss in rural population because the job prospects in tourism are much better. So, tourism can often be called the economic backbone of islands. Natural and agricultural areas suffer from this development. This paper shows the development on the islands of Elba and Malta, and addresses the effects and consequences of the land cover transformation that are not as obvious as urban sprawl, but are certainly just as meaningful.

Keywords: Land use; land cover transformation; Mediterranean islands

1. INTRODUCTION AND STUDY AREA

To observe land use transformation, it is important to know something about traditional land use. A good example for this was given by Hobbs et al., (1995) (Figure 1). There was a diversified land use system that uses the natural resources and possibilities, which start from sea fishing and ends at terraced farming on steep slopes. Even the villages or main villages were often not built immediately to the coast because of the danger of flooding or the mosquitoes from nearby swamps (e.g., the island of Elba up during the 1960s).

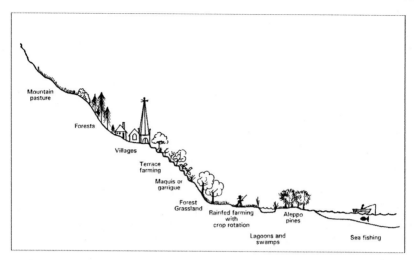

Fig. 1: Mediterranean land use system of the past. Compiled from Hobbs et al. (1995:21), altered

The ideal environment, shown in Figure 1, has many variations, especially in the case of forests. Wood has been used for construction purposes or as fuel has since Roman times in the Mediterranean, therefore, vast areas of that region are deforested. In some areas there was afforestation with pine or eucalyptus trees, and elsewhere the natural succession was able to reclaim areas in the form of maquis or garrigue. However, there was a significant share of agricultural areas in the valley floors or at the slopes in the form of field terraces. The scenic view of these terraces, e.g., in Cinque Terre (Italy), belongs to the cultural heritage of many Mediterranean countries. Apart from the share of the woodlands on the islands or the mainland, the land use system shown in Figure 1 fits nearly everywhere up to the middle of the last century. However, this system is rarely found in the landscape of today.

From that time, partly starting immediately after World War II, young people began to leave the agricultural milieu, mostly to earn money in the industrial areas of their or other countries, or to earn money in tourism. The migrations of the past century started in the 1960s. However, it was not the population loss (some reasons are given in Friedrich & Cyffka 2000) that led to land cover transformation. The process started at the other end, which can be seen in Figure 3.

Experiences in the Study of Land Cover Transformation

Fig. 2: Traditional land use on the island of Elba. The farmer is growing vine, vegetables, and other things for subsistence and selling. Even the flowers and prickly pears in the foreground of the photo are for the local market. Photo: Cyffka, March 1998

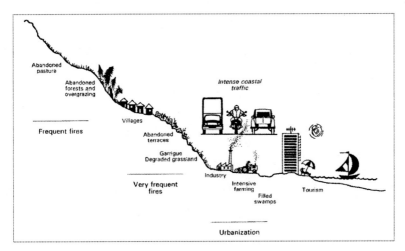

Fig. 3: Mediterranean land use system of today. From: Hobbs et al. (1995:21), altered

Industry and tourism were set up in the coastal regions of the islands and the mainland. Both served as an attraction for people from rural areas, but there were also negative aspects of the changes. Inheritance led to fragmentation of the fields, and often it was impossible for young farmers to feed their families because of the small size of their property. Therefore, many of them migrated into the industrial centers and to the coastal areas to earn their living. Many elderly people remained in the rural areas, and eventually, the fields became abandoned. If things were developing well from the sight of nature, the vegetation succession started to garrigue and maquis. In any case, there was a landscape transformation caused by land abandonment.

As mentioned before, the study area consisted of islands in the Mediterranean, Elba, and Malta. The Maltese Islands comprise six islands; mainland Malta is 246 km^2 and the island of Elba is 224 km^2 (Figure 4). At first glance there seems to be only a few differences between Elba and Malta, especially because both are situated in the central Mediterranean area. However, the island of Elba is an island of the state of Italy and able to rely on Italy's economic system. Malta is a state as an island, and with that completely responsible for its own economic development. Although the population of Elba is able to buy nearly everything from the Italian market, the Maltese people have to import every single fruit, vegetable, beverage and so on, that is not produced within their economy.

Fig. 4: Position of Elba and Malta on a map of Italy

More differences become noticeable when one begins to describe the landscape. Elba is often called the 'Green Island,' whereas Malta is barren in many parts. This is not only a fact of the degree of aridity the confrontation of meso- and xero-Mediterranean climate conditions, it is also the expression of the land use system that is preferred on the island and is described below. Although there is green cover on Elba, built by higher semi-natural maquis vegetation and forests (mainly pine plantations), the Maltese vegetation is a patchwork dominated by steppes and garrigue as a result from degradation. Also, the small maquis patches are degradated, and woodland is hardly present. The barren impression is strengthened by karstic limestone, which mainly looks light grey or yellow.

It is not possible or necessary to describe the islands here in detail; this is done by Schembri (1996) for Malta and e.g., Mori (1961) and Friedrich & Cyffka (2000) for Elba. Landscape impressions are shown in several photos in this paper.

The influences and the interconnection between the different landscape units that are the main targets of research are important aspects of this study. It is possible to identify the same landscape units, but only to a certain extent (Figure 5), on both islands.

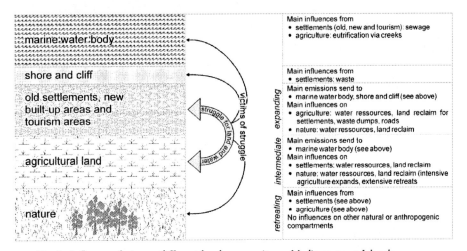

Fig. 5: Influences between different landscape units on Mediterranean Islands

Although there are influences from the marine water body, the shore, and pristine nature (as far as present), there is a struggle for land and water between the two other landscape units: settlements, especially tourism areas, and agricultural land. The basic problem is that the tourism areas are the center of concentration of nearly everything important and valuable. Therefore, it might be sensible to foster the extension of tourism and tourism areas, but this is not possible with a certain amount of land reclaim. It is only possible to get this land from agriculture, the next landscape unit in the general sequence. However, it is also necessary to feed the tourists from these few patches of arable land, which are the interconnection, as well as the interdependency, between the two areas. Normally agriculture would start to drive out nature, and this has happened to a certain extent, but there is also the case of land abandonment in agriculture, which possibly gave nature the chance to regain patches of land.

This partly vicious cycle should be investigated by several methods. The socio-economic investigations are part of a parallel work and cannot be shown here in depth. The center of this research is the development of nature, mainly vegetation and soil, under the conditions of land abandonment, and the interconnection of land abandonment with soil erosion.

2. EXPERIENCES AND DEVELOPMENT OF THE LAND COVER TRANSFORMATION ON ELBA

The above described migrations from Elba to other regions left grave traces in the agricultural landscape. The extent of the loss in the agricultural area is shown in Table 1.

Table 1: Development of the agricultural area on the islands of Elba and Corsica from 1800 to 1991. Source: Own investigations (Elba) and Lücke (1984)

Elba (223.5 km²)					
Year			1971	1981	1991
Used for agriculture (ha)			4.441	2.385	1.650
In % island area			19,9%	10,7%	7,4%
Corsica (8.700 km²)					
Year	1800	1900	1970	1980	
Used for agriculture (ha)	380.000	295.000	7.483	8.310	
In % island area	43,6%	33,9%	0,9%	1,0%	

It is not possible to compare Elba and Corsica because of their different size, but Table 1 shows that the development went in the same direction. On Elba there was a loss of more than 2.500 ha in only 20 years. Some of these abandoned parts were afforested with pines. This mixture from maquis and pine plantations leaves a landscape impression that gave Elba the epithet the 'Green Island.' The picture in Figure 6 illustrates the reason for the epithet.

Fig. 6: The towns of Poggio and Marciana Marina, Elba, embedded in a green landscape, build up by maquis and pine plantations. Photo: Cyffka, March 1998

Poggio can be seen in the center of the photo. It is an old town that is situated in a hilly area that can be compared with the situation in Figure 1. Left from Poggio, and not to be seen on this photo, there is Marciana Alta, which can be compared with Poggio in size, function, and historical condition. The twin town of Marciana Alta is Marciana Marina, to be seen at the coast in Figure 6. The land use system of today in Marciana Marina is depicted in Figure 3. The tourism facilities are located in this town, and are managed from Marciana Alta. There is a small harbor, a beach, hotels, restaurants, and so on. All three towns, Marciana Alta, Marciana Marina, and Poggio are surrounded by a mixture of pine plantations and higher maquis. Small agricultural patches can be seen in the left center of the photo. The rest is the result of land cover transformation.

Therefore, it is evident that the situation shown in Figure 2 is not the typical impression people get when visiting Elba. There are vast areas with pine plantations, exposed to the frequent Mediterranean fires, and with the same share maquis shrubs, partial woodlands, and part of the succession from agricultural land to the Mediterranean sclerophyll forest. The high amount of high maquis (average height 4-6 m) shows the time that has passed since land abandonment. It is the same with the pine plantations. In former times, there was a sequence of land use from the coast to the mountain tops that contains fishing, rain-fed, and terrace farming, maquis/garrigue and mountainous forests (Figure 1). The individual form depends on the relief. Nowadays the mountainous parts often are abandoned, especially the field terraces. The coastal parts are covered by traffic and tourism facilities, as well as urban sprawl. The old mountain villages remained in an ancient stage, often only inhabited by elderly people.

Since the beginning of the 20th century the mentioned land use change started with an increasing abandonment in agriculture. This development can be easily recognized by the following example: Up to the end of the 19th century, grapes were the most important agricultural product. In 1842, 32 million grapevines were counted. In 1961, about 120 years later, only 24 million grapevines exist (Mori, 1961). This negative trend continues, especially since the 1960s, when tourism has begun to develop into the main source of income.

But, what happened to the abandoned field terraces from the environmental point of view? A single and clear answer is not possible at this moment (Landi, 1999), but initial investigations indicate that the normal and natural succession from abandoned agricultural land via grassland to garrigue, maquis, and possibly to a Mediterranean oak forest often is not on its way. Many alien plants in this case often *Ampelodesmos mauritanicus*, a grass that came from Africa (Mingo et al., 1997), in combination with mainly man-made fires stop the natural development, and many areas remain in the stage of a grass steppe. This depends on many parameters, e.g., the aspect of the dryness of the slope that can be clearly proven by vegetation uptakes (Cyffka, 2005). On Elba, two research areas in the eastern part of the island were chosen, called 'Acquaviva' and 'Terranera'. It can easily be seen in these locations that the *Ampelodesmos mauritanicus* grass steppe is a bad stage of succession. The number of species is only 10 in such areas, which is an indication of considerable biodiversity problems (Cyffka, 2005).

However, this was not the only conspicuous fact in the results of the field research (Cyffka, 2005). Figure 7, the distribution of the vegetation units, clearly shows that the vegetation of the valley separated into two parts: woodland and grassland. The grassland slope is exposed to the south and is, therefore, much dryer than its opposite. This could be one reason for the more frequent fires. The following theory could be derived from the field research in this area: The entire valley, covered by a pine plantation (*Pinus pinea* and *P. pinaster*), burned down in many parts sometime within the past 25 years. The slope, with a southern aspect, was dryer and more affected. There was regeneration on the burnt patches on the northern and the southern slopes. Figure 7 shows many areas with maquis that are dominated or characterized by different leading species.

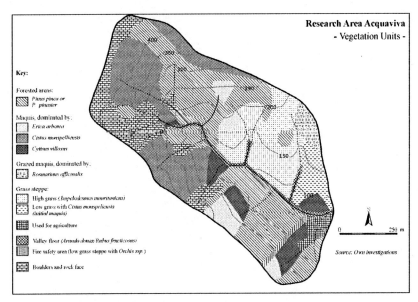

Fig. 7: Map of vegetation units in research area 'Acquaviva'

About 17 years after the first fire, the southern slope burnt down again. This fire destroyed the upcoming maquis and gave way for the massive invasion of *Ampelodesmos mauritanicus*. Since that time, the southern slope has developed completely different than the northern one. The grass rows quickly and forms eyries. Water washes the loose soil material easily between these eyries, and so erosion occurs because seeds and shoots are washed down to the valley floor, along with the soil.

Fig. 8: Ampelodesmos mauritanicus steppe in research area 'Acquaviva.' Photo: Cyffka, March 1998

The development of the Acquaviva Valley on the Island of Elba is not unique. In principle, it corresponds to the development of vegetation that is predicted for the Mediterranean area. However, it is interesting to see the examples shown in Figure 9. The usual succession goes from abandoned land via garrigue and maquis to the climax community, a Mediterranean sclerophyll forest. Without fire and grazing this might work, but the impact of man has not been evaluated thoroughly in this area. The right side of the diagram shows the path of fire, and the important branching after the stage of an open garrigue. This poses the question: Fire, often or seldom?

If there are frequent fires or regular fires it ends in a grass and herb community with soil erosion exactly at the stage of the southern slope in Acquaviva Valley.

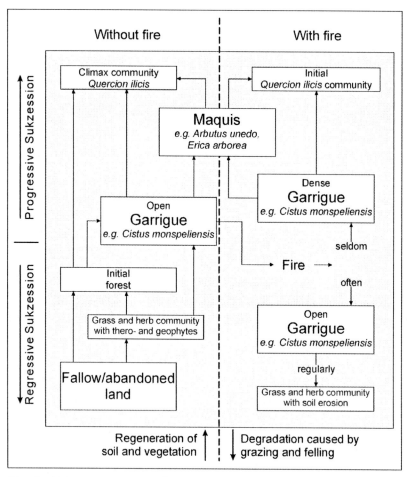

Fig. 9: *Anthropogenic triggered succession in maquis and garrigue. Compiled from: Burrichter (1961) and Lücke (1984), altered*

Therefore, land abandonment is able to influence land use transformation in many ways, if there are essential factors also in the direction of biodiversity. The question of Richter (1993): "Abandoned land in the Mediterranean—a chance for regaining nature-like areas?" is justified, but probably cannot be answered with 'yes' in any case.

3. EXPERIENCES AND DEVELOPMENT OF THE LAND COVER TRANSFORMATION ON MALTA

On Malta there is no population loss as on Elba; in fact, it is quite the reverse. However, there is a retreat of the population from agriculture, specifically from agriculture as the main occupation. Because of economical changes and rising prices, many inhabitants of the Maltese Island still have their small patch of land to grow vegetables after their main work and on weekends. Wide areas of the island are still used for agricultural purposes, but

Fig. 10: Agricultural landscape on Malta. Photo: Cyffka, October 2001

this does not compensate for the effects of a changing economy. In spite of a rising population, the share of employees in agriculture went down from nearly 10% in 1957 to less than 3% in 2002. During the same time, for example, in the service industries the number of employees went up from 12% to approximately 70% (cf. Figure 11)! All this took place while the population and the number of tourists were increasing.

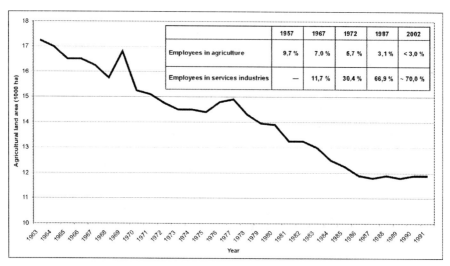

Fig. 11: Employees in agriculture and registered agricultural land with time. Compiled from different sources: Axiak et al. (1998), NSO (2003), Rainer Aschemeier, personal communication)

Figure 11 clearly shows the fall in agricultural land of nearly 32% in about 30 years. The number of employees falls from 9.7% to less than 3% in nearly the same period. This is the known system: mechanization enables fewer people to do more work. But what are the reasons for this change? In earlier times, Malta was an agricultural community. Agriculture was fostered by several of the conquerors of Malta. It was not until the 19th century that under British rule industry became more important. At that time, nearly the entire island was covered by agricultural land, starting from the top of the ridges over the slopes (terraced fields) down to the valley floors. The latter once had the best soil properties and were preferred.

The 1960s produced a complete change in the Maltese economy and agriculture. Industry and the services sector gained more share, especially tourism. As on Elba, it became more and more attractive to work in the cities, factories, shipyards, and hotels. Again land abandonment came along with these phenomena and triggered a complete change in the Maltese agriculture. Aside from the part-time farmers, there was and is a process of concentration in growing crops--vegetables as well as vines. The difficult-to-cultivate field terraces were abandoned and major farms on the valley floors emerge. In more recent times, additional new techniques, green houses, and plastic films are used to grow new products for the near European market that Malta joined in May 2004.

Experiences in the Study of Land Cover Transformation

Fig. 12: Modern agriculture vs. abandoned field terraces, Gnejna Valley, Northwest Malta. Photo: Cyffka, April 2003

Figure 12 illustrates the situation described above: the slopes are more or less abandoned. Intensive agriculture takes place on the valley floor and the plains (see Figure 12, left background), often in combination with greenhouses and plastic films. Modern Maltese farmers grow lettuce for the European market in plastic gutters with fluid culture medium in greenhouses. Strawberries grow under the plastic films during the winter months. There is no frost on Maltese Islands, but greenhouses and plastic films protect the fruits and vegetables against the salty sea spray.

The field terraces in the right middle of the photo (Figure 12) are mainly abandoned. Only a few patches of land are cultivated, though the slope gives a stable and intact impression because of the gentle terrace edges. They are naturally fixed by vegetation without rubble walls. A terraced slope of this type will grow over within a short period of time, and in this case, abandonment might be a change for nature to regain patches. In former times nearly every single piece of land in this area was used for agriculture. Obviously a major change has happened. These changes can be quantitatively proven by aerial photographs. It is possible to determine major changes by analyzing of aerial photos from 1957 to 1998. The result of this analysis is shown in Figure 13.

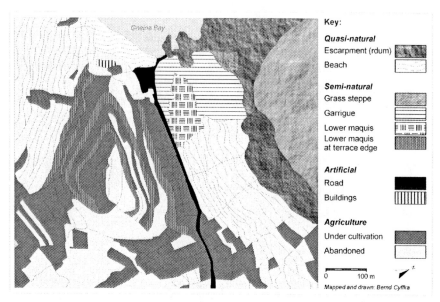

Fig. 13: Change of land use in Gnejna Valley, Northwest Malta. Source: Aerial photographs from August 17, 1957 and November 11, 1998

The huge amount of abandoned fields is very obvious. Nearly every field shown in Figure 13 as 'abandoned' was under cultivation in 1957. The land cover of the abandoned fields is variable and depends on the natural conditions. Although the geology is more or less the same, the aspect and inclination of the slopes is the steering factor. Southern aspects lead to more evaporation and water stress for the vegetation. These slopes are much dryer and generally covered by garrigue or grass steppe (Figure 14). This is somewhat the same as the island of Elba.

More humid conditions lead to thick green fuzz, the Maltese type of maquis. This large shrub layer often consists principally of Carob *(Ceratonia siliqua),* which was formerly planted by man (Schembri, 1997). From the view of nature this might be a progress in an often karstic landscape, but as far as biodiversity is concerned, it is a retrograde step. Development of this type of maquis often occurs out of fewer than eight species (own investigations in 2001) because the dense Carob shrub brings a lot of shadow to other plants.

Experiences in the Study of Land Cover Transformation 99

Fig. 14: Slope with abandoned field terraces in southern aspect under very dry conditions (in the foreground). Vegetation was grass steppe that had burnt down at time of taking the photo. In the middle of the photo, in the valley floor and on shallow slopes there are field terraces and greenhouses in cultivation. Photo: Cyffka, September 2004

Recent investigations not far from this area in a semi-natural maquis at a relatively inaccessible site at the foot of an inland cliff (the Maltese word for such areas is *rdum*) gave results of 15-20 species. Interestingly enough, Carob was missing in the shrub layer of this area completely and obviously 'replaced' by Lentisk (*Pistacia lentiscus*) and a Spurge (*Euphorbia dendroides*), which are more natural in the Maltese Islands than Carob.

The most impressive change has happened on the lower parts of the sunny southern slopes, mentioned above. In 1957, this terraced slope was built up by field terraces supported by rubble walls. In 1998 this slope was completely devastated. The rubble walls were broken down, and large patches of land were heavily eroded because of a lack of maintenance. Without regular care, the stones break out of the walls and fall down. Over time, a gap forms in the wall that becomes the first path for water erosion. And so, in the Maltese case, there is no invasion of alien plants as on Elba, but the hot and dry Maltese climate—especially on southern and steep slopes—prevent a quick and dense soil covering by vegetation. According to Aschemeier and Cyffka 2004, there is a chronological sequence of the collapse of terraces and rubble walls:

1. Abandonment of the field terrace
2. Continued abandonment leads to damage and gaps in the rubble walls
3. Eventually some parts of the rubble walls completely collapse (e.g., Figure 14)
4. During winter conditions (cloudbursts, flash floods), erosion of parts of the soil slides down onto the next terrace (depending on the size of the gap)
5. The gaps begin to form large gullies
6. When the gullies form a 'pass' or 'gate' additional soil material behind the rubble wall is eroded
7. Erosion spreads to the next lower terrace
8. The entire slop is endangered (Figure 15)

Fig. 15: The final stage of the Maltese erosion problem on abandoned field terraces—a completely devastated slope in Gnejna Valley. Photo: Cyffka, April 2003

A nearly unimaginable amount of soil loss results from the above-mentioned processes. For that reason, soil erosion is one of the biggest problems in the Maltese Islands. It must be stopped, not only to preserve the field terraces and rubble walls as a culture heritage, but also to prevent the soil from being washed down to the sea. The significance of this problem indicates that there is an urgent need for more research on this subject (Brown, 1991, Engelen, 1999).

On the other hand, not everything is on retreat in the Maltese Islands. Malta joined the EU in May 2004, and with this accession came new problems as well as new possibilities. The Maltese pre-accession negotiations went very well for the country, and therefore, many special or transitional rulings were established (cf. Aschemeier and Cyffka, 2004). Some regulations in this process of integration included the recommendation to grow more vines, and over the past several years there has been a major increase in wine-growers buying abandoned vineyards. Therefore, many of the previously abandoned fields, often used for wheat growing in the past, are now becoming wine-growing areas (Figure 16). Several farmers are waiting for a breakthrough in economy, but wine-growing is not without problems. It is obvious and quite normal for grapevines to grow on a field without any other vegetation covering the soil. Weeds are unwanted, so grapevines are an open cultivation. Therefore, these fields are prone to soil erosion as well. Under these terms, the Maltese agriculture continues to have a major erosion problem, regardless of whether one considers the abandoned fields or the newly formed ones.

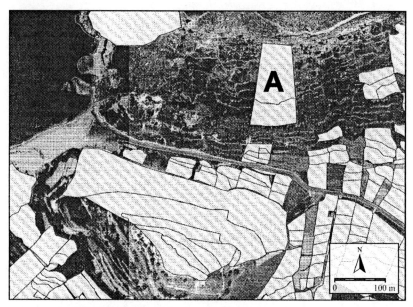

Fig. 16: Areas where land owners indicated to farm now or in the near future. Map redrawn from friendly communication with Maltese Environmental and Planning Authority, 2004

Agriculture on the valley floors is protected a bit better because of the accumulation of eroded sediments from the slopes, and at some places it is questionable whether a regaining of cultivated land for wine-growing or other purposes will be successful. The area marked with an 'A' in Figure 16 is exactly the same area shown in Figure 14 in the foreground. It is obvious when comparing these two illustrations that there are many problems to overcome before vine is growing can be totally successful.

Thus explains the dilemma in the agriculture of Malta. Abandonment, on the one hand, does not necessarily mean the regaining natural or at least nature-like areas. But, it is necessary, from both an environmental and a political perspective. After the integration in the European Union, there are not only possibilities for wine-growing, but the European law also demands more nature reserves in the Maltese Islands. The preliminary investigations of the author show that abandonment may also mean soil erosion and degradation - a very unwanted variant of land transformation. On the contrary, if they are starting an offensive to bring as many abandoned areas as possible into agriculture, there is again the risk of soil erosion and additionally economical risks and damages. No matter what is decided, land use change will provoke land cover transformations, and thus establishes the need to form an intelligent master plan for a sustainable land use development. And, regardless, the process of land abandonment must be stopped. There are many land use conflicts in other countries of the Mediterranean region (García-Ruiz and Lasanta, 1993), and land cover transformation seems to be one of the most urgent problems.

4. RECOMMENDATIONS ON THE BASIS OF THE EXPERIENCES

One of the main results of our research is that land use changes influence land cover transformations in different and often uncontrollable ways. For that reason, there is an urgent demand for more field research to find out what can be done to change it. It is necessary to set up large maps for landscape planning that show the state of the present land use and land cover, and it is also important to have the historical state of the landscape to identify the changes that have occurred over time. With this knowledge, one can start a process of planning to determine the concepts that are needed (e.g., for landscape development or landscape management). It is necessary to include as many aspects as possible into these concepts, such as vegetation, soil, agriculture, tourism, urban sprawl, etc. All this can be done by using modern methods such as GIS as a tool for landscape planning and remote sensing with high-resolution sensors (i.e., Ikonos) for noting the inventory or monitoring current trends.

Finally, it is necessary to have a goal or a 'plan as sustainable as possible or necessary.' Real sustainable development is impossible as long as an acting human society is involved, but mankind can get close to it!

REFERENCES

1. Axiak, V., V. Gauci, A. Mallia, E. Mallia, P.J. Schembri and A. Vella 1999. State of the Environment Report for Malta 1998, Valetta, Malta.
2. Aschemeier, R. and B. Cyffka 2004. Die maltesische Landwirtschaft an der Schwelle zum EU-Beitritt (Maltese agriculture on the threshold to Europe). Europa Regional, 122/2: 107-117.
3. Cyffka, B. 2005. Land Cover Changes by Land Use Changes on the Islands of Elba and Malta (Mediterranean Sea). – In: Milanova E., Y. Himiyama & I. Bicik (eds.) 2005. Understanding Land-Use and Land-Cover Change in Global and Regional Context, S. 187-205, New Delhi.
4. Brown, V.K. 1991. Early successional changes after land abandonment: the need for research. Options Méditerranéennes, 15: 97-101.
5. Burrichter E. 1961. Steineichenwald, Macchie und Garrigue auf Korsika (Mediterranean Oak Forest, maquis, and garrigue on Corsica). Ber. Geobot. Inst. ETH 32: 32-69.
6. Engelen, G. 1999. Desertification and land degradation in Mediterranean areas: from science to integrated policy making. In Conference Paper of the Workshop "The Impacts of Climate Change on the Mediterranean Area: Regional Scenarios and Vulnerability Assessment" in Venice, Italy, December 9th-10th.
7. Friedrich, I. and B. Cyffka 2000. Rezenter Landnutzungswandel auf der Insel Elba – Auswirkungen auf und Bedeutung für Mensch und Natur (Recent land-use change on the Island of Elba – effects on and significance for man and nature). Freiburger Geographische Hefte 60: 223-240.
8. García-Ruiz, J. M. and T. Lasanta 1993. Land-use conflicts as a result of land-use changes in the central Spanish Pyrenees. A review. Mountain Research and Development, 13: 213-223.
9. Hobbs, R.J., D.M. Richardson, and G.W. Davis 1995. Mediterranean-type ecosystems: opportunities and constraints for studying the function of biodiversity, pp. 1-42: In Davis, G.W. and D.M. Richardson. [eds.] 1995. Mediterranean-type ecosystems: the function of biodiversity, Springer, Berlin, Germany.
10. Landi, S. 1999. Flora e ambiente dell'Isola d'Elba. Rom, Italy.
11. Lücke, H. 1984. Macchie und Garrigue Korsikas (Maquis and garrigue of Corsica). Geoökodynamik 5: 147-182.
12. Mingo, A., A. Esposito and S. Mazzoleni 1997. Seedling establishment of three Mediterranean species in competition with adult plants of Brachypodium ramosum (L.) R. et S. In Proc. 40th Annual Symp. Intern. Ass. Veg. Sci. 1997, Ceske Budejovice.
13. Mori, A. 1961. Studi geografici sull'Isola d'Elba. Pisa, Italy.
14. NSO – National Statistics Office 2003. Population pressure and agricultural land use change. News Release No. 108/2003, Valletta, Malta.
15. Richter, M. 1993. Mediterranes Brachland – eine Chance für die Rückgewinnung naturnaher Standorte? (Mediterranean waste land – a chance for regaining nature-like areas?). Passauer Kontaktstudium Erdkunde, 3: 15-24.
16. Schembri, P.J. 1997. The Maltese Islands: climate, vegetation and landscape. GeoJournal 41: 115-125.

STUDY OF SOILS MODIFIED WITH STRUCTURE FORMING AGENTS

PIPEVA PETRANKA
University "Prof. Dr. A. Zlatarov" – Bourgas, Bulgaria

Abstract: Soil preservation and recultivation are basic guidelines of human scientific and practical activity in environmental protection. Continuous climate changes and improper use of soil lead to its destruction which is an important reason for wind and water erosion and decrease of its fertility.

One opportunity for overcoming these unfavorable changes is the use of structure forming agents in soils. Among them, synthetic polyelectrolytes recently attracted the majority of researchers' attention.

The present paper reports the results of a study on structure-forming agents synthesized at the "Prof. Dr. A. Zlatarov"- Universuty in Bourgas used with the two main types of soils existing in the mountain of Strandza: "Hromic luvisola" and Vertisols". The mechanical compositions of the spoils were studied by the dry and wet methods of Savinov using fractions with particle sizes of 10, 5, 2, 1 and 0.25 mm.

Microscopic observations showed an improved structure of the particles of the soils studied due to the versatility of functional groups present in the structure of the synthetic structure-forming agents used.

The physical and chemical properties of the soils studied were also improved. In presence of structure-forming agents, the contents of humus and total nitrogen were found to increase simultaneously with a decrease of pH of acidulous soils.

The synthesized structure-forming agents definitely facilitate the decrease of wind and water erosion of soils which is related to minimization of ecological hazards. They are neither toxic nor phytotoxic, which is an important prerequisite for their successful implementation. Last but not least, their positive effect on soil fertility should also be mentioned.

Keywords: soil, structure-forming agent, erosion, ecologic hazard, humus

1. INTRODUCTION

Soil preservation and recultivation are basic guidelines of human scientific and practical activity in environmental protection. Estimations of soil state provides lots of information for the process of taking the most appropriate and balanced ecological decisions related to the consequences for the environment and taking specific measures for its preservation and recultivation. Continuous climate changes and improper use of soil lead to its destruction which is an important reason for wind and water erosion and decrease of its fertility. Contemporary erosion is considered to be rapid and anthropogenic and started with the extermination of forests, extensive development of virgin soil, continuous increase of the intensity of use and mechanical treatment of soils, as well as uncontrolled livestock pasture.

To prevent these unfavorable consequences in agricultural practices, soil structure forming agents are used. Recently, the majority of them were synthetic polyelectrolytes known also with their common term "Crylliums" [1]. A number of special substances for forming the soil structure have been prepared in Russia under the name of "Series K" [2].

The present work reports the results obtained from a study of structure forming agents synthesized at the University "Prof. Dr. A. Zlatarov" – Bourgas used on two types of the most widely spread soils from the Strandja mountain – maroon forest soil "Hromic luvisola" and Smolnitsa "Vertisols".

2. MATERIALS AND METHODS

The water soluble polymers studied were prepared by hydrolytic modification of waste polyacrylonitrile fibers type "Bulana" obtained from an initial polymer referred to as Laxin [3] on a pilot installation at the University "Prof. Dr. A. Zlatarov" – Bourgas. The new polymers studied in the present work were obtained by hydroxymethylation and sulfomethylation of this polymer and were denoted as HMP and SHMP. They are homogeneous liquids of light-brown color, pH from 10 to 12, viscosity from 25 to 30 s, content of dry substance from 8 to 12% and content of bound nitrogen from 6.5 to 9%.

The main fragments of the Laxin macromolecule are as follows:

$$\begin{array}{llll}
\ldots\text{-CH}_2\text{-CH-}\ldots & \ldots\text{-CH}_2\text{-CH-}\ldots & \ldots\text{-CH}_2\text{-CH-}\ldots & \ldots\text{-CH}_2\text{-CH-}\ldots \\
\quad | & \quad | & \quad | & \quad | \\
\text{CONH}_2 & \text{COOH} & \text{COONa} & \\
\text{COONH}_4 & & &
\end{array}$$

Beside these fragments, the macromolecules of HMP and SHMP polymers contained the following fragments:

$$\begin{array}{ll}
\ldots\text{-CH}_2\text{-CH-}\ldots & \ldots\text{-CH}_2\text{-CH-}\ldots \\
\quad \text{C=O} & \quad \text{C=O} \\
\quad \text{NHCH}_2\text{OH} & \quad \text{NHCH}_2\text{SO}_3\text{Na} \\
\quad \text{HMP} & \quad \text{SHMP}
\end{array}$$

Au:

Method for investigation of the structure forming ability of the polymers synthesized

Samples of 250 mg air-dry soil taken from the surface layer of 0-20 cm depth was placed in wide plastic pans. The samples were then treated with 2.5 and 5 g water soluble polymer and 50 ml water. The polymer solution was introduced as thin stream under constant stirring of the soil. The reference sample was treated with 50 ml of water only, introduced by the same way.

After drying of the treated sample in air, the total amount of aggregates in the soil was determined by the method of Savinov [4]. For this purpose, the soil was placed on the upper sieve of a set of sieves with 10, 5, 2, 1 and 0.25 mm mesh. The fractions were weighed separately and their percentage in the total weight of the sample was calculated.

2.1 Determination of the water resistance of the structured aggregates

The experiments were carried out by the method of sieving the soil sample in aqueous medium. The water resistance of the structure aggregates was determined using 50 g samples of the fractions obtained by the dry method. These samples were flooded with water in a vessel for 1 h. Then, the second set of sieved was immersed in a special vessel. The samples were placed over them and the sieves were shaken 10 times up and down. The residues on each sieve were placed in heat resistant beakers and dried at 105°C to absolute dry state. After weighing, the contents of the fractions in the whole sample were calculated.

3. RESULTS AND DISCUSSION

Table 1 shows the results obtained from the experiments carried out by the dry method of Savinov for maroon forest soil. The structure forming agent known as KMC [5] was used as reference. The reference was used for soils treated with water only. The number in the code names of the samples corresponds to the amount of polymer introduced into the soil (2.5 or 5 g). As can be seen from the table, the polymers synthesized showed good structure forming properties for the type of soil used. With polymer HMP-5.0, for instance, the content of "crumb-grainy" fraction with particle size 5-2 mm increased up to 20.68% compared to 13.11% for the reference sample. Substantial increase was observed for the fraction 2-1 mm – 26.22% compared to 9.73% in the reference and for the fraction 1-0.25 mm – 15.12% compared to 5.39%. Simultaneously, the "lump" fraction with particle size >10 mm significantly decreased with both polymer concentrations to 13.73 and 9.87%, respectively, compared to 48.95% for the reference. KMC used as standard structure forming agent showed weaker effect on the "crumb-grainy" fraction and stronger on the "lump" fraction – 64.14% and 45.79%, respectively, compared to 48.95 of the reference sample.

The sulfohydroxymethylated polymer showed the best structure forming ability at both concentrations studied. The experiments with SHMP-2.5 polymer showed that the highest percentage had the fraction 2-1 mm – 31.58% vs. 9.73% for the reference and 11.66% with the standard substance KMC-2.5. These results were confirmed also by the microscopic observation. As can be seen on Fig. 1, larger soil aggregates were formed when SHMP polymer was used, compared to these of the reference sample (Fig. 2) and with the standard substance (Fig. 3). The most probable explanation of these observations is the versatility of the functional groups in polymers structures which induces better structuring of soil particles.

Table 2 shows the results obtained from the fraction analysis of soil "Smolnitsa" type treated with the different polymers and the standard substance. It should be noted that, in most cases, the content of the "lump" fraction >10 mm increased: 45.44% for HMP-1-5; 50.82% for HMP-1-2.5; 36.19% for SHMP-5; 46.77% for SHMP-2.5 compared to 42.27% for the reference.

For the "crump-grainy" fraction, slight increase was observed for fractions 10-5 mm and 5-2 mm (25.68% and 17.02% compared to 23.56% and 13.97% for the reference). For this type of soil, however, weaker structure forming effect showed both polymers synthesized as well as the standard KMC. It can be explained with some specific properties of the soil like heavey mechanical composition, unfavorable aqueous-physical properties and poor water permeability.

One of the main conditions and requirements to the synthetic structure forming agents for soils is their solubility in water and their transformation into water-insoluble form when introduced in the soil. Otherwise, the first rain will wash them away. Therefore, experiments were carried out to determine the water-resistant aggregates by the wet method of Savinov. The results obtained are presented on the histograms shown below. Obviously, the water resistant aggregates of fraction 5-2 mm preserved the highest content – 21.43% (Fig. 4) when the maroon forest soil was treated with polymer HMP-2.5. Without polymer, their content was only 7.21% and with standard KMC-2.5 – 9.53%. For fraction 2-1 mm, the water resistant aggregates in the soil treated with polymer SHMP-2.5 were 26.42% (Fig. 5). This was considered to be another confirmation that the new functional groups facilitated the formation of stronger chemical bonds between the soil particles and the polymer, thus reducing their solution and washing by the water. Polymer SHMP-5 turned out to be the best structure forming agent for the fraction 1-0.25 mm (Fig. 6). The content of water resistant aggregates was 11.44% compared to 1.08% for the reference sample and 2.81% with the standard KMC-5.

The studies carried out with polymeric structure forming agents showed that the chemical properties of the soils (humus, total nitrogen and pH) were also improved.

The results shown in Table 3 showed that the amount of humus in maroon forest soil treated with SHMP increased by 53% while for smolnitsa soil the increase was 40%. Similarly to humus content, the lowest amount of total nitrogen was found in the reference samples while the highest were

Table 1: Mechanical Composition of Maroon Forest Soil Hromic Luvisols with the Structure Formation, According to the Savinov's Dry Method

Sample Code	Dimension	Particle Size of the Fractions, mm						Total
	g, %	Lump fraction	Crumble-grain fraction				Powdery fraction	g, %
		> 10	10-5	5-2	2-1	1-0.25	< 0.25	
HMP-5.0	g	31.52	47.56	47.49	60.21	34.72	8.12	229.62
HMP-5.0	%	13.73	20.71	20.68	26.22	15.12	3.54	100.00
HMP-2.5	g	22.92	43.60	66.76	70.55	24.62	3.77	232.22
HMP-2.5	%	9.87	18.78	28.75	30.38	10.60	1.62	100.00
SHMP-5.0	g	21.11	41.83	53.83	71.50	36.32	7.20	231.79
SHMP-5.0	%	9.11	18.05	23.22	30.85	15.67	3.11	100.00
SHMP-2.5	g	26.82	32.80	57.10	73.16	35.00	6.78	321.66
SHMP-2.5	%	11.58	14.16	24.65	31.58	15.11	2.93	100.00
KMC-5.0	g	151.27	34.27	21.41	16.72	9.24	2.95	235.86
KMC-5.0	%	64.14	14.53	9.08	7.09	3.92	1.25	100.00
KMC-2.5	g	103.43	46.38	33.56	26.34	13.53	2.65	225.89
KMC-2.5	%	45.79	20.53	14.86	11.66	5.99	1.17	100.00
reference sample	g	109.4	50.86	29.90	22.20	12.30	3.50	228.16
reference sample	%	48.95	22.29	13.11	9.73	5.39	1.53	100.00

Table 2: Mechanical Composition of Maroon Forest Soil Hromic Luvisols with the Structure Formation, According to the Savinov's Dry Method

Sample Code	Dimension g, %	Particle Size of the Fractions, mm						Total g, %
		Lump fraction	Crumble-grain fraction				Powdery fraction	
		> 10	10-5	5-2	2-1	1-0.25	< 0.25	
HMP-5.0	g	93.06	47.88	27.26	22.20	11.77	2.19	204.34
HMP-5.0	%	45.54	23.43	13.34	10.86	5.76	1.07	100.00
HMP-2.5	g	106.91	50.12	26.67	17.96	7.32	1.40	210.39
HMP-2.5	%	50.82	23.86	12.68	8.54	3.48	0.67	100.00
SHMP-5.0	g	76.84	56.64	36.14	26.81	13.39	2.50	212.32
SHMP-5.0	%	36.19	26.68	17.02	12.63	6.31	1.18	100.00
SHMP-2.5	g	98.99	40.34	31.58	25.49	12.77	2.50	211.67
SHMP-2.5	%	46.77	19.26	14.92	12.04	6.03	1.18	100.00
KMC-5.0	g	88.38	37.68	28.88	27.50	15.72	3.20	201.36
KMC-5.0	%	44.89	18.71	14.34	13.66	7.81	1.59	100.00
KMC-2.5	g	94.22	45.73	29.47	24.20	12.58	2.66	208.86
KMC-2.5	%	45.11	21.90	14.11	11.59	60.02	1.27	100.00
Reference sample	g	87.42	48.80	28.88	26.25	13.08	2.38	206.81
reference sample	%	42.27	23.60	13.97	12.69	6.33	1.15	100.00

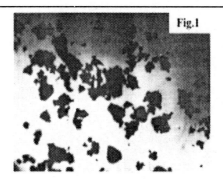

Fig.1. Microphotograph of maroon forest soil treated with SHMP-5

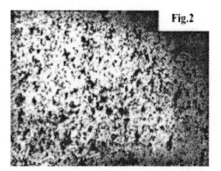

Fig.2. Microphotograph of maroon forest soil reference sample

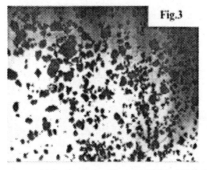

Fig.3. Microphotograph of maroon forest soil treated with standard used KMC

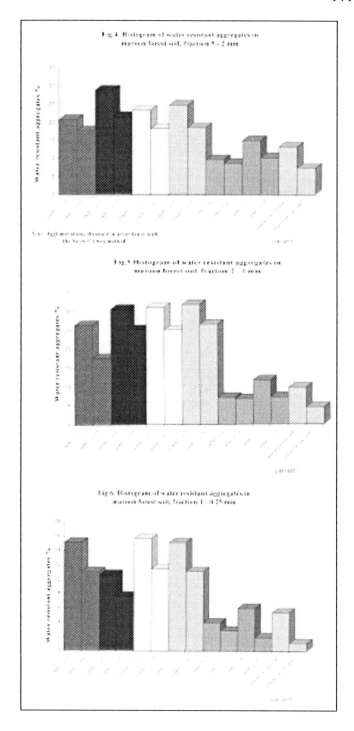

Table 3: Chemical composition of the soil studied

Variant	Humus, %	Total nitrogen, %	PH
Maroon forest treated with standard KMC	3.60	0.10	4.97
Maroon forest treated with HMP	4.04	0.12	6.05
Maroon forest treated with SHMP	4.30	0.13	6.12
Maroon forest reference	2.80	0.08	4.50
Smolnitsa treated with standard KMC	4.80	0.11	7.34
Smolnitsa treated with HMP	5.04	0.11	7.40
Smolnitsa treated with SHMP	5.10	0.13	7.42
Smolnitsa reference	3.85	0.09	7.36

observed in soils treated with SHMP polymer – 47% for maroon forest and 43% for smolnitsa soil.

It was interesting to study the effect of the structure forming agents on pH of the soil substrate. The maroon forest soil had highly acidulous reaction. The presence of polymers significantly changed the pH values – soils treated with KMC transformed from strong acidulous to weak acidulous. The smolnitsa had almost neutral reaction and the polymers did not affect it.

In conclusion, the results can be summarized as follows:
The structure forming abilities of two types of water soluble polymers synthesized from technologically asserted waste materials were studied. On the one hand, this would diminish the ecological hazard and on the other hand, the properties of the soils studied – maroon forest "Hromic luvisols" and smolnitsa "Vertisols" were improved.

It was found that the polymers studied significantly increased the contents of the agronomic most valuable structural aggregates and their water resistance which reduces the risks of water and wind erosion.
The strongest structure forming ability showed the sulfohydroxymethylated polymer (SHMP-5) in concentration of 2 mass% in the soil.

An additional effect of the water soluble polymers was the increased contents of humus and total nitrogen and reduction of pH of acidulous soils.

REFERENCES

1. Donov, V., Mountain soil science, Martilen Publ.house, Sofia, 1993.
2. Revut, I., I.Romanov, Soil science, №1, 1996.
3. Water soluble polymer Laxin, Industriasl standard 15 81935-88.
4. Gyurov, G., T.Totev, Soil science, Zemizdat publ.house, Sofia, 1990.
5. Madjikov U., A.Bektemirova, Structure forming ability of polyelectrolytes modified with KMC in soil dispersions, Uzb.Chem.Journal, №3, 1989.

PROBLEMS OF INSTABILITY IN AGRARIAN NATURE MANAGEMENT AND FOOD SAFETY IN LARGE COUNTRIES OF CENTRAL ASIA

B. KRASNOYAROVA, AND I. ORLOVA
Institute for Water and Environmental Problems (SB RAS), Russia

Abstract: The aim of the article is to analyze the problems of food safety and instability of agricultural nature management in four large countries of Central Asia, Kazakhstan, China, Mongolia and Russia, and in the remote boundary regions.

Keywords: food safety, agricultural nature management, agrarian nature management, foodstuff problem, instability, Central Asia regions, adaptation mechanisms

One of the latest definitions of food safety is formulated by the UN Commission on Human Rights in 1999 says: "by food safety is meant free access of all the people to the foodstuffs necessary for healthy and active life…" The concept of food safety of people incorporates steady physical and economic food accessibility as well as the safe and balanced food.

Food safety depends on various factors including the following: heavy dependence of national agrarian-industrial complex on environmental conditions and natural cataclysms, population, purchasing capacity, international activity of the country, economic development, labor productivity in agriculture, land resources reserve, etc.

At present many researchers agree in opinion that daily food rate should be as large as 2700-2800 kcal per capita; optimal rate is about 3000 kcal and not less than 3200-3300 kcal for people living in severe weather conditions. Scanty food allowance that causes physical degradation of personality comprises less than 1000 kcal a day.

Considering food safety on a global scale one can say that it provides only minimal consumption. It means that world foodstuff production is sufficient to meet minimal demands of people for healthy and active life. However, it is true only for the following conditions: equitable food distribution among the members of world community and uniform distribution of food among countries and continents. Since it hasn't been realized until now the provision of food safety in various countries and strata of society is different.

The industrialized countries produce and use more than 2/3 of world foodstuffs in value terms, even though their population accounts for less than

15% of the world population. From 1960 through 1990 the food self-sufficiency rate in the developed countries increased from 99 to 113%, and food consumption was 3300-3400 kcal per capita a day [2]. On the contrary, the majority of developing countries exhibit a very low rate in spite of the fact that average food consumption grew up to 2500 kcal in early 1990 as opposed to 2000 kcal in the middle of XX century. The worst situation is observed in African countries where food supply decreased up to 2040 kcal a day [3].

Our current interest is focused on Altai that is a geographical center of Eurasia. Here the peripheral regions of four large countries, Kazakhstan, China, Mongolia and Russia, which haven't even approached the national food safety, are located. The position of these countries in the world economy is given in *Table 1*.

Table 1: Position of Kazakhstan, China, Mongolia and Russia in world economy

Country	Population, mln.people	Area, th. km^2	Share in gross world product, %	Place according to GDP per head
Kazakhstan	15	2725	0,18	68
China	1272	9598	10,2	87
Mongolia	2	1567	0,01	119
Russia	145	17075	2,51	49

By gross national product per capita these countries lag behind the world average value that is $5120. In Kazakhstan it is $1350, in China - $890, in Mongolia - $400 and in Russia - $1750.

It should be noted that agriculture in these countries shares a reasonably large portion in gross domestic product (GDP). For instance, the world average index is 3.9%, in Mongolia it reaches 30.5%, in China it makes up 15.2%, in Kazakhstan – 9.0%, and in Russia – 6.8%. The number of employees in agriculture is rather high (in China – 60%, in Mongolia – 47% of able-bodied citizens).

Particular emphasis is placed on the data represented in Statistical reference book of World Bank [5] regarding small children suffering from undernourishment (*Fig. 1*).

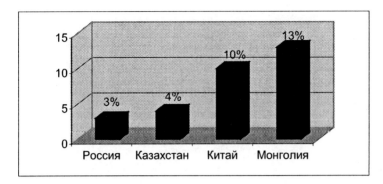

Fig. 1: Portion of small children (before 5 year-old) suffering from undernourishment

Let us consider main factors bringing the threat to the food safety in the countries mentioned.

In **China** the problem of food safety has several aspects. On the one hand, the economy is progressing rapidly. From 1978 through 1997 the GDP increased by a factor of 5.7 that corresponds to the annual growth by 9.6%. From 1995 to 2000 the annual increase of GDP made up 8.7%. Thus, for the last 20 years GDP in China increased by five times and real income of population – by 4 times [2].

In spite of the economic crisis in Asia countries the agrarian sector in China developed successfully and amounted to 30% of GNP. Recently China has been practically self-sufficient in agricultural products. Nowadays it exports rice, tea, sorghum, vegetables, eggs, meat and meat foods, fish, mushrooms, maize, peanut, honey, etc.

On the whole the ratio of export to import is rather favorable for the country and doesn't threaten the food safety (*Fig. 2*).

Plant cultivation accounts for two-third of the total agricultural production. China is typified by a large share of grain-crops (four-fifth of the area under crop), in particular, rice. Although cattle-breeding is of minor importance, the country is among the first in the world in cattle stock, namely pigs. China has a world lead in grain-crop growing (457 mln.t or 22.1% of the world production), rice (200.5 mln.t or 35% of the world production) and wheat growing (114.4 mln.t or 19.6% of the world production). It ranks second in maize growing (126.2 mln.t or 21% of the world production).

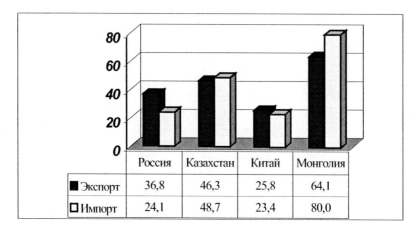

Fig. 2: Ratio of export to import of goods and services (% of GDP)

On the other hand, China is falling behind not only the highly developed countries but the developing ones in GDP rate per head (*Table 1*), labor productivity, etc. One of the main reasons is overpopulation. Due to the tight public policy of population control the demographic rate decreased from 2.77% in late 1960 to 1.1% nowadays. However, China is still the most overpopulated country in the world (1272 mln. people).

Clearly food consumption is ahead of its production. For instance, grain demand is 500 mln.t, while in 1998 its production was 392 mln.t and in 2003 it was as low as 322 mln.t [5].

Rapid economic growth and industrialization in China result in the reduction of plough-land that in its turn lead to the slowdown of foodstuffs production. The number of agricultural resources per capita is significantly less than the world average. For every Chinese there is only 0.09 ha of arable land that is 4 times less than the world average.

One more problem to be considered is desertification of the country. The process has affected 16% of the total territory. In the last decade due to sand encroachment 24 thousand villages of North and East China were left by the people. Average desertification rate is 3700 km^2/year, and in the past 50 years more than 50 th.ha of arable land were affected [5, 6].

Population of **Kazakhstan** is 15 mln. people and its increment is gradually decreasing (-0,7%). The largest part of arable land is occupied by pastures (188 mln.ha); cropland takes up about 40 mln.ha [7]. For each person there are 2.6 ha of arable land and more than 12 ha of pastures. Kazakhstan is distinguished by vast territory and various natural resources. The country is able to meet the foodstuffs demands, however, due to insufficient land use even the minimal required volume of production hasn't been attained.

Plant cultivation in Kazakhstan includes spring wheat (53% of total crops), rice, millet, cotton, sugar beet, sunflower, vegetables, melons and gourds, tobacco, etc. Gardening is widespread. Sheep and cattle breeding form the major portion in animal husbandry.

Recently the total grain crop in Kazakhstan has constituted 14-15 mln.t, potatoes – 1.5-2.0 mln.t, vegetables – 1.5 mln.t. Meat production (in deadweight) reaches 650 th.t, dairy farming - 3.5 mln.t [8]. The calculations performed [9] showed that domestic food production to sustain the food safety should be as follows: cereals - 13 mln.t, potatoes – 2.5 mln.t, vegetables and melons and gourds – 1.6 mln.t, meat (in deadweight) – 1.3 mln.t, milk – 6.5 mln.t, fish – 0.18 mln.t. Thus, the lack of domestic cattle-breeding production is evident.

Natural and weather conditions in Republic of Kazakhstan make the growing of tea, coffee and cacao beans, citrus plants, bananas, pineapples, etc. impossible, therefore preference should be given to the production of cereals, meat, milk, rice, maize, soy, fruits, sugar, fish.

The agrarian reform currently implemented in Kazakhstan has resulted in the reduction of farm production, its marketable value, increase in territorial distribution of production, and transformation of channels of products sales. Most of the products including 67% of cereals, 25% of meat and milk, 40% of wool are exchanged. Only 6% of cereals, 3% of cattle and poultry and 2% of wool are marketed through wholesale trade, commodity exchange trade fairs. The contract and other efficient ways of products realization are not popular due to non-payment, undercharge prices for raw materials, the delayed contract payments, that finally deteriorates the competitive position of agricultural producers. As a result, a negative tendency disturbing the balance in national food market, in particular, scanty supply and rapid rise in prices as compared to the growth of people cash income is observed. Import of food products makes up 43% [9].

As in China, desertification, cropland degradation and lack of nature protective technologies in agriculture bring the threat to food safety.

Mongolia is distinguished by extremely low level of social-economic development and homogeneous food market. The population of two million people increases further (1.1%), but doesn't have a marked effect on food situation.

Food safety in Mongolia is endangered mainly such factors as remoteness and difficulty of access, undeveloped transport communication with Asia countries, severe natural and climatic conditions, lack of export, namely of agricultural products. Except for meat, food products are imported from Russia and China.

On the one hand, import dependence undermines the economic safety of the country, adversely affects the government monetary reserves causing the growth of its foreign debt. On the other hand, import that forms the bulk of commodity foodstuffs and national food prices makes depressive effect on food markets.

The share of agriculture in GDP of Mongolia is one of the highest in the world (30.5%). Note that cattle breeding makes up 70% of agricultural gross output. In Mongolia livestock per head is the highest in the world (more that 100 heads of cattle per capita). The main cattle breeding branch is sheep breeding though cattle, horses, goats, camels, yaks are breed as well. Thus such an agricultural specialization results in exporting mostly cattle breeding raw materials like sheep and camel wool, hides and leather goods, knitted fabric, sheepskin coats, meat and meat products, biologicals, etc.

Animal proteins prevail in nourishment ration of Mongolian people that is unhealthy and insufficient for having active mode of life. As a result high sickness and death rate as well as low quality of life are observed.

Farming is traditionally undeveloped here. Arable lands constitute only 775 th.ha (0.4 ha per capita); as this takes place, 85% of arable lands are situated in the unsustainable farming zone. Wheat, oats, barley, potato, vegetables are grown but in small quantities.

Pastures occupy approximately 80% of the territory but actually 30-50% of them are in use. Areas nearby the administrative centers are overloaded with cattle and exposed to pasture digression; their productivity went down to 30-50%. On the contrary, the remote pastures are used insufficiently [6].

Russia holds rather diverse natural potential including rich land resources for agrarian sector development. On average 1.41 ha of agricultural lands including 0.86 ha of arable lands per capita is registered in Russia that exceeds world average indices.

The main prerequisite for food dependence of Russia is apparently lack of state support of national agricultural producers and excessive intrusion of import foodstuffs into the basket of goods that results in basic national products supplanting.

The Russian AIC peculiarity is in the leading role of agricultural production that makes up 48% of total AIC production, 68% of total industrial funds, 67% of agrarian sector employees [2].

Approximately 40% of gross agricultural output falls on plant cultivation, i.e. grain (wheat, rye, maize, barley, oat) growing; the sown area makes up 55%. Growing of groats (buckwheat, millet) and industrial crop (white beet, oil-bearing crops) is widely spread.

Animal husbandry constitutes 45% of gross agricultural output; cattle-breeding is in the lead (livestock reaches 22 mln. heads). Sheep and pig breeding as well as poultry keeping are widely spread also.

Consumption of national foodstuffs by rather big part of population (not less than 20%) at the level of minimal physiological standards is considered to be the lowest food safety that corresponds to 10-15% of imported products while 30-35% of imported food products realization is evidence of critical situation leading to independence loss.

The foreign commerce analysis of Russia shows that a share of foodstuffs and agricultural raw in the import structure is extremely high (23%), while in the export structure it is only 1.9% (*Table 2*).

In Russia the output of many foodstuffs is traditionally based on the use of raw stuff and materials purchased abroad by import, most and foremost it concerns the raw of tropical origin: tea, cacao-beans, coffee, spices, juices, fruits and vegetables. Import of product group mentioned above doesn't seriously threaten food safety of Russia.

Moreover, every year Russia has to import big volumes of raw (sugar) – 2.0-2.2 mln t; tobacco leafs- 20-25 th.t; wine materials-$20-25 mln; maize- 200-250 th.t, etc. High share of imported products (meat and meat products, milk and dairy products, oil, etc) that could be produced by the country itself means great under exploitation of the AIC potential [10].

Table 2: Export/import of foodstuffs and agricultural raw materials in the Russian Federation, $ US, bln (11)

Indices	1990	1991	1992-1993	1994-1998	1999-2002
Export, total	71,1	50,9	56,6	76,6	95,6
Including foodstuffs and agricultural raw materials	1,5	2,1	2,1	1,5	1,8
Import, total	81,8	44,5	43,7	45,7	38,0
Including foodstuffs and agricultural raw materials	16,6	12,4	9,8	11,9	8,8

By 2002 three groups of products dominated in the commodity composition of trade, namely: fish, fish and sea products (more than 50%), grain-crops (up to 40%), alcohol / soft drinks (about 5%) [11].

Recently food self-sufficiency increased greatly though significant differentiation by some food products and agricultural raw self-sufficiency is marked. For example, in 2001 the level of self-sufficiency with grain-crops was 120.8 %, potato-101.6%, eggs-98.4%, vegetables, melons and gourds- 93.4% to consumption volume. Worse situation turned out to be with self-sufficiency with meat and meat products (65%) as well as dairy products (88.9%).

The problems of low production and labor productivity in agrarian sector, obsolete technologies, irrational structure of lands, intensive degradation processes on agricultural lands and relative insularity of regional food markets are topical nowadays.

The problems of product safety are of great interest for the near boundary regions situated within the Altai Mountain Country that includes East-Kazakhstan oblast' of Kazakhstan. Xinxiang- Uigur Autonomous Region (SUAR) of China, Altai Krai and Republic of Altai of the Russian Federation, Bayan-Ulgy and Khovd aimaks of Mongolia of 591.6 th.km^2 and population - 4466 th. people (*Table 3*).

Table 3: *General characteristics of Altai Region*

Administrative region	Total area, th. km^2	Population, th. people	Agricultural specialization	Key export products
Altai Krai (Russia)	168,0	2642,5	Plant cultivation (grains, sunflower seeds, white beet, soy, flux, potato), Meat and milk cattle breeding, sheep breeding, bee keeping, fishing, fur farming	Flour, meat and meat products, milk and dairy products, butter, sugar, mixed fodder, sunflower seeds, oil, honey
Republic of Altai (Russia)	92,9	205,2	Cattle breeding, fur farming	Fluff, maral horns, wool, meat, leather goods, crude drugs, honey
Altai Okrug SUAR (China)	117,0	583,2	Plant cultivation (grains, sunflower seeds, white beet, legumes), cattle breeding (cattle and small cattle)	Meat and meat products, grain crop, potato, vegetables
East-Kazakhstan oblast' (Kazakhstan)	283,3	1533,0	Farming (grains, sunflower seeds), cattle breeding, fur farming, fishing	Grain crop, flour- and cereals products, meat, meat by-products, oil
Bayan-Ulgy aimak (Mongolia)	45,8	100,0	Meat cattle breeding, fishing	Meat and meat products, biopreparations, fish, wool, cattle skins, fluff
Khovd aimak (Mongolia)	76,0	90,0	Meat cattle breeding, Plant cultivation (potato, vegetables)	Meat and meat products wool, skins, fluff, fish

The near boundary regions mentioned above are referred to agrarian, typically peripheral ones except for East-Kazakhstan oblast'.

All these boundary territories are characterized by backwardness in life standard and infrastructure that is aggravated due to remote location and lack of transport communication. Most part of this region is economically undeveloped. Low level of industrialization and urbanization, extensive development of agrarian complex, low labor productivity in agriculture, irrational structure of agrarian lands and branches, obsolete management, poor technical and information basis can be referred to currently existing problems.

The Altai Krai ranked as an agrarian-industrial one traditionally plays the important role in the economy of Russia, has significant soil-land resources for development of agrarian production (*Table 3*).

Agricultural lands (11031 th ha) makes up 65% of total area; as this takes place, 64% (6709 th.ha) are used as arable lands. Soil cover of arable lands is represented by black earth and chestnut soils (88%). Moreover, the Altai Krai is distinguished by favorable agroclimatic resources, i.e. mean daily air temperature on the largest part of the territory makes up $+10^0$ C, the sum- $1800-2200^0$ or even more. Such climatic conditions make possible to grow grains, white beet, sunflower seeds, maize, and soy beans. First-class and solid wheat sorts which exceed the world standard by gluten content are the pride of Altai farmers. Cattle breeding plays the important role as well.

The Altai Krai has close economic collaboration with contiguous countries and neighboring regions. It exports agricultural products mostly of crop production type.

The agricultural lands of **Republic of Altai** on the contrary are very limited. Available lands of Republic includes 93 th.km^2 The most part of territories is unsuitable for agricultural development because of steep slopes, side-rocks, glaciers and other lands (31.2%), forest massifs and shrubs (49.3%). All in all there are only 19.2% of agricultural lands including pastures and hayfields (17.6%) while 1.6% falls on arable lands [12].

On the whole the Republic's economy is characterized as agrarian-raw one, 76 % of population work in agricultural sector. Agricultural production dominates in the structure of Gross Domestic Product (approximately 60%). The share of agro-industrial complex in the economy is twice higher than on average in Russia.

Severe and complicated natural-climatic conditions force the agrarians to specialize in cattle breeding. Its share in gross agricultural production is 82.4% while the share of plant growing is only 17.6%. Total livestock exceeds 760 th. heads.

Republic of Altai predominantly exports fluff, maral horns, wool, herbs and imports food products. Foreign trade turnover of near boundary countries constitutes 67.2%, most part of it falls on Mongolia and China [13].

East-Kazakhstan oblast' is one of the most industrially developed regions in Kazakhstan and in the Region under study though its agriculture is well developed as well. Agricultural lands occupy 5231 th.ha.

Major types of agrarian sector include plant growing (grain, sunflower seeds, forage crops, vegetables, melons and gourds) and cattle breeding (cattle and small cattle, horses). Maral breeding, bee keeping and fur farming are also developed. Cropping reaches 370.4 th t and livestock- 3.8 mln.heads [14].

The share of state agricultural enterprises makes up 13%, the rest falls on private ones.

Altai Okrug of SUAR occupies the territory of 117 th. km^2 with population of 600 th persons. Though arable lands constitutes 1067 th.ha only 200 th.ha are in use. The agriculture share in GDP reaches 60% [15]. Plant cultivation is specialized in growing wheat, oil-bearing crops, white beet, and beans. Cattle-breeding is developed as well.

China owns huge labor resources and it is one of the largest exporters of labor resources. Development of transnational cooperation in the field of labor services in the Altai Region has great potential and wide prospects.

It should be noted that China has advanced in realization of the project on economic upturn of its regions further than other near boundary countries of the Altai Region. The areas of economic development in Khabahei and Burchun districts are being established as well as a large wholesale market in Urumchi has been constructed. Modern infrastructure objects to serve the international business are being erected.

The population in **Mongolian part of the region** is mostly involved in cattle-breeding. Over the years after the cattle privatization was permitted its livestock considerably increased, and currently the overgrazing has turned out to be one of the grand problems to be solved for natural resources conservation.

Mongolia takes a great interest in meat export to Russia mainly for the external debt redemption. The problem solution is of strategic significance for economic integration of Siberian regions with Mongolia that affects primarily the interests of Altai and Krasnoyarsk Krais, Novosibirsk and Irkutsk oblasts. Mongolia, in its turn, is ready for meat and Siberian products exchange.

Bayan-Ulgy aimak is inhabited by 97 th. people, 97% of which are the Kazakhs. Cattle-breeding (1.4 mln. heads) ranks the first in aimak's economy. Annual meat supply of domestic and external markets exceeds 2000 tons, sheep wool – 850-900 tons, goat fluff – 130 tons and cattle skin – 280-300 th. pieces [16].

Khovd aimak is distinguished by multi-national population, the representatives of 19 nationalities can be found here. Cattle-breeding is the major trade. Total livestock reaches 2 mln., sheep and goats prevail.

At present the joint Russian and Mongolian companies produce more than 1000 tons of meat and equal amount of wool. The products are bartered for 2000 tons of flour, rice, 1000 tons of fuel and consumer commodities produced in Altai Krai [17]. On the other hand, fish is traditionally not very popular among the Mongolian people. Such high-quality fish as Mongolian grayling and osmane could be supplied to the Altai market. In its turn, Mongolia could be provided with mixed fodder the lack of which is observed in the country.

In general we can maintain positively that all countries and regions considered above have reached the limit of their extensive development. At present food products can be produced solely due to increase in productivity on the same by size areas.

Unsustainable character of agricultural nature management should be taken into account. Such an instability results from as objective as subjective factors of the natural environment changeability and social development nature.

Among the objective reasons let us emphasize the changing of general moisture of the territory that is manifested in climate cyclic recurrence, i.e. permanent fluctuation of heat and moisture ratio. These fluctuations bear as interseasonal as annual character and influence greatly the productivity and yield in the agrarian sector. In the regions under consideration this process is exhibited by progressive desertification of steppe areas and intermountain depressions, changes in vegetable communities, decrease in productivity of natural fodder land.

Another objective reason is the geographical location in the heart of Eurasia when natural component agrarian nature management instability is redoubled by climate peculiarity and altitude zoning in most boundary territories. As a result the severe natural-climatic conditions limit greatly the assortment of cultivated crops, breeds of livestock, etc. This predetermines uniformity of agrarian markets within the region under study, restricts commercial exchange of food products in the region, increase dependence on the imported goods, etc. It is only Altai Krai and partially East-Kazakhstan oblast that can supply the interregional market with crop production. The rest regions are able to satiate the food market with cattle-breeding products; however, it is impossible due to subsistence agriculture. Foodstuffs produced are homogeneous and have high prime cost.

Some of the objective reasons for instability arise from social-economic problems. The continental location of the region determines a great role of transport component in setting the prices on the imported foodstuffs as well as on the ones exported to the densely populated and economically developed regions that could provide the effective demand for ecologically clean and biological net products to be produced in mountain and submountain conditions. Since all Altai regions are situated in the outlying districts they are characterized by low level of development that implies low productivity, high prime cost of production and its low competitive capacity on the interregional market. It is immediately followed by the instability and insufficient adaptation of the agrarian sector to complicated geographical and social-economic conditions. Elaboration of possible adaptation methods for regional nature management systems must be based on the search for ways of best conformity of nature management types, correspondence of technological and territorial organization with natural-resource potential of the specific territory and social-economic conditions as well as perfection of nature protection and nature rehabilitation activities.

For example, while various technologies including intensive ones (even berries, fruits, soy, maize cultivation) can be used in the flat territory of Altai Krai, under more contrasting and severe conditions of the mountainous Altai Republic passive methods of nature management systems' adaptation (distant-pasture cattle breeding, trades, bee-keeping, maral breeding, ecological tourism, etc.) are more preferable. In west Mongolia the instability if agrarian nature management is associated with pastures rotation, veterinary activity, processing and distribution of cattle-breeding production.

To smooth away all these objective reasons of instability in agrarian nature management is possible by means of subjective mechanisms as follows:

- Introduction of economic-organizing and engineering economically and ecologically sound decisions on soil processing, selection of cultures and species, inculcation of landscape-adaptive systems for agriculture management and soil protective crop rotation, selection and regulation of stock-raising (qualitative and quantitative) load;

- Infrastructure construction implying the development of processing industries and services, transport and communication, distributive trades, etc.;

- Using of national and ethnic experience of nature management and culture of root people.

However, the implementation of these mechanisms calls for large financial investments in which the economically backward Altai boundary regions are limited. The quest for the unified development strategy, mutually beneficial cooperation can give impetus to a positive dynamics in the development of the countries mentioned. The model of the establishment of "Altai" Transboundary Biosphere Territory referred to in the paper by Yu.Vinokurov, B. Krasnoyarova, S.Surazakova can serve as a basis for such a cooperation among four countries in coordination of their national interests for biodiversity conservation and social-economic development of local population with the interests of partner countries and world community as a whole.

REFERENCES

1. Isachenko A.G. Ecological landscape capacity: its connection with global food product problem and assessment approaches. In: Izvestiya of RGS.2001.V.133. Issue 6. 1-18 pp.
2. Chernikov G.P., Chernikova D.A. World economy. M. Drofa publisher, 2003. 432p.
3. Pulyarkin V.A. Global food problem as a geographical phenomenon. In: Izvestiya of RAN. Geographical series 2000. N 2. 20-27 pp.
4. Countries and regions. 2003. World Bank Statistical Reference Book. M: Ves' Mir publisher, 2004. 240 p.
5. Vlasova O. Extremely hungry billion. Expert.2004.N37. 50-58 pp.
6. Romanova E.P., et al. World natural resources. M.:MSU publisher, 1993. 304p
7. Sultanov O.S., et al. On land property in Kazakhstan Republic. In: Regional problems of social-economic development of AIC. Barnaul. ASAU publisher, 2003. V.2. Issue 7. 192-197 pp.
8. Tkach A. On food safety of CIS countries. In: International agricultural journal. 2001.N3. 56-59 pp.
9. Espolov A.T. Provision for food products' competitiveness in Kazakhstan markets. In: International agricultural journal. 2004.N3. 46-47 pp.
10. Borisenko E.N. Food safety of Russia: problems and prospects. M.:"Economy" pbl. 1997. 349p.
11. Korovkin V., Lenchevsky I. Foreign commerce of Russia at the turn of century: agrofood aspect. In: International agricultural journal.2004.N2. 39-44 pp.
12. Isachenko A.G. Ecological landscape capacity: its connection with global food product problem and assessment approaches. In: Izvestiya of RGS.2001.V.133. Issue 6. 1-18 pp.
13. Chernikov G.P., Chernikova D.A. World economy. M. Drofa publisher, 2003. 432p.
14. Pulyarkin V.A. Global food problem as a geographical phenomenon.In: Izvestiya of RAN. Geographical series 2000. N 2. 20-27 pp.
15. Countries and regions. 2003. World Bank Statistical Reference Book. M: Ves' Mir publisher, 2004.240 p.
16. Vlasova O. Extremely hungry billion. Expert.2004.N37. 50-58 pp.
17. Romanova E.P., et al. World natural resources. M.:MSU publisher, 1993. 304p.
18. Sultanov O.S., et al. On land property in Kazakhstan Republic. In: Regional problems of social-economic development of AIC. Barnaul. ASAU publisher, 2003. V.2. Issue 7. 192-197 pp.
19. Tkach A. On food safety of CIS countries. In: International agricultural journal. 2001.N3. 56-59 pp.
20. Espolov A.T. Provision for food products' competitiveness in Kazakhstan markets. In: International agricultural journal. 2004.N3. 46-47 pp.
21. Borisenko E.N. Food safety of Russia: problems and prospects .M.:"Economy" pbl. 1997. 349p.
22. Korovkin V., Lenchevsky I. Foreign commerce of Russia at the turn of century: agrofood aspect. In: International agricultural journal.2004.N2. 39-44 pp.

SUSTAINABLE LAND USE AS A BASIS FOR A HEALTHY NUTRITION AND A CORNER STONE FOR REGIONAL DEVELOPMENT

A. MEIER-PLOEGER

Kassel University, Faculty of Organic Agricultural Sciences, Department for Organic Food Quality and Food Culture, Witzenhausen, Germany

Abstract: Nutrition is the most intensive land use and influences human health directly. Traditional food processing, which has been done in regional units or on farms, increase the added value in the region and develop a regional culture for special food products and recepies. This regional culture is a treasure which makes people to be proud of their culture and landscape. By changing food habits towards more convenience products, the globalisation in food processing and trade influences todays knowledge about food. Young people might have to learn again more about food production and the link to regional land use and landscape. The paper presents some examples for Germany.

Keywords: regional development, sustainability, nutrition, processing, added value

Centuries ago, the purpose of farming was the production of food for the population of the country. The region, especially the capacity of the soil and climatic conditions, influenced the agrobiodiversity and the amount of what could be planted and harvested. Hilly areas were regions for raising sheep, goats and cows and special breeds were selected according to vegetation and the climatic conditions (Figure 1).

Typical landscapes, reflecting the agricultural practice, developed over centuries (FERNANDEZ- ARMESTO, 2001). Nutrition habits were different between the different regions of a country, based on the treasures of farmproducts within the different seasons. Typical recepies of the regions and the seasons were developed in households and in small scale processing companies. To preserve seasonal products, different processing techniques had been developed (preservation with sugar, salt, vinegar, fermentation, heat, drying). Food production, processing and preparation represented cultural habits which were taught and practised by children, too (MONTANARI, 1993;

Figure 1: Treasures of the region: UNESCO Biosphere Territory Rhön, Germany (special breed for sheep and cow)

BRAUDEL, 1985). There was a natural understanding of rhythm of nature, agri –culture, food and landscape and a deep trust in good food for a good health (CONFORD, 2001 p 130 ff). At the same time for farmers and regional processors added value stayed in the region thus protecting the social structure in general, making people proud of their region (Figure 2).

Fig. 2: Value of the region: UNESCO Biosphere Territory Rhön, Germany
(What we use in a sustainable way we can pass to our children)

The advantages of a sustainable landuse can be summarized as follows:
- Protection of hilly areas as a basis for animal feed and biodiversity in general
- Avoidance of import of animal-feed (e.g. soy beans)
- Protection of region specific breed e.g. Rhöner Fleckvieh in Germany
- High milk and meat quality (herbs, polyunsaturated fatty acids, taste)
- Quality management in smaller cycles leads to consumers trust in produce
- Increase in income by 30% for farmers in the region of Biosphere Territory Rhön, Germany
- Creating a typical landscape for tourism

- Protection of smaller food processing companies and restaurants
- Protection of structures in villages and towns

In the 19th century in Central Europe, parallel to the increasing industrialisation in general, the food market industrialised as well. Caused by the separation of working place and home (living) in bigger cities, the need for convenience products arose.

Preprepared soups (e.g. vegetables, or meat) as well as breakfast cereals were the first products which were introduced to a growing market. By this, "global" recepies took over regional recepies. The amount of frozen food consumed or food derived from heated glass house made peoples food habits free from seasonal conditions. As FERNANDEZ- ARMESTO described (p. 242 f) the cooling/ freezing techniques made it possible already in 1851 to send food for longer distance. The first shipment from Australia to London with frozen food was in 1880. The American market (spending in 1959 already 2.7 Billion dollars on frozen food, including half a billion on ready prepared meals of the "heat-and-serve" variety) splashed to Europe (e.g. products from General Foods Company). Food processing took more and more place in large companies on European and international level. The advertisement: "You can rely on us – it tastes always the same" expresses this global und unseasonal demand. The trust in a farmer and his/her products changed towards trust in a company and its Quality Management System (QM) as well as the trust in legislative requirements for a healthy food from government (CONFORD, 2001 p 130 ff; 146 ff). The new generation of convenience products in the global market makes food not only independent from region and season but also from agricultural products. This technique – called food design – allows companies to buy the cheapest carbohydrates (extracted from yams or potatoes e.g. as a basis for a soup) or fat or protein. Not the raw material and its structure counts but the ingredient which can be "fortified" with vitamins, minerals, flavour, colouring additives, processing auxiliaries (Figure 3). We can state that food today is independent from region, season and agriculture.

- Ingredience: carbohydrates, protein, fat
- Independend from agricultural production
- High energy need for intensifying food processing & packageing
- No typical product of a region but typical for a special brand

Fig. 3: Food Industry today- Food Design: 5 minute soup

At the same time, family structures have changed. In industrialized countries less children are born, men and women living alone (singel) increased and only one to two generations are living together. More and more, both partners in a family are going out for work. These socio demografic changes lead to more processed food, more convenience products and a deminishing knowledge of food processing and preparation (RÜTZLER; 2003). The contact to food and food preparation for the young generation in cities is often linked to advertisements in journals and TV Direct contact to nature in general (e.g. walking, playing in the forest, contact to farmers) decreased tremendously by using computers or watching TV after school. The advertisements in TV and journals are nearly always for highly processed food and snacks, rich in sugar and fat leading to an increasing obesity of the population in general (DGE, 2004). Politicians are warning that the increasing number of obese children will cause high social costs for e.g. diabetes treatment, coronary heart diseases and cancer. 20 % of the German children are already obese; nutrition depending diseases cost already 71 Billion Euros per year. Healthy eating and drinking, knowledge about food and preparation of meals has now to be learned in kindergarten and schools (MEIER-PLOEGER, 2002; 2003). For this training, LUDE (2001) suggested to include the experience in nature as shown in figure 4. Young people should learn about the beauty of nature in the different seasons, about how nature approach our senses (e.g. listening to birds, feeling the surface of wood), nature as the natural supplier of food (e.g. wild mushrooms), nature as a possibility to care for (nisting birds) or a place to relax (sporting) and about nature and farming (landscape, risk of pollution by agricultural use of pestizides). The increasing of time pupils spend in front of the TV or computer reduced the time that could be spent outside in nature and thus reduces the experience with nature (RINK, 2002, SCHUSTER & LANTERMANN, 2002).

Today food quality as seen by consumers includes not only desirable and undesirable nutrients in foods but social and ecological parameters of the production as well. Working conditions (e.g. fair trade) as well as energy used for production on farms and processing as well as packaging in companies are under consideration. "From farm to folk" or "from stable to table" means that each step in the food production chain is analyzed in regard to material flux, energy, emission and waste. The methodology (e.g. eco balance sheets, ecological footprint or life cycle assessment) has been improved in the last years and the results give the consumers the possibility to choose according to their understanding of a high quality product. Figure 4 presents some data made in Switzerland for tomatoes (glasshouse production).

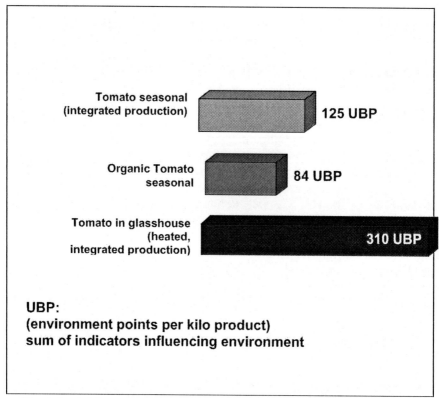

Fig. 4: Food and environment - Tomato production (: Epp 2002)

More and more farmers worldwide try to produce food in an ecologically sound way. In the EEC regulation 2092/91 the principles of an organic food production and processing are characterized (YUSSEFI, WILLER 2003):

- working in "closed cycles": fertilization (manure, clover, rotation)
  Animal feed (max 20% import)

- supporting "inner cycles": soil fertility – soil life
  Immune system of plants
  (Secondary compounds)

  Animal husbandry and health

A report published by the German Government (Federal Research Institute; Tauscher et al 2003) stated clearly the ecological advantages of organic farming as seen in Table 1.

Table. 1: Sustainability of organic agriculture (Tauscher et al., 2003)

Parameters (organic vs. conventional)	
• biodiversity	+
• landscape	=
• protection of soil and soil functions	=
• protection of water quality	+
• reduced eutrophication	+
• reduced acidity in soil	+
• reduced climatic relevant gases	+
• protection of recources in general	+
• reduced human toxicity	+

But sustainability cannot be carried and practised by the farmers and processors alone. It should be a value carried by all participants in the market (CONFORD, 2001 p.210 ff). Consumers have to be aware of environmental and social problems which are linked to global food market. Politicians should support environmentally sound production methods in industry in general and scientists should put more emphasis to invest sustainability in the food chain so that data can be used by industry, politics and households.

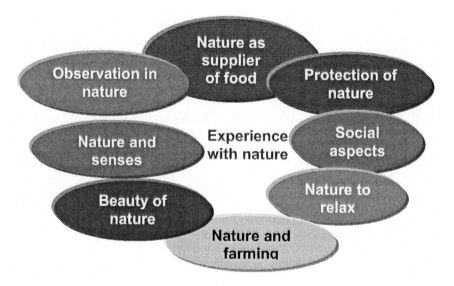

Fig. 5: Awareness of nature as a precondition for a healthy nutrition (LUDE, 2001)

REFERENCES

1. Birch, I (1980): Effects of Peer Models Food Choices and Eating Behaviours on Preschoolers' Food Preferences. In: Child Development 51, 489-496.
2. Braudel, F., 1985: Der Alltag. Sozialgeschichte des 15. – 18. Jahrhunderts. (Civilisation materielle, economie et capitalisme XV – XVIII siecle; Librairie Armand Colin, Paris, 1979) Deutschsprachige Ausgabe, Kindler Verlag, München, ISBN 3 463 40025 1; 671 pages.
3. Conford, F., 2001: The Origins of the Organic Movement. Bell& Bain, Glasgow ISBN 0-86315-336-4 287 pages.
4. DGE (German society for nutrition; ed), 2004: Ernährungsbericht 2004; Bonn; ISBN 3 88749 183 1 ; 484 pages.

5. Fernandez-Amesto, F., 2001: Food, A History. Pan Macmillan Ltd., London, Basingstoke, Oxford. ISBN 0 333901746 287 pages.
6. Lude, A. (2001): Naturerfahrung und Naturschutzbewusstsein. – Innsbruck.
7. Meier-Ploeger, A.,2003: Kulturlandschaft geniessen - Natur im Kontext der Ernährungskultur in: Erdmann, K.-H., Schnell, C. (Hg): Zukunftsfaktor Natur - Blickpunkt Mensch, p 257-272, Münster-Hiltrup, Landwirtschaftsverlag.
8. Meier-Ploeger, A., 2002: Klasse statt Masse..., Sinnesschulungen von Kindern und Jugendlichen als Maßnahme zur Steigerung des Qualitätsbewusstseins bei Lebensmitteln. Der kritische Agrarbericht 2002. p 309-313.
9. Montanari, M:, 1993: Der Hunger und der Überfluss. Kulturgeschichte der Ernährung in Europa. C.H. Beck, München, ISBN 3 406 37702 5; 251 pages.
10. Rink, D. (2002): Naturbilder und Naturvorstellungen sozialer Gruppen. Konzepte, Befunde und Fragestellungen. - In: Erdmann, K.-H. & Schell, C. (ed.) (2002): Naturschutz und gesellschaftliches Handeln. – Bonn- Bad Godesberg, p 23- 41.
11. Schuster, K. & LANTERMANN, E.-D. (2002): Naturschutzkommunikation und Lebensstile. - In: Erdmann, K.-H. & Schell, C. (ed.) (2002): Naturschutz und gesellschaftliches Handeln. – Bonn- Bad Godesberg, p 79 – 93.
12. Tauscher et al, 2003 : Bewertung von Lebensmitteln verschiedener Produktionsverfahren. Statusbericht. Münster.
13. Yussefi, M. & Willer,H., 2003 (ed),: The World of Organic Agriculture. Statistics and Future Prospects 2003. International Federation of Organic Agriculture Movements (IFOAM), Tholey-Theley ISBN: 3 934055 22 2 ; 127 pages.

CHAPTER 3
SUSTAINABLE DEVELOPMENT IN MOUNTAIN AND STEPPE REGIONS

PROBLEMS OF SUSTAINABLE DEVELOPMENT OF MOUNTAINOUS REGIONS OF TAJIKISTAN

H.M. MUHABBATOV.[1] AND H.U. UMAROV[2]

1: *Prof., Head of Department of Regional Economy Institute of Geography, Academy of Sciences, Republic of Tajikistan*

2: *Prof., Head of Department of Macroeconomy Institute of Economy Research under the Ministry of Economy and Trade, Republic of Tajikistan*

Keywords: Sustainable Development, Economic Circulation, Growth of Productivity, Natural Resources, Agriculture Crops, Row Materials, Transformation, Recession, Animal Husbandry, Ecological and Financial Institutions, Agricultural Reorganization, International Assistance

At present stage, i.e. in the third millennium humanity is at transitional stage of civilization's development. The transition is extremely uneven all over the world, and especially in mountainous regions. It aggravates already existing contradictions and complicates the problem of transition to sustainable development of nature and society.

From the strategic point of view, at present Tajikistan faces the problem of gradual transition to sustainable economic development. This problem is bound up not only with growing globalization of all spheres of the public life, but with those restrictions, which have arisen in the Soviet period, in the period of rapid development of large-scale industry and public sector in agriculture.

In accordance with standard nutrition norms, for the sustenance of 6.25 million people, it is necessary to produce 5.62 million tons of grain. But in 2002, only 700.1 thousand tons of grain was produced in Tajikistan, i.e. 2.5 times less than the necessary volume of production (we mean both food and fodder grain). Imported grain and the grain (including flour) received as the humanitarian aid, make about 600 thousand tons. The volumes of own production and received grain make up no more than 1.3 million tons, i.e. 1.6 times less than normative requirements. For production of normative volumes of grain at productivity of 30 centners from 1 hectare, 1,873 thousand hectares of irrigated arable land are required. However, the areas under crops make no more than 810 thousand hectares. In our country there are hardly any virgin tracts of land left for economic circulation. The dry-farming lands, which can be brought under cultivation, also remain less and less. Thus qualitative indices of such lands are deteriorating. Year after year natural fertility of soil is decreasing owning to reduction of volumes of imported mineral and organic fertilizers. Crop rotation and volumes of all chemical elements necessary for plants-nitrogen, phosphorus, potassium, etc. are also reduced.

Cereals yield of the last decade (1990s) has decreased from 14.3 to 13.1 centners per hectare, raw cotton from 27.6 to 19.1 centners, olive crops from 5.4 to 3.4 centners, tobacco – from 26.6 to 13.8, vegetables – from 193 to 117, corn and green forages – from 215 to 130.1, perennial herbs – from 67.7 to 39.4 centners, etc.[1]

In case of inertial development, the situation can worsen and would lead to further aggravation of food problem. For the improvement of situation, i.e. growth of productivity of agricultural crops, it is necessary to increase the use of mineral fertilizers and pesticides. In both cases population can lose and the natural potential will be damaged.

The situation in mountainous regions becomes more and more unmanageable. In mountains, destruction of natural resources goes at high rates, rare species of flora and fauna are on the brink of disappearance. The processes of deserting get large scales. Once woody slopes of mountains turn into bare deserts, deprived of any vegetation. The reason of this phenomenon is demographic boom in mountainous areas of Tajikistan, where the rate of annual natural increase of population for the last 40 years (right up to 2000) had made no less than 3.3%. A rapidly growing population of mountains, due to conducting a traditional way of life, absence of the organized maintenance of fuel and electricity, the low level of incomes from public sector, mainly use natural resources. Thus, any attention was not given to the fact that in some cases the results of massed pressure on natural resources might have irreversible nature. During the Soviet period, in all mountain systems of our country valuable relics of nature such as woods of juniper, pistachio, almond were ruthlessly cut down for some demographic and economic reasons (e.g. necessity of fulfilling plans). For the same demographic reason, the number of domestic animals has repeatedly increased and the processes of overgrazing on the permanent winter and spring-summer pastures resulted in reduction of natural fodder resources of mountains. In consequence, the high pressure on soil has sharply increased the scales of erosion of pastures and their

---

[1] See: Agriculture of Republic of Tajikistan. – Dushanbe, 2003. – p.37-38

productivity has decreased. Fertile layers, of soil on the slopes are washed down by precipitations and soil formation is practically brought to nothing. In the nearest future, southern slopes of Turkistan ridge, eastern slopes of Karatau and Sorsarak will turn into solid mountainous deserts with rare vegetation. The number of wild animals and birds are reduced hundreds of times. After the disintegration of the USSR because of discontinuance of coal delivery, irregular electricity supply to mountainous regions, deteriorations in forage and food grain supply, demographic pressure on mountain ecosystems has repeatedly grown. In essence, the destruction of biological resources is going on in mountain areas. Deterioration of life-support systems in mountain areas, among which biological resources began to play a determining role during the last years, led to growing scales of labor migration of population. Due to this, there could be more negative phenomenon – depopulation of mountains. As for the development of cities and industry, in recent years owing to deindustrialization and deurbanization the levels of water, air and ground pollution have been reduced. For the period from 1991 to 2000 the emissions of harmful substances into the atmosphere were reduced from 100.5 thousand tons to 30.8 thousand tons, the emissions of solid substances – from 28.7 to 5.3 thousand tons, the gaseous and liquid emissions – from 71.1 to 25.5 thousand tons. The most appreciable was the reduction of sulphureuos anhydride – from 17.0 to 1.2 thousand tons, oxide of nitrogen – from 7.1 to 0.6 thousand tons, hydrocarbon – from 2.1 to 0.8 thousand tons.[2]

As it is well known, an economic growth can be observed in our country since 1997. There is no doubt that this tendency will carry not only long-term, but increasing character. Unfortunately, the country doesn't have precise conception of restoration of the industry until now. A technocratic approach to restoration processes in the industry can increase the risk of instability of development in our country.

Year by year the manufacture of non-ferrous metals, on the basis of local mining raw materials will demand the increase in ore output counting upon unit of final effect, manufacture of nitric fertilizers demand more and more expenditure in transition to exploitation of domestic deposits of natural gas, manufacture of cotton will demand the expansion of the ground area and planting monocultures, because of deterioration of domestic fertility of ground, etc. But the most dangerous is that the expansion of production costs will lead to the growth of waste products, polluting the environment and causing damage to qualitative aspects of population's well-being.

---

[2] Statistical annual of Tajikistan. – Dushanbe, 2003. – p.140

A conclusion might be the planning of a national economy, overcoming of transformation recession, and an acceleration of economic growth. Thus, the applications of non-conventional approaches, which under condition of successful realization are capable to provide sustainable development of economy, become necessary.

The basis of sustainable economy "should be the new structural principle meaning transition from irreversible exhaustion of natural resources to renewed energy sources and continuous repeated using and processing of salvage". This definition, given by scientists L. Brown and K. Flovin is concretized by examples of such order: "In sustainable economy volume of fish catch should not exceed of the volume of stable increase of fish resources, speed of pumping water out of underground water layers – speed of water regeneration, rates of soil erosion – natural rates of soil formation, rates of deforestation – rates of planting green plantations, and emissions of carbon – ability of nature to tie contained in atmosphere C^o".[3]

There is no doubt about validity of such conception of economic stability. Its application in condition of Tajikistan, where the demographic boom is not ceased yet, and process of destruction of natural resources proceeds, is connected with great difficulties.

Hence, there is a conclusion that a variant of development, which can find the expression in cause – and effect – relationship, such as "restoration of economy, acceleration in the rates of economic growth, acceleration of birthrates and a natural increase of the population", is inadmissible. Logically, this variant proceeds from the logic of interrelations between changes in economy and in demographic development. If present transformation recession assists falling of birthrate, then beginning economic growth and increase of population's prosperity will lead to the increase in population. Such interrelation expresses prevalence of traditional type of reproduction of population in Tajikistan. Hence, follows a conclusion about necessity of radical change in the described interrelation.

It is necessary to achieve full change in proportions of dynamics between economy and the population. The idea of such change is that the revival of economy, its restoration, increase in rates of economic growth, should be accompanied by deceleration of population's increase and formation of slightly extended reproduction of population. It means transition of reproduction from traditional to modern type.

---

[3] See: Sustainable economy. The office of regional programs. – Vienna, 2001. – p.15

Measures on restoration of industry must be taken on the basis of the corresponding conception. There is no such conception for the present, but in view of sustainable development it is possible to speak about its certain aspects.

First of all, it is necessary to approach branches of industry differently, as the real potential of sustainable development and economic prospects of separate branches differ mainly. The engineering and metal-working industries can be exposed to technical updating and restructuring, taking into account that namely these branches can form the technical basis for stable development of power engineering and agriculture. According to the economic plan the given branches will be able to perform two more functions: import substitution and replenishment of domestic market with technical goods. They also could produce the equipment for drop and other progressive kinds of irrigation, means of mechanization, hand and electric implements for agricultural, storage and transport works. The branches in question can be specialized in production of a complete set of equipment for all types of small and micro hydroelectric power stations with various capacities (from 0.5 up to 3000 kw) so that by 2015 to electrify completely the rural areas, manufactures, constructions, agriculture and transport.

Besides, the surplus of electric power in mountainous region of Tajikistan would provide ample opportunity for taking large-scale nature protecting and restaurating measures. So it is necessary to develop wood restoration works in all mountainous regions, because during the last years for the lack of coal, black oil, liquefied gas and electric power the scales of wood felling have repeatedly increased. For the last 10 years wood restoration works were seriously slowed down.

The volumes of wood restoration from 1991 to 2002 were reduced from 4.0 to 2.2 thousand hectares, i.e. two times. In 1991 planting and sowing of woods were carried out on the area equal to 3.9 thousand hectares, in 2002 – on the area of 1.7 thousand hectares. A natural wood restoration was carried out on the area of 1.7 and 0.6 thousand hectares.[4] These works are scanty in comparison with required scales of wood restoration. It should be mentioned that mountain territories make up 93.0% of all the territory of country, whereas wood area makes up 401 thousand hectares or 3.0% of all the territory of Tajikistan.

---

[4] Statistical annual of Tajikistan. – Dushanbe, 2003. – p. 140

It is necessary to start a large-scale creation of building, energy and biological plantations simultaneously with wood restoration works. The first are necessary for production of building wood, the second – for fire – wood, the third – for increase of fodder resources and supply of green fertilizers. It is impossible to create such woods without developing power engineering on a large scale. But the destructive processes in nature will assume irreversible character, if such woods and plantations are not created.

The seedlings on the given plantations will be irrigated by means of electric pumps and small hydroelectric power stations. Irrigation works will continue until roots reach a water-bearing layer of ground.

The same equipment is needed for realization of restoration works on the summer pastures which are in very erosive condition.

As regards other branches of industry, each of them has its own specificity of development. The mining industry can develop with due regard for strategic demands of world market for lead, zinc, tin tungsten, antimony, mercury and etc.

Demand for some elements (for example lead and mercury) will decrease, for others (tin and tungsten) it will possibility increase. In any case, it is necessary to use technologies causing no harm to society and population. The introduction of such technologies (especially biochemical processes) requires a complete reconstruction of mining and processing industries. Our country has no sufficient financial resources for realization of reconstructions. Foreign investments should be accompanied by transfer of pure technologies, i.e. those completely preventing pollution of environment and deterioration of people's health.

The agriculture of Tajikistan should make a step-by-step transition to sustainable development, depending on favorable technical, economic and social preconditions. A sharp reduction in volumes of mineral fertilizers and chemical weed and pest killers will lead to irreversible results. In recent years, reduction in volumes of agricultural deliveries caused serious damage to the country's economy. In 1980, the supply of mineral fertilizers made up 235.5 kg counting upon 1 hectare of areas under crops. In 2000 accordingly such deliveries made up 114.0 kg, i.e. were reduced more than 2 times. In the same year nitric fertilizer deliveries were reduced 1.8 times. For the last 10 years there almost was no supply of phosphate, potash and combined fertilizers. For the above-mentioned period there was sharp drop in the level of crop yield and animal husbandry (owning to reduction in fodder production). Accordingly, there was decrease in volumes of organic fertilizers, counting upon a unit of areas under crops. All that decreases and reductions were caused by crisis in fodder production, cattle breeding and also growing use of manure as wood-fire.

At present Tajik authorities cannot approve the offers of international ecological and financial institutions as to reduce gradually volumes of using mineral fertilizers and pesticides, go over to using green and organic fertilizers, and apply biological methods of plant protection. Such agricultural reorganization demands time and cannot be realized without international assistance. The contrary approach may lead to further intensification of crisis in Tajikistan, where 85% of the population lives in poverty. Such approach may transform the country into a zone of social disasters.

Therefore, with the purpose of ensuring food safety it is necessary to use alternative methods of "descent and braking".

At first it is necessary to increase the volumes of delivery of mineral fertilizers and chemical weed and pest killers, and then to start their gradual reduction to a limit that can provide ecologically safe production of agricultural products. In parallel with these quantitative shifts, it is necessary to emphasize on realization of qualitative ones such as improving the quality of mineral fertilizers, their assimilation by plants, and irrigation of croplands. It is not less important to introduce the crop rotation and realize scientifically-grounded actions on prevention of soil erosion. In additional fertile tracts of land, which are exposed to water erosion, should be transformed to green plantations. The means for such actions should be allotted by the state budget and external humanitarian aid.

Crucial preconditions for sustainable economic development are:
a) technology and information transfer;
b) international financial aid;
c) effectively functioning institutions;
d) a favorable public opinion.

There is no doubt, that industrial countries should support developing countries. They should share with information on methods and technologies of effective nature management, ecologically pure industrial systems and ways of transition to sustainable economic development. The agriculture of Tajikistan requires transfer of knowledge and technologies necessary for restoration of forest tracts in the mountainous regions, and also technologies for substituting chemical methods of plants protection biological ones. If these technologies are not transferred to developing countries, then existing demographic tendencies will aggravate ecological crisis within this decade.

The same situation takes place regarding financing of processes of transition to steady economic growth. Tajikistan as well as overwhelming majority of developing countries has not necessary financial resources for timely realization of such transition. Therefore, industrial countries and international financial organizations should help developing countries with financial resources, sufficient for their timely transition to steady economic growth.

Tajikistan has a special need for ecologically effective technologies on transformation of coal into gaseous fuel for production of automobile accumulators, solar batteries, the equipment for wind power stations, automobile and tractor engines of internal combustion, engines functioning on fuel cells, for production of biopower installations both for separate families microdistricts and separate village.

REFERENCES

1. Agriculture of Republic of Tajikistan. – Dushanbe, 2003. – p. 37-38.
2. Statistical annual of Tajikistan. – Dushanbe, 2003. – p. 140.
3. Sustainable economy. The office of regional programs. – Vienna, 2001. – p. 15.
4. Statistical annual of Tajikistan. – Dushanbe, 2003. – p. 140.

IMPLICATION OF ENVIRONMENTAL LAW FOR MOUNTAIN PROTECTION IN CHINA

JIEBIN ZHANG
Xinjiang Institute of Ecology and Geography, Chinese Academy of Sciences, China

Abstract: The paper examines the general provisions relevant to mountain protection in China environmental law in terms of substantive rules, procedural rules and institutional mechanisms. It also reviews some implementation regulations and procedures, and other basic laws relevant to the mountain protection. The results reveal that the China environmental law contains most substantive norms relevant to mountain protection. However, it lacks of a comprehensive legal instrument and relevant procedural norms necessary to support the integrated mountain protection and management. As for the institutional mechanisms, it adopts the administration by various departments rather than by a unified administration institution. Such mechanisms are inappropriate to the required integrated mountain protection and management. These findings suggest that the legal framework of environmental protection and management has been established and environmental law can generally apply to the mountain protection and management. However, a specific regulation should be formulated in light of integrated mountain protection and management. In particular, the regulation should be formulated in consideration of good governance and by integrated approach, emphasizing the issues with which China is recently confronted and should deal.

1. INTRODUCTION

China is a mountainous country with large population depending on the mountains. Mountains are an important source of water, energy and biological diversity in China. Also, they are a source of such key resources as minerals, forest products and agricultural products and of recreation. During the 1950's and 1960's, China accelerated its development of mountains guided by the then policies and five-year plans for national social-economic development. Mountain ecosystems were rapidly changing towards to degradation. The major issues, on the one hand, were accelerated soil erosion, landslides and rapid loss of habitat and genetic diversity. There, on the other hand, poverty was widespread among mountain population. Chinese government recognized these issues as early as the beginning of the 1970's, and adopted a series of policies to prevent such degradation and tackle the poverty issues. However, with a new country-wide accelerated development beginning at early 1980's, most mountain areas have been experiencing environmental degradation again. Hence, the coordinated management of mountain resources and socio-economic development is imperative in accordance with the policy and legislation gradually established in China in particular the environmental law, in order to protect and reserve the mountain ecosystem and sustain human livings.

With the reestablishment of China's legislation in the late 1970's the Environmental Protection Law of the People's Republic of China was adopted in 1989. It has laid the foundation for good environmental protection and management in China. Furthermore, China has formulated a lot of regulations and procedures to embody this law. In despite of this environmental protection law, China also enacted other basic laws in light of natural resources development and management, which also contain provisions relevant to the specific subjects in the environmental protection, especially the ecosystem protection. Although China has not formulated any specific law and regulations concerning mountains, it considers that the existing environmental and resources legal system could apply to the mountain protection and management. Based on the checklist of existing legal system, the paper is aimed to examine what are the substantive and procedural rules provided in the environmental and resources laws and how are they relevant to integrated mountain protection. At the same time, the paper also examines the issues of existing environmental protection laws and regulations as well as institutional mechanisms, in order to provide some necessary recommendations.

2. CHINA'S ENVIRONMENTAL LEGISLATIVE SYSTEM AND RESULTS

2.1 Legislative system

In China, the term "legislation" means specially defined activities by a specially appointed organ to work out, recognize and change laws and regulations by using its designated rights, following certain procedures and applying the necessary technique. China's current environmental legislation system has basically three levels, e.g. by the National People's Congress and its Standing Committee, the State Council and its relevant departments and the

Provincial People's Congress and its Standing Committee. As provided in the Constitution, the National People's Congress and its Standing Committee are entitled to exercise the legislative power of the State and enact basic laws. The State Council and its relevant departments are entitled to enact administrative rules and regulations and issue decisions and orders. The provincial People's Congress and its Standing Committee formulate and adopt the local regulations. Figure1 shows in detail.

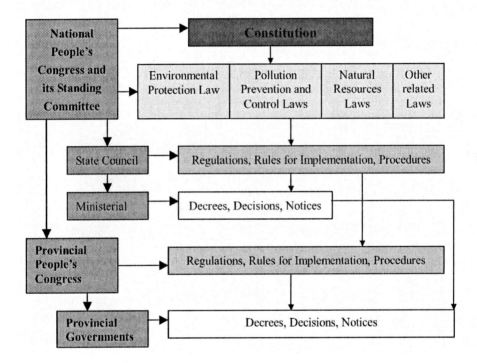

Fig. 1: China's legislation structure and environmental and resources law legislation

2.2 Results

Since the 1980's, China has gradually enacted and formulated a series of laws and regulations by its three legislative bodies to govern and regulate the activities in environment protection and resources management. The following provides a checklist of the major laws and regulations relating to the mountain protection and management.

Basic Environmental Laws

- Environmental Protection Law of the People's Republic of China (adopted: 1989, effective: 1989)
- Law of the People's Republic of China on Prevention and Control of Water Pollution (adopted: 1984, amended: 1996, newly effective: 1996)
- Law of the People's Republic of China on the Environmental Impact Assessment (adopted: 2002, effective: 2003)

Basic Resources Laws

- Forestry Law of the People's Republic of China (adopted: 1985, amended: 2002, newly effective: 2003)
- Grassland Law of the People's Republic of China (adopted: 1985, amended: 2002, newly effective: 2002)
- Law of the People's Republic of China on the Protection of Wildlife (adopted: 1988, effective: 1989)
- Fisheries Law of the People's Republic of China (adopted: 1986, effective: 1986)
- Water Law of the People's Republic of China (adopted: 1988, amended: 2002, newly effective: 2002)
- Law of the People's Republic of China on Water and Soil Conservation (adopted: 1991, effective: 1991)
- Mineral Resources Law of the People's Republic of China (adopted: 1986, amended: 1996, newly effective: 1996)
- Land Administration Law of the People's Republic of China (adopted: 1986, amended: 1998, newly effective: 1999)
- Law of the People's Republic of China on Desert Prevention and Transformation (adopted: 2001, effective: 2002)

Other Relevant Laws

- Law of the People's Republic of China on the Protection of Cultural Relics (adopted: 1982, effective: 1982)

- Agriculture Law of the People's Republic of China (adopted: 1993, effective: 1993)
- Administrative Punishment Law of the People's Republic of China (adopted: 1996, effective: 1996)
- Law of the People's Republic of China on Administrative Permission (adopted: 2003, effective: 2004)

Important Regulations

- Regulations on Restoring Farmland to Forest (2002)
- Regulations of the People's Republic of China on Natural Reserves (1994)
- Regulations of the People's Republic of China on Wild Plants Protection (1996)
- Regulations for the Implementation of the People's Republic of China on the Protection of Terrestrial Wildlife (1992)
- Interim Regulations on Administration of Scenic spots (1985)
- Procedures for the Management of Forests and Wild Animals Species Nature Reserves (1985)

3. GENERAL NORMS RELATING TO MOUNTAIN PROTECTION

3.1 Substantive norms

The substantive norms mean those that establish rights and obligations concerning the environment and management. The review of existing environmental and resources law system and other relevant laws outlines the following norms, which can be defined as substantive: (1) coordinated development and environmental protection (2) obligation to environmental protection and environmental damage prevention (3) prevention and control of pollution and "polluter pays" (4) reasonable utilization and protection of natural resources (5) protection and preservation of historical and cultural heritage and natural scenic spots.

As the cornerstone law in the field, the environmental protection law generally provides those norms under its general provisions. In particular, it provides coordinated development and environmental protection under article 1. Under article 6, the law provides that "all units and individuals shall have the obligation to protect the environment and shall have the right to report on or file charges against units or individuals that cause pollution or damage to the environment." The norm of prevention and control of pollution and "polluter pays" is highlighted through its Chapter Four entitled Prevention and Control of Environment and Public Hazards and explicated in the series of pollution prevention and control laws concerning water, atmosphere, solid waste, noise and radiative materials. The norm of reasonable utilization and protection of natural resources is provided in its Chapter three entitled Protection and Improvement of the Environment and explicated in the series of resources laws. Also, this chapter provides the norm of protection and preservation of historical and cultural heritage and natural scenic spots and it is detailed in the Law on the Protection of Cultural Relics, especially in Article 2 which provides that "the state shall place under its protection, the cultural relics of historical, artistic or scientific value."

3.2 Procedural norms

As opposed to "substantive norms", the procedural norms prescribe the method or process for the implementing or the enforcement of substantive norms, or anything dealing with it, principally with procedures which the unit and individuals must follow. Following is the checklist of some major norms: environmental impact assessment, environmental protection planning, "Santongshi", environment monitoring and reporting, pollution discharge reporting and registering, fact-finding, collection of pollution discharge fees, limited period for pollution reduction and control, timely notification of an accident or any other exigency, natural reserve establishment, administrative and criminal punishments.

The environmental protection law refers to all these procedural norms, however, it emphasizes those pollution prevention procedures. In particular, the environmental impact assessment and "Santongshi" are emphasized in the relatively detailed provisions, of which "Santongshi" means that the installations for the prevention and control of pollution at a construction project must be designed, built and commissioned together with the principal part of the project. These two procedural norms have been further detailed in the recently adopted Law of the People's Republic of China on the Environmental Impact and 1998 Regulations on Environmental Protection and

Management of Construction Projects. Other norms relating to pollution have been well developed in the relevant pollution laws and regulations, rather than the environmental protection law. Although the law refers to the function of nature reserve in environment protection, there are no provisions provided for the procedures. The natural reserve establishment procedures are detailed in the 1994 Regulations of the People's Republic of China on Natural Reserves and resources laws concerning forest and wildlife rather than the environmental protection law.

4. MAJOR ISSUES AND RECOMMENDATIONS

The environmental protection law is the cornerstone law in the field and generally applicable to all activities in environment protection and management. However, integrated mountain protection and management are not defined in the scope of this environmental law. In addition, in application of this law, the issues rise in the ecosystem protection and prevention. The law highlights the pollution prevention and control rather than the ecosystem protection and prevention. Most substantive and procedural rules for ecological environment protection and management are provided in the individual natural resources laws and regulations, and other relevant laws and regulations. This is disadvantageous to the mountain protection and management which require adopting an integrated approach and an integrated legal regime. In order to supply this gap, the author strongly recommends formulating and adopting the Regulations for Mountain Protection and Management, based on present environmental and resources legislation. The Regulations should clearly define the scope and incorporate all existing substantive and procedural rules along with the detailed explanations. At the same time, the Regulations should also add two new substantive norms—top-priority of ecosystem protection and the obligation not to cause transboundary environmental damages, and three procedural norms—prior notification and consultation, information sharing and exchange, cooperation and dispute settlement.

It has been recognized by the central government, that the fragmental responsibilities have impeded the integrated environmental protection and management in China. The sectional management has been unified to environmental protection agencies after the adoption of the environmental protection law and successive institutional reform of the State Council all over the country of China. As provided in the environmental protection law, the competent department of environmental protection administration under the State Council shall conduct unified supervision and management of the environmental protection work throughout the country. The competent departments of environmental protection administration of the local people's governments at or above the county level shall conduct unified supervision and management of the environmental protection work within areas under their jurisdiction. The competent administrative departments of land, minerals, forestry, agriculture and water conservancy of the people's governments at or above the county level shall, in accordance with the provisions of relevant laws, conduct supervision and management of the protection of natural resources. However, the practice reveals that it is difficult to define a competent authority for the integrated mountain protection and management. Also, principle and subordinate relation between the various authorities is not clearly defined. In fact, it does not greatly change the administration by various departments into a unified administration institution. Mountain protection and management is artificially segmented. Such mechanisms are inappropriate to the required integrated mountain protection and management. How can the effective integration be ensured between the competent stakeholders? How to avoid the long-standing action on its own will of each authority and take the effective coordinated action, in particular the mountain ecosystem protection, is still a challenge in China. Given such uncertainty, the author suggests that a mountain committee should be established under the existing environmental protection commission of the State Council. The committee should be legally defined in the envisioned Regulations and mainly exert coordinative functions between the departments under the State Council and provincial governments. Also, the participatory approach should be adopted to include all stakeholders in decision-making. Figure 2 shows the detailed suggestions.

5. CONCLUSION

Issues of mountain ecosystem degradation and human poverty have been recognized by the Chinese central government and various governments at lower levels since early 1970's. They have adopted a series of policies to prevent such effects. Legally, governments consider that the existing environmental and resources legal system could apply to the mountain protection and management. Based on the checklist and review of the existing environmental legal system, it is found that the system is disadvantageous to the mountain protection and management, which require adopting an integrated approach and an integrated legal regime. In particular, institutional mechanisms are inappropriate to the required integrated mountain protection and management. In order to supply this gap and protect the ecosystem along

Implication of Environmental Law for Mountain Protection in China 157

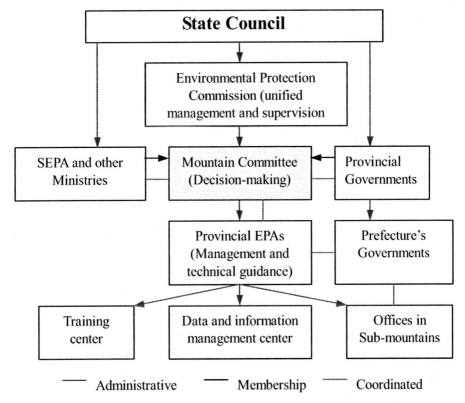

Fig. 2: Mountain Committee and its functions

with sustaining the human livings in the mountains, the author strongly recommends formulating and adopting the Regulations for Mountain Protection and Management, based on present environmental and resources legislation and internationally accepted norms specific to environment protection. The Regulations should also legally define the institutional framework such as envisioned Mountain Committee under the State Council. In the same time, the Regulations should emphasize on providing the effective mechanisms for information sharing and exchange, participatory approach in decision-making, and cooperation and dispute settlement.

REFERENCES

1. Department of Policies, Laws and Regulations of State Environmental Protection Administration (SEPA), Collections of Laws and Regulations of Environmental Protection in China (1982-1997), Chemical Industry Press, Beijing, 1997.
2. Department of Policies, Laws and Regulations of SEPA, Collections of Laws and Regulations of Environmental Protection in China (1997-2001), Chemical Industry Press, Beijing, 2001.
3. Department of Policies, Laws and Regulations of SEPA, Collections of Laws and Regulations of Environmental Protection in China (2001-2002), Chemical Industry Press, Beijing, 2002.
4. Department of Policies, Laws and Regulations of SEPA, Collections of Laws and Regulations of Environmental Protection in China (2003-2004), Chemical Industry Press, Beijing, 2004.

ACTUAL ECOLOGICAL SITUATIONS IN THE TERRITORY OF MOUNTAIN REGIONS AND BIODIVERSITY PROBLEMS (THE CASE OF GEORGIA)

Z. TATASHIDZE, I. BONDYREV & E. TSERETELI
Vakhushti Bagrationi Institute of Geography, Tbilisi, Georgia

Abstract: The complicated social-economic and political situation has brought about an abrupt deterioration of the natural environment in Georgia for the last decades. At present, 4.5 million ha of the territory of Georgia is under the threat of dangerous elemental processes. As a result of activation of the exogenous processes, stipulated by the same factors, 5% of arable lands are withdrawn yearly from agricultural turnover, and 50% is affected by erosion. Under the impact of anthropogenic factors, about 100-130 t/ha of fertile soil layer of arable lands is annually washed-out in Eastern Georgia and about 150-160 t/ha in Western Georgia. The total number of mudflow basins in Georgia exceeds 2.7 thousand. The peculiarity of mountain relief and climate has stipulated not only the presence of various landscapes, but also a high level of biodiversity within the organic world of Georgia. A successful solution of the problem of balanced development and the conservation of biodiversity depends on two main factors - influence of elemental-destructive processes and anthropogenic impact, which are directly bound with each other.

Keywords: ecologic; climate; seismic activity; floods; glaciers; landslides; mudflows; biodiversity; sustainable development; Georgia.

1. CHANGE OF CLIMATE

Climate warming all over the planet, as it is ambiguously estimated both by criminologists and scientists of allied scientific disciplines, is characterized by the unevenness of its spatial-provisional dynamics. The best example to this is the territory of Georgia, western part and highlands of which have a trend towards warming, weakened by the influence of the Black Sea. The situation is quite different in Eastern Georgia, within which a constant rise of air temperatures is registered, on an average of 0.006° C per annum, and a reduction of rainfall during the summer (12-15%) [28].

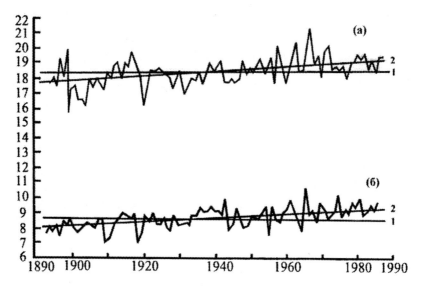

Fig. 1: Change of mean minimum (a) and maximum air temperatures in the city of Tbilisi for the last 100 years. 1- norm, 2- trend [14]

In our opinion, the change of climate in Eastern Georgia and desertification processes caused by these changes are largely of anthropogenic nature.

2. SEISMIC ACTIVITY

The territory of Georgia, on the whole, is situated in a magnitude of 7, and its mountain regions in an 8-9-magnitude zone. Herewith, the area of discharge is timed to neotectonic upstanding blocks. The energy of stress release comes to about 80 bars ($8 \cdot 106$ din/cm^2) [13,32]. An earthquake in the Racha-Imereti region on the 29.04.1991, covered the territory at the joint of the east block of Okriba-Khreiti tectonic zone and the Gagra-Javakheti eastern segment. The epicenter was a tectonic block situated on the south slope of the Racha ridge, an area of about 7,8 thousand $km/^2$ On the whole, more than 700 populated areas happened to be in the earthquake zone.[4,9,13] More than thousand inhabitants of this region perished and suffered damages. About 100 thousand inhabitants were deprived of their dwelling houses. A powerful stone avalanche, formed as a result of earthquake, completely destroyed the village of Khakhieti (Sachkhere region). The general material damage was estimated to be 10 billion dollars. 80% of blocks of houses and buildings situated in the epicenters of earthquakes, which rocked Georgia in 1991-1992, were distorted and unfit to reconstruction. 20 thousand new landslides and collapses were formed simultaneously [13,20].

At present, we face the period of sharp seismic activity that threatens the country with new cataclysms.

Herewith it should be noted that more than 25% of housing resources of large cities of Georgia, where more than 80% of the population is concentrated, are under emergency conditions. Hence, we may conclude that Georgia as well as other countries of the Caucasus is absolutely unprepared for disastrous earthquakes, consequences of which could be tragic for the large cities.

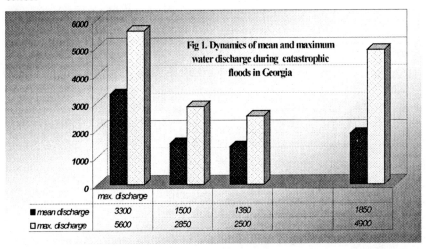

Fig. 2: Dynamics of mean and maximum water discharge during catastrophic floods in Georgia [10,25,26,31]

3. HIGH WATER AND FLOODS

There are about 32198 rivers in the territory of Georgia, 75550 km of total extent (out of which 61% - is in Western Georgia) and 600 lakes. General mean annual river discharge is 55km^3, module of discharge is – 24.2 l/sec · km^2.

Disastrous floods are bound either with sharp warming after snowy winters and intensive snow thawing or with the outburst of dammed lakes in the nival-glacial zone.

Thus, as a result of the disastrous flood on the Rioni River during 1811-1812, the number of the population in Imereti was reduced to 30-35%. The data of later years show that a rise of the water level in the Rioni River first of all was stipulated by the enumerated above reasons that took place in 1839 and reached 9,6m, in 1911 - 2-3m, in 1922 - 2,8m (water discharge was 2420 m^3/sec), but in 1968 - 7,7m (water discharge was 5220 m^3/sec). As a result of floods in 1982, a territory of 130 km^2 was inundated and damage was estimated to 12 million US dollars [29-33].

A flood in Western Georgia in January 1987, has caused the inundation of about 200km^2, damaged 3.2 thousands and completely destroyed 2 thousand dwelling houses and 650 public construction works, 1.5 thousand hydro-technical structures, 16,5km railway tracks, 1.3 thousand kilometers of highways; demolished 1.1 thousand kilometers of power transmission lines and 0.7 thousand kilometers of communication lines and more than 16 thousand people were evacuated. The total damage caused was about 300 million US dollars.

Repeated disastrous floods and freshets were noted in the Kura River basin. Heavy rains and hail in 1968 resulted in landslides in the Borjomi gorge, which blocked the Kura River valley. The outburst of the bulkhead has brought a catastrophic water drain with a discharge of 2450 m^3/sec. Freshets in 1982-1992 also had disastrous nature and inflicted great damage to the economy of Georgia.

As a result of violent earthquakes, block landslides, colossal mudflows (including glacial), glacier shifting, etc., retaining lakes are occasionally formed on mountain rivers. Some of them have already existed for a long time (Ritsa, Amtkeli, Kvedrula), but the time of their existence varies from a few days (Devdoraki, Terek, et al.,) and some months.

Thus, the burst of glacial lakes in the valleys of Devdoraki, Amali and Gveleti glaciers (the Terek River basin), repeatedly brought about disastrous floods. A similar situation took place at the burst of retaining lakes in the Irony River upper course, as well as in the basins of the Gordjomi, Atsgara, Skhaltba, Patsa, Khakhieti, and Dzhruchula et al., rivers [17].

In South and South-East Georgia, large floods were noted on the Paravani and Khrami River basins. The activity of high waters exceeds the indices of bygone years on the shallow water rivers of this area about 2-3 times.

4. GLACIAL AND PERIGLACIAL PROCESSES AND PHENOMENA

In the mountains of Georgia, there are registered about 786 glaciers, with a total area of about 550km^2 [15,16,24]. Out of them, 82.5% is in the Kodori, Inguri, Rioni, and Terek Rivers' upper courses.

For the last 150 years, significant glacier retreat (0.8-1,7 km)[1] and shrinking of 16% of their area have been observed. Since the middle 1940s of the 20 th century [24], the glaciological situation has been characterized by a sharp reduction of glacial areas, with the simultaneous increasing in their number, on account of the disintegration of united glaciers into separate smaller ones, though at the same time separate movements took place.

At some transshipping sections of Georgian highways, the material damage due to elemental-destructive processes as snow avalanches and collapses effected that 50% from the total freight turned over to motor transport.

About 31% of the territory of Georgia is subjected to avalanches (18% - in Eastern and 13% - in Western Georgia). In the Terek, Argun, Assa River basins, more than 70% of the territory is subjected to avalanches; in the Bzibi, Kodori, Chkhalta and Inguri Rivers - about 50%, and up to 25% in the Khobi, Alazani and Iori River basins [18].

5. LANDSLIDE AND MUDFLOW PROCESSES

At present, about 4.5 million ha of the territory of Georgia is under the threat of elemental natural processes. As a result of active exogenous processes, stipulated by these factors, 5% of arable lands are withdrawn yearly from agricultural turnover, and about 50% are damaged through erosion. Under the influence of anthropogenic factors from farmlands of Eastern Georgia, about 100-130 ha of soil layer is washed down, in Western Georgia approximately 150-160 t/ha a year. During torrential rains, these indices rise 2-4 times. On the whole, the territories in Georgia, affected by elemental processes, had been constantly increasing and reached 4.7 million ha by 1996.

Out of 5 thousand landslide basins registered in the Caucasus, more than half – 2.7 thousand - are in Georgia. Large mudflow hearths are registered in the Inguri, Tskhenistskali, Rioni, Alazani, Bolshoi Liakhvi, Aragvi, Terek et al., river basins, where more than 60% of the country's population is

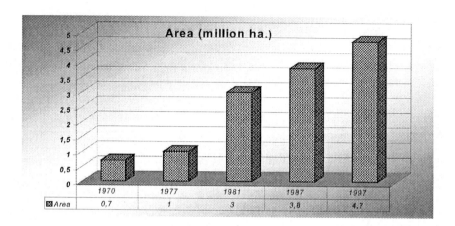

Fig. 3: Area of the territory of Georgia subjected to dangerous geodynamic processes and phenomena [3,5,6,729,31]

concentrated. For the last 220 years, about 37 mudflow occurrences of disastrous nature were registered in Georgia. A single evacuation of mudflow material quite often exceeds 1 million. Thus, the volume of mudflow on the Duruji River on 14.09.1961 was 1,2 millionm^3 [10], whereas a disastrous mudflow on the Terek River in August 1967 evacuated some million m^3 mudflow material. The most dangerous is the Duruji mudflow, which threatens to destroy the town of Kvareli. For the last 100 years, this small borough was subjected to destroying 15 times, and its present-day existence is endangered because of missing required means and facilities for the reconstruction of destroyed mudflow-resisting buildings. And, as an example of such catastrophe, there should be noted a grandiose disastrous mudflow in the Carmadon River valley in the North Caucasus, killing 113 people.

Under the conditions of the Atachara mountain regions, the amount of evacuated material under the influence of gully erosion on the slopes with the gradient up to 150 is 02-0m^3/km^2, but on the slopes with the gradient over 200 it is 0,05-0,69 m^3/km^2. In Abkhazia these indices are accordingly 0,2-6,0 and 0,07-0,38 m^3/km^2, but in Imereti they are 0,02-0,6 and 0,03-0,7 m^3/km^2 [21,25,26].

The events of the summer 2004, (continuous and abundant precipitations), brought an excessive activation of geodynamic processes in the mountains of the Greater Caucasus and created an extremely complicated geo-ecological situation in this region. Not only that landslides and mudflows, occurring along the Georgian Military Road (20 June), caused its closing for several days in the area of Mleta. Also, the development of the probably most powerful mudflow basins in the Caucasus had to fight with manifestations that the activity is impossible by means of technical or any other facilities, because of their extent (Photo 1 and 2).

All this practically means that the village of Mleta and its population is destined to be relocated to a safer place.

Photo 1: 1 km Extent Debris Cone at the Village of Mleta

The damage from torrential rains which took place in the Svaneti region was estimated at 2 million US dollars. In Mestia region, 80% of hayfields were destroyed, 100% sowings of potatoes, 5 dwelling houses (65 became useless), and defensive dams in the Mleta-chala River valley were ruined as well as all the bridges and separate parts of highways. Svaneti region is completely deprived of electric power, since the majority of power transmission line supports were completely swept away [12,23].

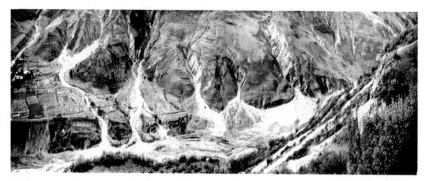

Photo 2: General View of Mleta Mudflow Region

6. NEGATIVE ATMOSPHERIC PROCESSES AND PHENOMENA

Hail should be noted amongst such phenomena, which not only inflicts enormous damage to agriculture but also is one of the reasons causing manifestation of mudflows.

Photo 3: Consequences of elemental processes in Svaneti Region

In Georgia, the length of hailstorms varies in broad ranges, depending on geographical positions and heights above SL. Thus 2-10 minute hailstorms are characteristic of Western Georgia regions and Shida Kartli, 15-20 minute storms - for the regions of Kakheti and Lower Kartli. The maximum length is noted in Kazbegi region and at Kakheti Caucasus, where it lasts 1-1,2 hours.

In the territory of Eastern Georgia, one of the negative phenomena of hazardous nature is drought. 134 dry months (12 %) were registered in Georgia in 1900 - 1990, during which the period air temperature rose by 4-50° C in comparison with the mean perennial; the relative air humidity was 30% lower, and the wind velocity reached 5-10m/sec. Particularly arid were the years 1921, 1946, 1963, 1981 and 1989. Herewith, severe droughts (<100mm precipitations a year) were 1-3%, and strong droughts (<150mm) 40%.

A severe drought was noted in the summer of 2000, when the whole territory of Eastern Georgia was declared to a zone of ecological disaster. Only 20% of grains and about 5% of sunflower were harvested. For neutralization of these phenomena, irrigation systems of the country consume annually 9 to 12% of the rivers' run-off. 3.608 million m^3 fresh water is spent on irrigation, waste accounts for 50-65%.

The irrigation systems in Eastern Georgia are almost ruined and require urgent reconstruction works which is hindered by the absence of means.

The annual damage inflicted to the country by these phenomena during the years of their background manifestation, was estimated on an average at 250-300 million dollars, during the years of their extreme activity (repeated every 3-5 or 8-11 years) -at 0.6-1,0 billion US dollars. Only during the years of 1987-1992, this number reached an astronomical value - 10 billion dollars.

All this once more testifies the interconnection among these processes and phenomena and requires a thoughtful approach to their study as well as the development of practical actions for an elimination of the consequences of natural catastrophes.

7. BIOLOGICAL AND LANDSCAPE VARIETY

There are 97 types of diverse landscapes in the territory of the country. Consequently, for every 10 000 km^2, there occur 14 types of natural landscapes on an average. Analysis of information shows that most diverse natural landscape-types are situated in the Shida Kartli region- 36 types for 10 000 km^2. Then comes the region of Gombori ridge - 35, the mountain part of Atchara-Trialeti ridge ranks third - 34 types [6,7]. This is stipulated by the fact that, only in these regions, an abrupt transition from humid or dry subtropical plains to subalpine and alpine landscapes of the Atchara-Trialeti range in the south slope of the Eastern Caucasus takes place, and, additionally, the climate dividing effect from the Gombori Range (with slightly developed subalpine landscapes), rising above plains surrounding it. Consequently, the defining role belongs to the relief and climate.

However, this relationship involves not only the presence of definite correlation dependencies from certain relief parameters, but also landscape biodiversity of the relief structure itself. Under the structure we signify the lithologic-tectogene base of the relief, as well as the nature of its genesis. Herewith, the more diversified the relief forms from a structural point of view on the territory under study are, the more diversified they are according to their origin and forms of relief, the higher is the level of landscape biodiversity in the region.

Landscapes, as other open type dynamic systems, represent classical homestas, able to maintain a definite regime of structure forming processes, under the conditions of even very significant, but not prolonged changes (natural or anthropogenic) of the natural environment.

Peculiarities of a mountain relief and climate have stipulated not only a variety of landscapes but also a high level of biodiversity of the organic kingdom of Georgia. 7 thousand types of different plants and about 8 thousand types of mushrooms grow in the territory of the country.

The animal kingdom of Georgia is presented by 109 types of mammals, 322 - birds, 52 - reptiles, 13 - amphibians, more than 120 -freshwater, transitional and sea fish, 111 - crustacea, 52 - millipedes, about 1300 - arachnidas (spider like), insects - more than 10 000, worms - 1546, simplest parasites - 235, etc. Thereby, out of 2667 described and studied major types of mammals inhabiting the planet, 4.1% of mammals, 3.8 % -birds, 0.7% reptiles, 0.3% amphibians, 0.6% -fishes are represented in Georgia, (an area which forms only 0.05% of the whole land). Thus, if the relation between the earth's surface land area and the one of Georgia is 2129:1, worldwide biodiversity and biodiversity of animal kingdom of Georgia relation on an average is 24:1. This index is two orders higher than the index of relation of the areas and indicates the high level of biodiversity of the national territory.

If we compare the number of types included in the "Red Book" of IUCN (International Union for the Conservation of Nature) with the number of corresponding types enlisted into the "Red Book" of Georgia - the types protected in the reserves of the country, then it will turn out that 9.8% mammal types is protected in Georgia (from all types of mammals registered and protected by IUCN), 11.0% - birds, 14% -reptiles, 13.8%- amphibians and about 0.5% - fish.

Herewith, most areas of protected territories of the country are in hypsometric zones of 2600-3500 m altitude- 8.67% and 1400-1800m - 4%. All this testifies that the protecting conditions for the animal kingdom in Georgia (representatives) is at a sufficiently high level and suggests successful decision of the biodiversity protection problem in the country.

Especially should be stressed the fact that many ecological problems in Georgia are stipulated by aggravated economic and social conditions of the population (felling of wood, damaged communication lines). The aggravation of the ecological situation in its turn strains the social-economic situation in the country. Thus we have to face some kind of exclusive circle.
See below, do you really want to repeat exactly the same part?

8. REGIONS OF ECOLOGICAL DISASTER

Specially should be noted the completely degraded landscapes in the vicinity of the towns of Chiatura and Madneuli, concentrating groups of mines contaminated by chemical compounds of the territory adjoining to the town of Zestafoni, ruined banks and communications at the mouth of the Inguri River, the areas southward from the town of Poti up to the village of Kobuleti. Poisoned waters of the Mashavera (below the village of Kazreti), the Debeda (below Alaverdi Concentrating Group of Mines), the Jejora (below the village of Kvaisa), the Kura (below the city of Tbilisi) rivers, etc. should be attributed to their number. It is necessary to introduce the differentiation of so called "red zones" i.e. - ecologically disaster areas. In the territory of Georgia such highly congested and ecologically rather sensitive ("explosive") mountain regions are Imereti, Atchara et al.

9. SUSTAINABLE DEVELOPMENT

The complicated social-economic and political situation has brought an abrupt deterioration of the natural environment in Georgia for the last decades. All over the territory of the country, forests, wind protecting forest stripes and forest-parks have been significantly damaged. Fauna resources are quickly destroyed due to uncontrolled fishery, hunting and poaching. The sanitary condition of cities has become extremely poor. Significant changes of the natural environment in the areas of natural disaster and armed conflicts have been observed. State funding of ecological and nature conservation programs has been considerably reduced. All this requires taking prompt actions for stabilization, and consequent recultivation and reconstruction of the country's natural landscapes.

To tackle the problems of balanced development and the conservation of biodiversity, constant consideration of two main factors is necessary: influence of anthropogenic and elemental-destructive processes, which are directly bound with each other.

Especially should be stressed the fact that many ecological problems in Georgia are stipulated by aggravated economic and social conditions of the population (felling of food, damaged line communications). The aggravation of ecological situation in its turn strains the social-economic situation in the country. Thus we have to face some kind of exclusive circle.

10. CONCLUSIONS

- At the present stage of research, it is difficult to define precisely to what extent climatic fluctuations affect natural disastrous phenomena and deteriorate geo-ecological situation in the country, and what percentage should be attributed to anthropogenic factor. Based on very rough calculations, the role of the latter in particular regions could be estimated by the values of 40 to 85%.
- The activation of elemental-destructive processes is triggered by numerous cycles of space-planetary characteristics imposed upon geological substratum, relief and landscape, surplus anthropogenic factor.
- One of the acute problems is the arrangement and complex use of stationary, field, cosmic and computer methods and approaches. The work for creation computer database has already begun for every type of natural hazards to compile a catalogue of manifestation of particular facts (a catalogue for over 8 thousand manifestations of various types of geodynamic processes is available), which gives a possibility to make calculations of possible dates and places of their possible activation.
- A complex study of geodynamic processes, forecast of their development tendencies in close contact with engineer-economic activity is impossible without well organized monitoring. The absence of such monitoring costs to the government too much and financial loss evaluated exceeds 100 times the sum required for the study of the

condition of processes and gaining short-term information about their change.
- In spite of the fact that a great part of the Georgian population resides in the high-risk zone of natural and technogene processes, most of them have no idea of the possible danger. That is why it is necessary to teach them how to behave in extreme conditions and provide them with necessary information.
- Unfortunately, we are too slow in understanding the unity of man and nature. It is necessary to concentrate all intellectual resources of our society in search for quite different ways of tackling the problem and look for an alternative decision based on the fact of lifetime granted to us by the fortune.

REFERENCES

1. Abdushelishvili K.L., Gagua B.P., Kerimov A.A., et al. Dangerous hydro-meteorological processes at the Caucasus, L. Gidrometeoizdat, 1980, 288 p.
2. Bondyrev I.B. Aerospace study and the problem of ring structures, Tbilisi, Techinform, 1991, 68 p.
3. Bondyrev I. V. Main problems of study and development of high mountain regions in Georgia, Tbilisi: GruzNIINTI, 1987, 72 p.
4. Bondyrev I.V. A new vision of some problems of Geomorphology of Georgia, Tbilisi:, Poligraf, 2000, 72 p.s
5. Bondyrev I.V., Maisuradze G.M. The essay of study and peculiarities of spatial distribution of the frozen-glacial relief forms, beyond the Caucasus border. "Quaternary system of Georgia", (XI Intern. Congr. Quater. Moscow, 1982), Tbilisi, Metsniereba, 1982, 74-88 p.
6. Bondyrev I.V., Seperteladze Z.Kh. Structure of relief as one of the defining factors of landscape diversity, Intern. Conference "Ecological Geomorphology", Belgrade: BelGu, 2000, 28-29 p.
7. Bondyrev I.V., Seperteladze Z.Kh. Protected territories and national parks of Georgia-reliable basis of natural environment and its rational use, Conference held in Georgia, "Mountains - wealth of our country (problems of sustainable development)", Tbilisi-Mtskheta- Gudauri: Metsniereba, 2000, 86-89 p.
8. Bondyrev I.V., Sulkhanishvili G.S. Experimental study of mechanics of frost weathering of mountain rocks. Trudi GPI, ser."Hydro geology and eng.geol." 12 (354), 1989, 19-25 p.
9. Bondyrev I.V., Tatashidze Z.K., Singh V.P., Tsereteli E.D., Yilmaz A.Impediments to the Sustainable Development of the Caucasus-Pontdes Region //"New Global Development" // Journal of International & Comparative Social Welfare, Twentieth Anniversary Special, 2004, v.XX, No 1, 33-48.
10. Vladimirov L.A., Water balance of Georgia, Tbilisi, Metsniereba, 1974, 184 p.

11. Gvindadze D. Prospects of sustainable development, renewable sources of energy, UNESKO and Georgia, problems of ecology. v.2, Tbilisi, Pub. Tech. University, 2000, 70-90 p.
12. Jaoshvili V.Sh. Impact of elemental natural processes upon demographic processes. Sc. Session "Enviroment and elemental natural destructive processes", Tbilisi, Metsniereba, 1994, 12-13 p.
13. Jibladze E.A., Butikashvili N.K., Tsereteli N.S. Seismotectonic and tectonic deformations in the area of Ratcha earthquake, 1991, Theses of sc. session "Environment and elemental natural destructive processes, Tbilisi, Metsniereba, 1994, 43 p.
14. Gzirishvili T., Beritashvili B. ,Svanidze G. et al. The first national information on climate change in Georgia. Tbilisi, Metsniereba, 1994, 43 p.
15. Gobejishvili P.G. Present day glaciers of Georgia and evolution of glaciation in the Eurasia mountains in late pleistocene and Holocene. Doctor's synopsis of thesis, Tbilisi, Inst. Geography Georgian Acad. Sc., 1995, 66 p.
16. Gobejishvili R.G. Elemental destructive processes on the south slope of the Kavkasioni, Sc. session "Environment and elemental natural destructive processes", Tbilisi, Metsniereba, 1994, 50-51 p.
17. Gobechia G. Tsereteli E. Some dangerous geological processes in Georgia and role of water factor in their formation, Tbilisi, KEPS, 1992, 18 p.
18. Gobechia G.,Tsereteli E., Paroxysm of elemental natural processes and their primary geoecological complications, Metsniereba da teknika, 1998, No1-3, 72-76 p .
19. Kiknadze A.G. On the regulation and management of beach formation processes at the Black Sea coast of Georgia. Sc. session "Environment and elemental natural destructive processes, " Tbilisi, Metsniereba, 1994, 22-23 p.
20. Lobzhanidze G. On some consequences of Ratcha earthquake, Theses Sc. Session dedicated to 70 anniversary of foundation A. Janelidze Inst. Geology, Tbilisi, TGU, 1995, 87 p.
21. Mdinaradze .L A, Tsereteli E.D., Meliksed-beg D.A. et al. General lay-out of counter erosional measures in Georgian SSR in 1981-1990 till 200, Tbilisi, Sabchota Sakartvelo, 1987, 728 p.
22. Nakhutsrishvili G. Present day ways of changing and protection of vegetative cover of Georgian mountains, Seminar-meeting "On measures of rational nature use and protection of environment", Tbilisi, Tsodna, 1987, 406 p.
23. Neidze V.E Ecological disasters in Georgia: Socio-geographical aspects of research, Sc. session "Environment and elemental destructive processes" Tbilisi, Metsniereba, 1944, 17-19 p.
24. Panov V.D. Evolution of present day glaciation in the Caucasus, SPB: Gidrometeoizdat, 1993, 432 p.
25. Svanidze G.G., Tsomaya V.Sh. Catastrophic flooding in Georgia and method of calculation their maximum discharge, Sc. session "Environment and elemental-destructive .processes", Tbilisi, 1944, 35-37 p.
26. Svanidze G., Kaldani L., Tsomaya V. Glaciological and hydrological effect upon trans-shipping points of the roads in mountain regions of Georgia, 5th conference dedicated to 120 anniversary from the birth of I.Javakhishvili, Tbilisi, TGU, 1996, 95-102 p.

27. Tatashidze Z.K. Scientific preconditions of passing to sustainable development in mountain regions of Georgia,"Geographical aspects of the problem of passing to sustainable development in CIS countries ", Kiev- M. NAN Ukraine, 11999, 122-128 p.
28. Tatashidze Z.K., Bondyrev I.V., Mumladze D.G. Tendencies of climate change in Georgia and their ecological and socio-economic consequences, in book " Global and regional changes of climate and their natural and socio-economic consequences", M., GEOC, 2000, 224-229 p.
29. Tatashidze Z.K., Tsereteli E.D., Bondyrev I.V. et al. Conditions and prospects of study elemental destructive processes in Georgia, 5 th conference dedicated to 120 anniversary from the birth I. Javakhishvili, Tbilisi, TGU, 1996, 47-49 p.
30. Tatashidze Z.K., Tsereteli E.D., Khazaradze R.D. Activation of natural hazards in Georgia and reasons of their formation. Sc. session "Environment and elemental destructive processes", Tbilisi, Metsniereba, 1994. 10-11 p.
31. Tsomaya V., Saneblidze L., Kenkebashvili C. Hydrometeorological peculiarities of freshets in high mountain zone of the Caucasus under the present day conditions of climate warming. Hydromet. Transactions,, Tbilisi, 1998, v.101, 160-164 p.
32. Tuliani L.I. Seismicity and seismic danger (on the basis of thermodynamic and geological parameters of tectonosphere) M., Nauchni Mir, 1999, 216 p.
33. Tsereteli E.D., Tsereteli D.D. Geological conditions giving rise to mudflows in Georgia. Tbilisi, Metsniereba, 1985, 182 p.

NEW WAYS AND NEW FORMS OF LIMITED AND CONTROLLED NATURE MANAGEMENT IN THE STEPPE REGION OF NORTHERN EURASIA

A. A. CHIBILYOV
Institute of Steppe, Orenburg, Russia

Abstract: The analysis of nature reservations framework in Northern Eurasia steppe zone is done. New approaches to the forming of a conserved nature territories network in the steppe zone are offered. The necessity of organizing a natural park to the Przhevalsky horse re-acclimatization in Orenburg region is substantiated.

Keywords: nature reservation, steppe ecosystems, pasture-grounds, environmental framework.

The last 20th century has made the Eurasian steppes the most anthropogenic landscape on the planet. The current structure of the steppe land supply in different regions of the Newly Independent States indicates that the natural ecosystems have been completely replaced by the anthropogenic agroecosystems, resulting in an entirely new biota. The main drastic changes in steppe landscapes in Ukraine and southern European Russia had occurred in the 19th century, whereas the rest of the Eurasian steppes that lies south of the Don was finally transformed agriculturally in 1930s, and then in 1950-1960s. Despite the fact that in the course of history it was dangerous ecological state of steppes, in the late 1890s-early 1910s, that gave rise to the environmental movement in Russia, throughout the entire 20th century, Soviet scientists have failed to form a representative network of the steppe landscape reserves and to solve in practice issues concerning the optimization of steppe ecosystem conservation (Mordkovich *et al.*, 1997, Chibilev, 2000).

By the end of the 20th century 100 reserves had been established, the reserves in the taiga region (the Russian name for boreal forests) numbering 19, in the forest-steppe region and mixed forest region – 15, and in elevation zones areas – 53. In the steppe region and in the dry steppes going into the semi-desert as well, only 4 reserves functioned. One should pay attention to the fact that the area of state reserves in the tundra and forest-tundra is 5.24 % and in the steppe region is only 0.11 % of the native zones' total area. The zonal representative coefficient calculated by A. Nikoliskiy and V. Rumjantsev (2001) for the state reserves of the Russian Federation is 2.36 for tundra and forest-tundra, 1.69 for desert, but only 0.09 for the steppe and semi-desert. Provided that the zonal representative coefficient equals 1 (the norm), the total area of zonal steppe reserves in Russia should be increased by more than 10 times. Establishment of the biosphere reserves in steppe and forest-steppe areas well-developed economically involves serious difficulties. They result mainly from contradictions between the scientific views on environmental protection and the economic interests of different sectors of the national economy. The absence of steppe reserves in most of the physiographic provinces is nowadays the major unsolved problem in reserve management and studies of the country. As the hands-on experience indicates, it may be solved by following the principle of "cluster ecosystems" (Gusev, 1988, Chibilev, 1992). The central-Black-Earth state biosphere reserve established by V. Alehin in 1930 is one case in point. Another example is the steppe reserves in Ukraine. In that region there are 4 reserves ("Luganskiy", "Ukrainian steppe" reserve, "Askaniya Nova", and "Chernomorskiy") with 7 steppe standards protected (Schelyag-Sosonko et al., 1987). These 7 steppe areas form a peculiar landscape profile of steppes in the longitudinal direction from the northern (meadow), rich in herbs and wild grasses steppes through the southern poor in herbs steppes consisting of Fescue and feather grass, to the sandy littoral zones along the Black Sea coast.

Despite the territory limitation of every reserve area in Ukraine (five of seven are only 200 to 1000 ha), they are most representative series for the whole Eurasian steppe region. The creation of similar complexes of protected steppe areas in the Russian Federation and Kazakhstan is the urgent task in our days. The more anthropogenic level of landscapes the higher the need is for development of nature reserves network well-grounded scientifically. But for all that biosphere reserves should be regarded as ecological representative landscapes in the given region rather than unique corners and nooks of landscapes.

Given that it is impossible for large reserves to be created within the agricultural steppe and forest-steppe zones of the country, the proposal was given to form an unbroken grid consisting of small and medium in size natural complexes to be protected (Chibilev, 1986, 1987). A basis for the creation such a network is supposed to be the representation of natural landscapes: objects should be typical for a region and zone and at the same time be unique in general, they should be endangered, and the projected areas should be also of some value for wildlife conservation providing refuge for various living organisms. Small (up to 1,000 ha) and middle-sized (1,000-10,000 ha) standard territories may be included in one reserve. If the location of such isolated areas has the ecological justification and landscape substantiation, there is a possibility not only for the establishment of a biosphere reserve in the region but also the creation of a territorial base to carry on research and ecological monitoring. In addition, only a small amount of land is removed from agricultural land and industrial use, thus it is possible to avoid the greater expense of maintaining several reserves.

During formation of the spatial structure of an agricultural landscape in the steppe region, the well-known Landscape Polarization Principle took on special significance (Rodoman, 1974) that territories used and protected intensively should be as far apart as possible. Application of this principle enables construction of rational and optimum models of the today's steppe landscape. One of the most interesting ideas deserving particular attention is the formation of a landscape-ecological framework in the steppe provinces that enables ecologically valuable landscapes to be uniformly distributed alongside axes of any types in terms of present structure of environmentally homogeneous grounds. The system of areas proposed for protection such as natural monuments, protected landscapes, species refuges, and reserves is supposed to be developed alongside axes of this framework. In projecting a reserve network and other protected nature territories the Principle of Natural-migratory Channels should be used which allows combining both present and projected nature reserves and refuges with the aid of different conservation zones – "natural geographic windows", "environmental transit corridors" and "buffer strips".

The boundaries of the unmanaged standard areas in steppe reserves and their buffer zones is supposed to secure relative ecological autonomy of landscapes reserved, i.e. to safeguard their independence against anthropogenic and other random actions. This is especially important in open (a plain steppe landscape with transit flows of matter and energy) landscape. The real way of development of protected network in the steppe region is the creation of the largest possible number of small reserves. In order to preserve at least some of the biota, first of all vegetation, invertebrates, small mammals and specific steppe ornitofauna, when organizing steppe reserves consideration must be given to their scattered and mosaic location and configuration. Permanent, cluster biosphere reserves can be used as landscape etalons when studying environmental conditions. They make it possible to assess present natural diversity, the damage caused to natural resources, and biological efficiency for ecosystem principal categories in each economic region. The more than century-old experience in the organization of steppe reserves and refuges in Northern Eurasia allows making the following conclusions being dictated by necessity to conserve the steppe etalons within the bounds of protected areas.

1. When projecting steppe reserves, on the basis of the ecosystems current state evaluation, a recovery (reanimation) period from 2 to 5 years should be set. For this period, throughout the protected territory an absolutely closed regime is in force as well as the conservation grazing that eliminates any summer grazing camps (summer ranch houses) to be located in the buffer zone.

2. After the recovery period the protection land organization is carried out for each area reserved according to which areas with complete protection regime are detached as well as areas with regulated kinds of the economic use, imitating different scenarios of the steppe landscape development throughout a historic period.

3. There are active and passive protection regimes in the steppe reserves. The former provide absolute protection from interference in the current landscape dynamics and structure. The active protection regime stimulates the recovery of the zoological complex including wild ungulate animals (e.g., saigas (*Saiga tartarica*) or free pasturing of domestic horses). This regime promotes recovery of populations of typical steppe animals such as Marmots (*Marmota bobac Muller, 1776*), Great Bustards (*Otis tarda*), Little Bustards (*Tetrax tetrax*), Steppe Eagles (*Aquila rapax*) and other species.

4. In steppe reserves where recovery of the entire zoological complex is an impossible task, after the reanimation period the protection regime with the selective grazing should be gradually introduced. In this case, also, certain areas are selected in which no pasturing will be allowed, and time periods ("critical phenological windows") are determined in which the steppe will remain completely undisturbed.

5. The protection land organization can provide other regimes for scientific purposes: the pasturing of different domestic animals for a limited time

period and the selective horse-drawn hay – mowing mainly for the reserve's own needs.

The great potential of maintaining and optimizing the biodiversity of steppe ecosystems can be found in the conservation zone of steppe reserves. Here active biotechnical practices can be carried out that promote formation of the characteristic steppe biota. The experience of the state reserve "Orenburgskiy" (1989-2001) consisting of four protected areas, has shown that it is in the conservation zone that ideal conditions are created to form steppe biodiversity.

From the running experience of the state reserve "Orenburgskiy" and other reserves in Ukraine, the Central Chernozem area (the Central-Black-Earth District) and the Volga region (Povolzhie), the conclusion can be drawn that the existing protection system in the country does not solve all the issues on conserving landscape and biological diversity in the steppe region. When working out the Regional action plan on conserving landscape and biological diversity in the Orenburg region, we have included fundamentally new forms of protection of the steppe landscape etalons. The first and most widespread form of protection is the etalon (standard) steppe areas. Approximately 40 steppe areas (50-1.500 ha) have been isolated and designated as natural monuments with no withdrawal of lands from land users. Various land use regimes with controlled grazing including limited hay- mowing are provided for the steppe natural monuments.

Another form of steppe reservation is the pastoral reserves (the pasturing of ungulate animals do not go against the protection regime in the steppe). These are sufficiently large steppe areas (1.500-15.000 ha) that are also owned by former land users. In these areas, in agreement with the land users, a special pastoral regime is set with the removal of every summer grazing camp from the steppe sites creating higher and more unique biodiversity.

Clearly Russian residents would not be able to experience the spiritual interaction with the steppe nature without creating a steppe national park. The Askaniya Nova reserved in 1898 became in its time the prototype of such a park. As USSR not exists, Russia lost its opportunity to use the Askania-Nova as its own steppe reservation. This is why Institute of steppe substantiated the expediency to organize the Steppe Park and biostation "Orenburgskaja Tarpania". A plot of 15000 ha is selected from the national land reserve to organize it there. It is situated in the country between Ural and Ilek rivers in the subzone of herb-cereal steppes on southern black soils. The reacclimatizing of Przhevalsky horse, the aboriginal inhabitant of Eurasian steppes, is planned within its territory. Unlike a reserve, a steppe park solves more then problems of landscape and biota diversities conservation. It satisfies also human request of rest, tourism, spiritual self-development and scientific knowledge. It serves also the enlightenment. The steppe park is planned to become more than an area, where nowadays typical inhabitants of steppe can be observed. It is planned also to become an opportunity to restore populations of wild hoofed animals, which are lost components of steppe ecosystem.

Measurements intended for the representative steppe reservations development and for the steppe landscape ecological restoration correspond national and regional interests of states of whole the steppe belt of Northern Eurasia: from Hungary and Moldova to Mongolia and North-Eastern China.

REFERENCES

1. Gusev A.A. Organizing principles of a biosphere reserves. // Geological and geographical research and environmental problems of especially conserved territories in Urals and Siberia. Cheljabinsk, 1988.
2. Mordkovich V.G., Giljarov A.M., Tishkov A.A., Balandin S.A. The destiny of steppes. – Novosibirsk: "Mangazel", 1997. – 208 p.
3. Nikolskiy A.A., Rumjantsev V.J. The zonal representative quality of nature the reserve system of Russian Federation. // Topical problems of ecology and nature use. – Moscow, Russian university of nations friendship editorial board, 2001. Issue 2.
4. Rodoman B.B. The polarizing of a landscape as means of conservation of biosphere and recreation resources. // Resources, environment, spreading. – Moscow. – 1974.
5. Chibilyov A.A. On principles of the forming of conserved nature territories network. // Geographical problems of reserve theory: Theses of report on the Union-wide scientific conference. Samarkand, - 1986.
6. Chibilyov A.A. On the landscape and ecological grounds of the conserver nature territories network development // General and regional problems of landscape geography of USSR. Voronezh, - 1987.
7. Chibilyov A.A. The ecological optimizing of steppe landscapes. – Sverdlovsk: Urals branch of USSR AS, 1992. – 172 p.
8. Chibilyov A.A. Recent problems of steppe science // Questions of steppe science. – Orenburg: RAS Urals branch, 2000. – P. 5-8.
9. Sheljag-Sosonko J.R. et al. The Green Book of Ukrainian SSR. – Kiev: Naukova dumka, 1987.

CHAPTER 4

BEYOND BOARDERS: TRANSBOUNDARY BIOSPHERE RESERVE "ALTAI" AS AN APPROACH FOR REGIONAL DEVELOPMENT

SUSTAINABLE LAND USE CONVERGENCE IN BORDER AREA IN CENTRAL EUROPE

J. KOLEJKA AND D. MAREK
Mendel University of Agriculture and Forestry, Faculty of Forestry and Wood Technology, Institute of Geoinformation Technologies, Brno, Czech Republic

Abstract: The digital landscape model consisting of natural landscape map, DEM and three historical land use maps (1829-40, 1935-38, 2002) has been applied to identify general and local tendencies in the land use development under different natural conditions in the mountainous area on the Czech-German border. Regardless to different starting points, and economic and social causes driving the land use changes, the natural background plays the decisive role in the final detail allocation of human activities in the study territory and supports the convergence of land use pattern in the present time.

Keywords: natural landscape units, historical land use changes, GIS, pattern convergence.

1. INTRODUCTION

Mountain regions represent areas with well developed vertical zoning of geoms. The vertical ordering is typical for classes of present landscapes reflecting natural conditions. Differences between the ways of the landscape utilising in the same natural mountain environments can be caused by different economic and social features of regions. If natural, economic and social parameters of regions are similar, also the landscape utilising has to be similar regardless of their geographic location. Such similarity has to govern the landscape view of both areas as well. Due to different human conditions, diverse and impact printing ethnographically based cultural landscapes have developed. The running globalisation raised and equalises economic and social differences between regions of the world at certain scale. Such development touches mountain regions imminently. Results of this process can consequently manage a theoretical convergence of structures and view of presently different cultural landscapes in nearby future. The loss of identity of typical cultural landscapes is a real threat of mountain regions of Altai, Alps, Ands, Carpathians or other mountains in the world. The protection of mountain regions has to include the preservation of typical landscape views at least there where it is typically developed. The results of a long-year research carried out in the border mountain area of Šumava at the Czech-German boundary can serve as an example of such threat.

2. MATERIALS USED FOR STUDY OF LAND USE DYNAMICS

The study of historical land use changes represents a traditional branch of landscape ecological research not only in Czech and Slovak science. The identification of changes originated during the time using various cartographic or statistical data is the common aim of the most examples (Bičík, 1998, Štěpánek, 2002, Lipský, 1996 and many others). The data processing runs usually on the background of certain administrative unit or map sheet. An effort to describe the relationship between the land use change and natural background ends in very general explanation, because the comparison is related to general description of climatic or terrain features without any detail location of site where land use change was detected. The relationship to geology, land form (slope), humidity, soil is not identified. The knowledge about the relationship between land use change and natural background parameters of the site is important for understanding the cause and role of the change. The research results of such kind are still very rare (Kolejka, 1987, Olah, 2002).

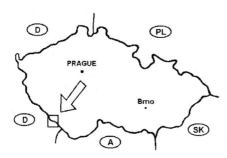

Fig. 1: Position of research areas

An extended research has been carried out in the framework of the project No. 5 01 308 of the Czech-German Future Foundation under the title „The border community" in two areas on the both sides of the border. The study area consists of two sections: Strážný (13 sq km) on the Czech side and Philipsreuth (10 sq km) on German side of the border (fig. 1).

Every territory contains a couple of cadastral areas with similar natural parameters. Common border line between them is about 6 km long. A digital landscape model (Kolejka, Plšek, Pokorný, 2003) was compiled for both the areas using available cartographic data, aerial imagery and field survey data.

All the data was integrated from the viewpoint of size, scale, cartographic projection, resolution, format and logical relationships between natural parameters into the natural landscape units (geosystems) and landscape utilising in 1829-40, 1935-38 and 2002 finally in the resolution common in maps at the scale of 1:10 000. It is not only a multilayer set of topographic or thematic data (e.g. as German Landschaftsmodell, see Zölitz-Möller, 2002), but it is a 3D presentation of a real landscape segment. Such kind of digital landscape model consists of at least three multiparameter information layers (fig. 2): natural background, human impacts and development limits as well as of the digital elevation model. The natural background layer, present and historical land use layers and DEM were applied for the study of historical land use changes

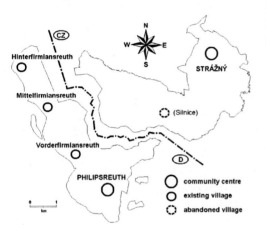

Fig. 2: Study areas on the Czech-German border

3. METHODS APPLIED

All kind of collected geospatial data sets were compiled in a digital form and stored in GIS database. Software MicroStation and ArcView were used for its digitising, SW MicroStation and ArcView, GIS ArcView v.3.11 and its extensions Geospatial Analyst and 3D Analyst were applied as the main processing tools.

The integrated data layer "natural background" (fig. 3) as a map of natural landscape has been compiled by overlaying analytic (component) data layers and their manual integration into one layer. Every polygon of the integrated data layer is labelled with information about bioclimate, geology, soil type and kind, and humidity conditions.

The integrated data layer "land use in 2002" originated using the on-screen digitising over the black-and-white aerial orthophoto taken in 2000 by GEODIS BRNO Ltd. (in S-JTSK co-ordinate system for Czech part of study territory, in Gauss-Krueger system for German territory). The aerial data was verified in the field in 2002. This map contains the information about the land utilising (land cover) as well as about the present quality of land use presented by the ecological stability coefficient (from 0=min. to 5=max.). This coefficient value shows the similarity ration between present and potential natural land cover. There are 21 land use forms distinguished in the map here.

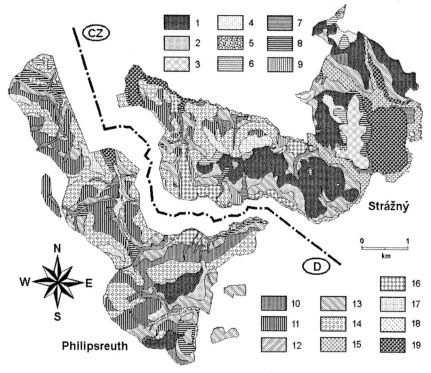

Fig. 3: Integrated layer "natural background" shows net of natural landscape units

The land use map for years 1935 (Czech part) and 1938 (German part) represents the period before the World War II. At that time, the German population dominated on the both sides of the border. Germans were resettled in 1946 according to the decision of the Potsdam Allies.

Conference concluded by USA, USSR and UK. The Czech map is based on the digitised cadastral map from 1935 using St.-Stephan projection and georeferenced in Kokeš system. Local map corrections were conducted using land use map 2002 to adapt as much as possible lines from the newer map to get the best map overlap. The map for the German territory was provided by the German research partner from the University of Regensburg and later adjusted to project use. There were 8 land use forms represented in the map.

The years 1829 (German part) and 1840 (Czech part) are documented in maps proceeded from different sources. The German map was provided in the digital form by the project partner from the University of Regensburg and finalised for research needs. The map of the Czech part of the study territory was digitized on screen over digitised map sheets of the Austrian Second Military Survey carried out at the original scale of 1:28 800. Selected map sheets were initially georectified in MATKART system (developed by Prof. B. Veverka, Czech Technical University of Prague). The data digitising in GIS ArcView was supported by the program Base App (developed by J. Pokorný, Prograf Brno) specifying the position of raster data of the Second Survey shifting it into the optimal position corresponding with lines of the land use map 1935. The shifted raster data was repeatedly opened in ArcView where the digitising continued. If possible, many lines were incorporated into the land use map 1840 from the younger one. There were 7 land use forms represented in the map.

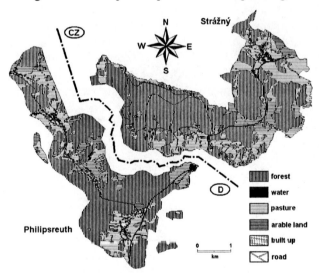

Fig. 4: Land use in 2002

The set of three historical land use maps required qualitative generalising to get an uniform comparative basis. Because of the legend of the land use map 1829-1840 was the simplest one, the others for 1935-38 and 2002 (fig. 4) were modified to show the same land use classes. The map simplification was carried out in ArcView.

The geoprocessing tools were applied to calculate the ration of any land use class in every natural landscape unit in absolute and relative values for three time periods. Similarly, the average relative ration was counted for every of 19 natural landscape unit classes (table 1). Since the forests, meadows (with pastures),arable land and built-up areas were extremely dominating land use classes (more than 90 %) in any natural landscape unit class, the attention was focused on these land use forms only to detect development trends (fig. 5a, b).

The statistical data set consisting of average values of three dominant land use forms (forest, meadow, arable land) in every class of natural landscape units calculated for any of three time periods was created in ArcView separately for the Czech and German part of the territory. It was converted from DBF format into XLS one used in MS Excel. Every class of natural landscape units on the Czech and German territory was described with a data matrix involving 4 x 3 elements and with an appropriate diagram. These matrixes and diagrams document development trends of four land use forms on the territory of individual classes of natural landscape units. This way, 18 matrixes and diagrams were compiled for the Czech territory (class no. 18 is missing here) and 15 ones for German territory (classes no. 4, 5, 9 and 12 absent here). Pairs of graphs representing the same classes of natural landscape units on the both sides of the border were integrated into one graph (fig. 6). Although the visual observation of diagrams provides certain identification of development regularities in individual landscape units, the more exact classification needed the application of methods of numerical taxonomy, namely the cluster analysis. The statistical Unistat 5.5 system was applied for these purposes.

Table 1: Parameters of identified natural landscape unit classes

No.	CLIMATE	FOREST STAGE	GEOLOGY	SOIL TYPE	SOIL MECHANICS	HUMIDITY
1	cold	6	crystalline	dystric cambisols	loamy-sandy	normal
2	cold	6	crystalline	gleyic cambisols	sandy-loamy to clayic-loamy	fresh
3	cold	6	crystalline	eutric cambisols	sandy-loamy to loamy-sandy	normal
4	cold	6	crystalline	podzols	loamy-sandy	normal
5	cold	6	peat	organosols	biodetritic	wet
6	cold	6	crystalline	rankers	stony	dry
7	very cold	7	(deluvio)fluvial deposits	fluvisols	loamy-clayic to clayic	humid
8	very cold	7	slope deposits	gleysols	loamy-clayic to clayic	humid
9	very cold	7	peat	gleysols	biodetritic	humid
10	very cold	7	crystalline	dystric cambisols	loamy-sandy	normal
11	very cold	7	crystalline	eutric cambisols	sandy-loamy to loamy-sandy	normal
12	very cold	7	(deluvio)fluvial deposits	pseudogleysols	clayic-loamy to loamy-clayic	moist
13	very cold	7	slope deposits	pseudogleysols	clayic-loamy to loamy-clayic	moist
14	very cold	7	crystalline	podzols	loamy-sandy	normal
15	very cold	7	(deluvio)fluvial deposits	organosols	biodetritic	wet
16	very cold	7	peat	organosols	biodetritic	wet
17	very cold	7	crystalline	rankers	stony	dry
18	cool	8	(deluvio)fluvial deposits	organosols	biodetritic	wet
19	cool	8	peat	organosols	biodetritic	wet

(Legend for forest stages: 6 – fir-beech, 7 – spruce-beech-fir, 8 – spruce)

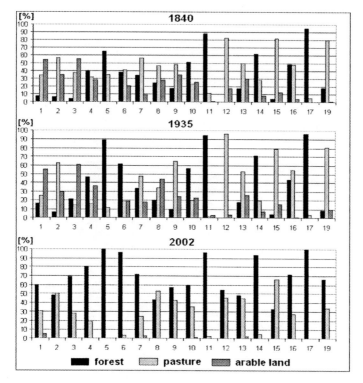

Fig. 5a: Ratio of studied land use forms in natural landscape units in German territory

The classification input data consisted of values of the index of changes (I_C). Its individual values were calculated using formula (1):

$$I_C = (B-A)/\text{average of }(A,B), \qquad (1)$$

where:
B - % ratio of individual land use form on the territory of given natural landscape unit class in consequent period,
A - % ratio of same individual land use form on the territory of given natural landscape unit class in previous period.

Sustainable Land Use Convergence in Border Area in Central Europe

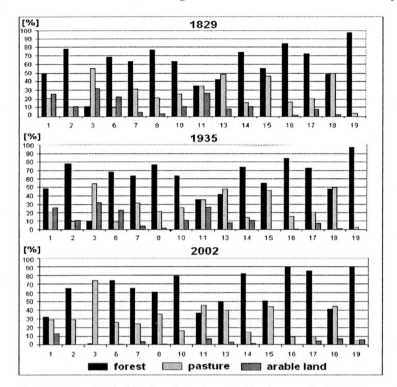

Fig. 5b: Ratio of studied land use forms in natural landscape units in German territory

Experimenting with various cluster analysis methods offered results saying that the useful (applicable and explainable) products provides the method of the furthest neighbour only. The final dendrogram (fig. 7) was used as a starting point for the explanation and description of identified land use development trends in this border area.

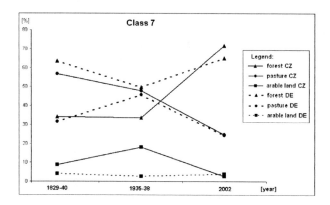

Fig. 6: Example of graph presenting chronologically ordered land use data in one class of nature landscape units

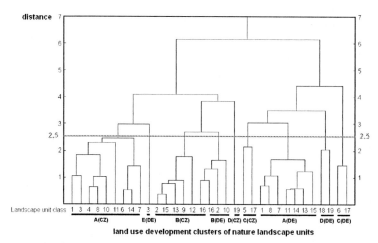

Fig. 7: Furthest neighbour dendrogram with iteration stopped on level 2.5 showing nature landscape units clustered into land use development classes

4. RESULTS AND DISCUSSION

The natural environments of both areas are relatively similar. They are located nearby the main European watershed in Šumava Mts. in the territory with numerous long ridges and wide flat valley in the elevation from 800 to 1139 m a. s. l. The average temperatures range from -3 to -5 °C in January and 14 - 16 °C in July. The annual precipitation amounts vary from 1000 to 1200 mm. The individual classes of natural landscape units (see tab. 1) are presented differently on the sides of the border, but some of the most common expressed similar ratios, e.g. classes no. 13 and 14 (fig. 8).

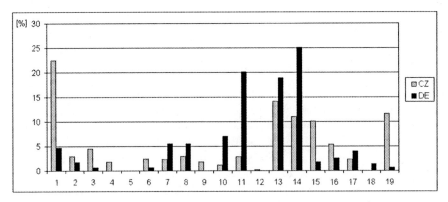

Fig. 8: Relative presence of nature landscape unit classes in national areas

The area development of forests, meadow, arable land and built-up areas in Strážný and Philipsreuth communities indicates, that regardless to similar natural conditions, the starting land use in the beginning of 19. Century was different probably thanks to different economic orientation or specialising of population of border communities. The forestry activities were dominant at this time but they cannot manage full support especially on the Czech side of the border. Other activities, namely the agriculture, were also important. The agriculture played the complementary role on the German side only. Such situation continued until the World War II. The post war political and economical situations on the Czech side were completely changed. The German population was displaced and a strong border zone was created. Economic activities were deeply limited here. Silnice village was abandoned. Old individual farms spread in a wide belt along the border crumbled. Only the buildings of the border crossing point and customs survived. The population

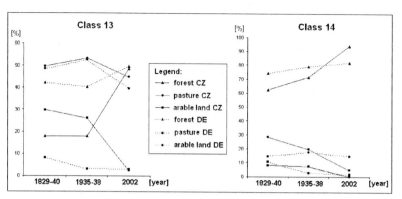

Fig. 9: Example of similar land use development trends in natural landscape units classes common on both sides of the border

number declined and political barriers caused fast reduction of arable land demand and the forested areas expanded. The population drain on the German side into the hinterland was less visible probably thanks to original smaller agricultural orientation of the territory. The employment restructuring supported the recreational role of all settlements and especially Mittelfirmiansreuth, what was transformed into winter ski centre with an appropriate technical equipment and infrastructure. Similar development is visible on the Czech village Strážný after the year 1989, when the tourist and entertainment infrastructure was developed. Since this time, the kiosk shopping was widely introduced. The present extension of arable land is negligible on the both sides of the border. The Czech part of the territory is protected in Šumava National Park and Šumava Landscape Park with exception of Strážný intravilan.

It is eminent that the development on the Czech side was completely different and totally independent on the development on the German side. Nonetheless the visual convergence of landscape originated, especially whet talking about the territorial distribution of forests, meadows and pastures in analogue natural environments and location. A similar land use pattern originated on the both sides in similar natural landscape units as well (fig. 9, see class no. 13-wet units and no. 14-podsolised unit widely presented on the both sides of the border). In opposite, the natural landscape unit classes with extremely different ratios in the both territories contrast with another land use pattern.

It was detected, that individual classes of natural landscape units "used" either very similar or very different strategies to get similar end effect therefore the similar land utilising separately on the Czech and German border sides (fig. 10).

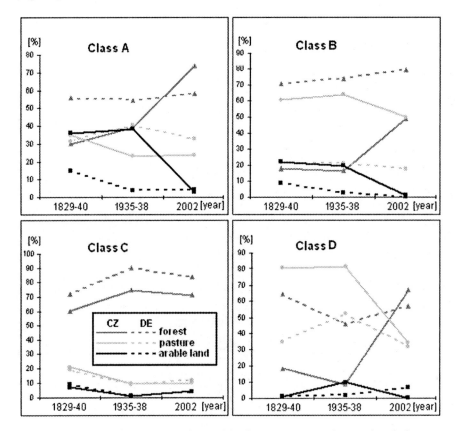

Fig. 10: Typical graphs of groups of natural landscape units according to identified development trends

The group A of the natural landscape units, represented on both border sides by wet valley bottoms and stony slopes, is characterised by antithetic development accompanied with dramatic arable land decline and forest area growth on the Czech side, while the changes on the German side are in sum fractional. Nonetheless such development caused the convergence of land use patterns. The group B with similar ratio of very different units from the humidity viewpoint exhibits slower development on the German side, while the same process runs faster on the Czech side. Finally, the land use patterns remain different. The German side is predominantly forested; the Czech one indicates the balanced forest/meadow ratio.

Units of class C characterised with extremely stony or wet areas show the same development on the both sides of the border accompanied with small fluctuations in some land use forms ratio. Dramatically and diametrically different changes are typical for class D on Czech and German territory finally leading to almost perfect converge of land utilising characterised by small dominance of forest above meadows. The group E originated on the German side only and the forest ratio decline at permanent its low ratio is typical, while the growth of meadows and arable land drop out is visible.

Similar natural conditions are followed with similar land use patterns. If some natural landscape unit has no equivalent beyond the border, the land use development trend is unique (fig. 11). The above mentioned regularities can be explained using a hypothesis, that a "general rational" land utilising is being constituted respecting the territory marginalising, gradual depopulation, agriculture decline and growing role of recreation regardless to governing social-economic system and previous life standards.

It will be also very interesting to compare some results reached in the inland research and identify if similar behaviour of natural landscape units and its utilising occurs there as well even under different economic pressure. The integrated exhibition of the nature, population density and economic climate should be expressed in similar land use pattern. This fact can be very important for the future landscape maintenance. Some classes of the cultural landscape can be directly endangered in the relationship to the convergence of the life style and the economic level of population in different regions of the European Union.

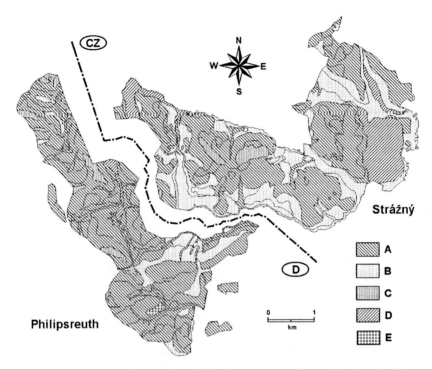

Fig. 11: Territorial distribution of natural landscape units with same land use development trends (classes A, C and D represent land use convergence)

REFERENCES

1. Bičík, I., 1998. Land Use in the Czech Republic 1845-1948-1990. Methodology, Interpretation, Contexts. Acta Universitatis Carolinae – Geographica, Vol. 32, pp. 247-255.
2. Kolejka, J., 1987. Landscape-historical Synthesis. Materials, Methods and Results. Ecology, Vol. 6, No. 1, pp. 51-62.
3. Kolejka, J., Plšek, P, Pokorný, J., 2003. Digital Landscape Model – an integrated tool for area management. In: Proceedings The 3rd International Symposium on Digital Earth „Digital Earth – Information Resources for Global Sustainability", Masaryk University, Brno, pp. 155-165.
4. Lipský, Z., 1996. The land use changes and their environmental consequences in the Czech Republic. In: Ecological and landscape consequences of land use changes in Europe. Tilburg, pp. 350-360.

5. Olah, B., 2003. Vývoj vyuzitia krajiny Podpol'ania. Starostlivost' o kultúrnu krajinu prechodnej zóny Biosférickej rezervácie Pol'ana. Vedecké stúdie No. 1, 2003B, Technická univerzita vo Zvolene, 110p.
6. Štěpánek, V., 2002. Czech frontier in the 20th century: major political shift in changing land use structure. In: Land Use / Land Cover Changes in the Period of Globalization. Proceedings of the IGU-LUCC International Conference, Charles University, Prague, pp. 110-115.
7. Zölitz-Möller, R., 2002. Geobasisdaten für die Planung? Standort – Zeitschrift für Angewandte Geographie. No. 3, pp. 110-114.

SUSTAINABLE DEVELOPMENT BEYOND ADMINISTRATIVE BOUNDARIES

Case study: Rhön biosphere reserve, Germany

DR. D. POKORNY
Rhön Biosphere Reserve, Bavarian Administration Unit, Germany

Abstract: Based on the practical experience in the Rhön biosphere reserve (=biosphere territory), which covers parts of three federal states in Germany, mechanisms and strategies for sustainable development beyond administrative boundaries are demonstrated and examples for trilateral cooperation are given. Based on this, recommendations for the further development of the Altai region as a transboundary biosphere territory are derived. Success or failure of a biosphere territory depends especially on the local people, as they need to decide on and take action for their own sustainable future on the regional level.

Keywords: biosphere reserve, biosphere territory, cooperation, organisational structures, local participation, success factors and strategies, regional marketing.

INTRODUCTION TO THE RHÖN BIOSPHERE RESERVE

The Rhön is situated in the centre of Germany, 150 km east of Frankfurt (see map) and was designated by UNESCO as a biosphere reserve in 1991. It is one of Germany's 14 biosphere reserves and covers an area of 1850 km². The population is 136.000 with a population density of 80 inhabitants per km² which, in Germany, is considered as a rural area. People live in small villages or towns which belong to 90 municipalities. They are part of five districts within three federal states or "Länder": Bayern (Bavaria), Hessen and Thüringen (Thuringia).

As biosphere reserves in Germany are within the jurisdiction of the individual "Länder" governments, the Rhön is a biosphere reserve consisting of three independent units.

Therefore, cooperation beyond administrative boundaries is the precondition for the functioning of the Rhön biosphere reserve.

To this aspect, the Rhön is very similar to a "transboundary biosphere reserve" and will in this paper be referred to as a "trilateral" biosphere reserve.

The Rhön is a low mountain area with rolling hills and highlands reaching a maximum altitude of 950 m above sea level. The climate is both sub-continental and sub-oceanic. The mean annual temperature ranges from 5°C in the highlands to 8°C in the lower areas. The mean annual precipitation follows a gradient ranging from 1000mm in the west to 550mm in the east.

Until the German reunification in 1990, the Rhön had been divided by the Iron Curtain for 40 years. The Thuringian part of the Rhön belonged to the former German Democratic Republic. Hence there are major structural differences in land use on both sides of the former Iron Curtain.

The Rhön is a picturesque, cultural landscape with a land use history which can be traced back to the 9th century. Although pristine or natural ecosystems have gone since long, the area has a high anthropogenic biodiversity, bound to low impact agricultural and forest management activities.

Managed forests (mainly mixed forests or broadleaf beech-woodlands dominated by *Fagus sylvatica)* cover more than 40% of the area. 30% are used as grassland (both pastures and meadows) and 22% are ploughed land, 8% are settlements, roads or other infrastructure. Due to the harsh climate and the poor soils, large parts of the Rhön can be regarded as marginal agricultural areas.

Sustainable Development Beyond Administrative Boundaries   201

Ecologically, however, the cultural landscape is very rich, consisting of precious habitats for a range of endangered and rare animal and plant species, which are protected both on the regional, national and European level (*e.g. Ciconia nigra, Milvus milvus, Crex crex,* or *Bythinella compressa, which is an endemic species*).

Biodiversity in the Rhön is particularly depending on extensive grassland management which may be bound to disappear in future as more and more farms give up due to economic difficulties. Conservation goals in the Rhön can only be reached by close cooperation between farmers, the nature conservation authorities and the agricultural authorities which provide grant schemes and programmes for adapted management.

Generally speaking, the task of the biosphere reserve is the conservation of abiotic resources and of biodiversity, which needs to be integrated in a sustainable regional development. In addition, environmental education which in more advanced steps should more widely be regarded as "education for sustainable development", needs to accompany this process, as well as applied research and integrated environmental monitoring.

Success or failure of the Rhön biosphere reserve depends especially on the local people, on their commitment, creativity and their preparedness to take economic risks to try out new ways. They also need to decide about their own sustainable future on the regional level and take action and responsibility accordingly.

The following chapters will show trilateral coordination in the Rhön biosphere reserve works in practice:

1. STRATEGIES AND MECHANISMS TO MAKE A BIOSPHERE RESERVE WORK BEYOND ADMINISTRATIVE BOUNDARIES

1.1 Common Visions and goals – process and product

Sustainable regional development depends on a consensus of all stakeholders and interest groups, on clear visions and goals. In a transboundary context, this is all the more important. It is crucial to make local people feel committed to put the goals into action afterwards.

In the Rhön biosphere reserve, visions and goals were elaborated as part of a so called "Management Framework", which was worked out trilaterally, beyond the three federal states. An independent landscape planning office was contracted for this task.

The "Management Framework" was embedded in an informal landscape planning process on the regional level. It was aimed at working out the region's strengths and weaknesses, threats and potentials.

As a first step it started with the inventory and analysis of the region's ecological situation, land use, and economy.

On this basis, goals for all major land use and economic sectors were discussed, negotiated and agreed upon: agriculture and forestry, aquaculture, sport fishery and hunting, mining, industry and handcrafts, settlement development, nature conservation, tourism and recreation etc. Also goals and measures for applied research and environmental education were elaborated.

Local participation was vital in this process. All stakeholders (munici-palities, districts, non-governmental organisations, government adminis-trations) in the three "Länder" were involved in round table discussions and were invited to make suggestions and interventions to the draft versions of the Management Framework.

The process took three years and was important by itself, resulting in:
- making local people think about the future of their region in a comprehensive context
- creating a platform for networking, as stakeholders from the different parts of the Rhön had met and discussed about their common future for the first time

- providing a common thematic basis for the sustainable development in the three parts of the Rhön area.

The product was
- a comprehensive informal management framework, including goals and objectives and a list of actions on how to reach these goals, including
- a zoning system, according to UNESCO guidelines, linking conservation, and development functions to specific areas within the biosphere reserve.

The zonation of the Rhön biosphere reserve clearly reflects the context of a traditional cultural landscape: The core zone of the Rhön covers only 2% of the area, consisting of woodland and bog ecosystems which are not pristine ecosystems. But they have been legally protected and set aside for natural succession and are not exploited any more by forestry or agriculture. The core zones in the long run are meant to be new "wilderness areas". 40% of the biosphere reserve belongs to the buffer zone (which is in the Rhön called "maintenance zone" with regards to the cultural landscape).

Not only should it buffer the core from harmful influences but furthermore it has a function of its own: It consists of anthropogenic semi-natural grassland or extensively managed forest ecosystems. In the Rhön, the buffer zone has the highest biodiversity, as it is based on land use of a low intensification level. The most important areas of the buffer zone are legally protected as managed nature reserves, which in the context of a cultural landscape means, that they are grazed or mowed according to conservation goals.

In the Rhön biosphere reserve it has proven to be successful that
- the transboundary Management Framework was prepared by an external private consultant who not only had profound expertise in landscape planning, but played the role of a moderator, mediator and coordinator in this process,
- the trilateral Management Framework itself is a profound but informal and not legally binding document, as this made the agreement and con-sensual cooperation beyond state boundaries much easier. Each unit of the biosphere reserve however commits itself to the same overall goals which have been agreed upon.

As a result of the planning process, the most important goal and vision of the Rhön biosphere reserve, which had been agreed upon by all stakeholders, was to conserve the Rhön cultural landscape with its high biodiversity of species and habitats for the sake of both people and nature. This was to be done in the context of a sustainable regional development.

It has furthermore been recognised, that traditional land use on a low level of intensification (meaning low input of fertiliser or pesticides, low grazing intensity) has created today's high biodiversity. Agriculture was identified as the key factor.

With more and more farmers giving up due to economic reason, the major threat to the cultural landscape will be fallow land or afforestation. On the other hand, farming intensity on the remaining more fertile plots may increase potentially endangering abiotic resources.

This would alter the ecosystems and destroy both the ecological and aesthetic value of the landscape and will thus negatively influence tourism business, which is bound to the beauty of the landscape.

Countermeasures, aimed at an increase of cultural landscape values were identified:

- strengthening the local farmers by processing and direct marketing of regional products; thus increasing the farmers' benefit,
- enhancing eco-tourism
- the linkage of both.

Specific quality products of the Rhön biosphere reserve are beef (traditional landrace breeds), Rhön sheep (traditional breed), Rhön apple (traditional varieties), Rhön brown trout (*Salmo trutta*), traditional farm bread made in wood stoves or, as to the non-food sector, Rhön beech timber (*Fagus sylvatica*). Main eco-tourism activities are hiking, cycling and mountain biking, cross country skiing and horse trekking, plane sailing and paragliding.
A short version of the Management Framework can be downloaded at: www.biosphaerenreservat-rhoen.de

The goals of the Management Framework were intended to be implemented in different stages of official planning procedures (regional planning, municipal planning, sector oriented land use planning for agriculture, forestry etc.). This, however, has only partly happened until now.

Furthermore, each biosphere reserve unit has to find its own appropriate measures to put the goals into action. Also every unit sets its own priorities, time frame and financial resources for implementing the goals.

Sustainable Development Beyond Administrative Boundaries

1.2 Networking and trilateral cooperation on different levels

Implementing the biosphere reserve goals is vitally depending on networks and on cooperation on all levels, both horizontally (e.g. cooperation among farmers or cooperation among restaurants) and vertically (e.g. cooperation of farmers with gastronomy or the cooperation of hunters with non-governmental nature conservation organisations).

In a transboundary or trilateral context, cooperation becomes even more important and can become quite complex. Trilateral cooperation is dealt with on different levels and in different sectors and is described by different examples:

1.2.1 Examples of trilateral cooperation in the private business sector

Cooperation on the private sector concerns e.g.:

- Organic farming: The enterprise "Rhönhöfe" is a joint venture of several organic farms in the three biosphere reserve units, ranging from small family farms of less than 20ha each, up to formerly collective farms of more than 2000 ha. The "Rhönhöfe" market their produce in farm stores, farmers markets and offer a delivery service for households.

- Gastronomy: "From the Rhön – for the Rhön" is a trilateral co-operation between 10 restaurants and a number of farmers in the region. The restaurants changed their menu to regional cuisine and have a commitment to buy a certain percentage of their produce on regional farms. A similar initiative is "Charming Rhön", consisting of 60 restaurants. Their standards are less strict, yet they promote regional dishes and local produce as well.

- Timber: The "Rhön wood processing initiative" is a small group of forest enterprises, sawmills and carpenters which work together. The initiative promotes the processing of timber, mainly beech wood, for high quality

solid wood furniture, wooden staircases, floors and others. Recently, a trilateral cooperation of further wood enterprises was established covering additional sectors, such as wood energy production, wooden arts and wooden/ wood-frame houses.

1.2.2 Examples of trilateral cooperation in the institutional sector

1.2.2.1 Cooperation between the biosphere reserve administration units

For each of its parts, the Rhön biosphere reserve has an administration unit. The three administration units are public institutions of each federal state, belonging to different administrative levels: to the district council (for Hessen) and to the regional council (for Bavaria); in Thuringia they are directly linked to the ministerial level. Number and qualification of staff as well as tasks of each biosphere reserve administration unit is within the jurisdiction of each state and varies accordingly.

Altogether the three units consist of 10 professional academic staff.

The units work independently on implementing the biosphere reserve goals which had been agreed upon in the "Management Framework". All issues of trilateral importance are discussed regularly and in depth by the head of the three administration units who meet once a month. In a three years' rotation system one unit takes the lead for common issues and projects of trilateral importance in terms of managerial or secretarial business.

Trilateral projects are financed out of a trilateral budget which is based on a trilateral administration treaty. One example is the common Geographical Information System (GIS), which is financed by the three units. Another example of trilateral cooperation is public relations work, which is done by a freelance journalist who works for the three biosphere reserve administration units and prepares and distributes press releases on all relevant issues and activities of the biosphere reserve.

A trilateral advisory committee, consisting of local and outside experts who represent agriculture, tourism or industry, advises and accompanies the biosphere reserve administration units.

It has no decision - making function.

Research is coordinated and organized trilaterally between the three units in terms of defining research needs and avoiding overlapping inventories. Research projects which are financed trilaterally are dealt with by the trilateral research coordinator.

Still every biosphere reserve unit is independent in carrying out research projects and allocating budgets. Recently a trilateral "Open Research Forum" was put in place, which is a discussion platform of both biosphere reserve administration units and local stakeholders in order to suggest applied research topics which could be beneficial to the region.

Activities of environmental education are coordinated in a similar way.

There are several interpretation centres in the biosphere reserves, whose issues of exhibitions and activities are coordinated trilaterally in order to clearly profile the different interpretation centres and avoid overlaps.

Yet all centres are run within the jurisdiction and the budget of the individual states.

1.2.2.2 Cooperation of districts: The Rhön Regional Work Group

The biosphere reserve covers parts of five districts. Due to the initiative of the biosphere reserve administrations, the five district councils took part in a nationwide competition, which was called "regions of future". The fact that the Rhön was among the proud winners raised the interest of the local politicians in the biosphere reserve, as they realized that the biosphere reserve would be an asset to the profile of the Rhön region.

The organisational structure, which had been set up during the joint project, was the basis of what is now the Rhön Regional Work Group. It is committed to sustainability and to the goals of the biosphere reserve, but the area concerned goes even beyond the biosphere reserve boundaries, thus it is linking the biosphere reserve with adjacent areas.

Three subgroups focus on regional labelling, tourism and public relations. The workgroup runs its own internet platform at: www.rhoen.de

The creation of the Rhön Regional Work Group can be considered as the breakthrough of shifting responsibility for the development of the Rhön biosphere reserve from government administration level to the political level (to the local people respectively to their representatives).

The biosphere reserve process which (historically speaking) had been driven by a "top down" process for many years, only since recently is backed by a vital "bottom up" process.

Examples for joint trilateral projects

- Based on funding by the European Union LIFE programme, two major nature conservation projects with a total funding of 15,4Mio € could be accomplished beyond Länder boundaries from 1993 to 2001. Each of the Rhön biosphere reserve administration units together with stakeholder groups set up a range of measures to be taken within the programme. A joint project coordinator was employed to manage the trilateral project accordingly.

- A common label for the Rhön has been developed beyond state boundaries. This „Rhön Regional Label" is meant to make the Rhön better known and support the region and its economy. It can be used trilaterally by municipalities or institutions of the five Rhön districts. A quality label for organic agricultural products will advertise regional products and help the customer to identify them.

- A trilateral project on hiking, cycling and mountain biking tourism is in process, aiming at improving and harmonizing the trails signs beyond state boundaries. These show striking differences because they had been set up by different institutions. Another task is to identify hiking trails and tours in the Rhön which may be suitable for national or international tourism marketing.

1.2.3 Examples of trilateral cooperation in the non-governmental sector

Besides cooperation on the institutional level there is a close cooperation also on the level of non-governmental organisations, for example:
- Trilateral cooperation for Rhön apple orchards

Fruit orchards are a typical feature of the Rhön landscape, may it be around the villages or in the countryside. Apart from their aesthetic values, they provide important habitats for wild plant and animal species. As fruit from traditionally managed orchards had however become less competitive on the markets, traditional orchards decreased dramatically throughout the country in the last 20 years.

In 1996 a transboundary apple initiative for the conservation of orchards was founded by locals, which set up a joint apple project in co-operation with the biosphere administrations.

With the financial help of the EU-LEADER programme and of a private enterprise, an extensive inventory of the genetic potential of fruit varieties was carried out as a transboundary research project. Every land owner who had fruit orchards was invited to have fruit specimen identified by hired scientists. Thus 170 apple varieties, 38 pear and 12 plum varieties were found in the Rhön.

Different private enterprises joined the initiative trying to find best marketing potentials for the different varieties. By now, a wide range of organic apple products is sold on the regional and national markets, including Rhön apple juice, a mixed drink of beer and apple juice, apple chips, apple champagne, cider, vinegar, mustard, jam etc.

Apple fairs, an apple interpretation trail and an in situ conservation site for rare apple tree varieties and the publication of a recipe book for regional apple dishes support the marketing strategies.

As a result, the price for apples from traditional orchards has in-creased four times compared to the beginning of the 1990. By now, land owners are motivated to invest in organic apple orchards as there is already of shortcoming of fruit in the region.

- Trilateral cooperation for species conservation

The black grouse *(Tetrao tetrix)*, which is a large ground breeding bird, used to be quite common in the Rhön until the 1970ies. Since then, the population dropped significantly. It seems to be due to various factors such as climate change, afforestation, changes in landscape management and increasing tourism.

Therefore conservation measures are run by the public sector such as the designation of large nature conservation sites or control and guidance of visitors.

Non-governmental organisations also got involved in the black grouse protection: A transboundary work group consisting of members of the local hunters association, non-governmental nature conservation organisations, district and region nature conservation administrations and others meets regularly and discusses hands-on conservation measures for this bird species. They raised money from both governmental sources and foundations to run a black grouse monitoring and research project and trilaterally tailor conservation and land management measures to the protection goals.

2. STRUCTURES AND DRIVING FORCES: WHO MAKES THE BIOSPHERE RESERVE WORK?

Biosphere reserves depend on cooperation in many aspects. They deliver action at the local level, are voluntary rather than regulatory, involve the co-ordination of numerous stakeholders rather than

single agency management and have multiple benefits in terms of conservation goals, economic values and social well being – for example job opportunities.

They bring together disparate groups into new partnerships, are forward looking, yet seek to match action with conservation of landscape qualities and quality of life, and they solve occurring conflicts through approaches that focus on win-win solutions.

The most important driving force are the local stakeholders who are proud on their region, have a strong regional identity and in depth practical knowledge of the region and its peculiarities. People's creativity and belief in the region's potential, their willingness to co-operate beyond spatial, jurisdictional or professional boundaries for the sake of both people and nature are the most important assets and driving forces to make a biosphere reserve work.

Tasks of the biosphere reserve coordinators concern the integration of biosphere reserve's visions and goals in regional conservation and development concepts. Biosphere reserve coordinators motivate and bring partners together, moderate and mediate, coordinate projects and set thematic and spatial priorities, assist in searching money for model projects and accompany projects, e. g. through applied research.

In a transboundary or multilateral context as in the Rhön biosphere reserve, this becomes quite complex but it is all the more necessary especially in order to overcome "virtual boundaries" which often exist in people's minds.

There is no set organisational structure designed to coordinate individual BRs. Each reserve designs a tailored organisational structure based on its own unique situation. The task is to bring together public responsibility for a biosphere reserve ("top down" in terms of financial and human resources) and local initiative ("bottom up" in terms of ideas and activity). The challenge for BRs is to find the proper balance between the "top down" and "bottom up" approaches, so that both appropriate resource inputs from government and volunteer passion with connection to local communities can be maximised. Coordination rather than management techniques seem to be the right tool since biosphere reserves have a non-regulatory and consensus mandate.

The biosphere reserve coordination can thus be part of a public administration, or it can be a new or existing not-for profit organisation, or a combination of both. Usually a steering committee (representing government institutions and local stakeholders), a technical advisory committee (with representatives of science and advisors from different agencies) and a technical coordination office which does the operational work are needed.

In a transboundary context, a coordination group representing all countries involved is needed in order to harmonize and agree upon projects and measures which are of transboundary importance.

3. CONCLUSIONS AND RECOMMENDATIONS FOR THE ALTAI REGION

Based on the experience from the trilateral Rhön biosphere reserve and backed by the experience of other transboundary biosphere reserves (UNESCO 2000), it can be recommended for the countries in the Altai region to

- identify local and national partners of a transboundary work group in order to define strengths and weaknesses, threats and potentials of the Altai region and define those which are of transboundary importance for the planned biosphere territory as key issues for co-operation.
- agree upon common visions and goals in a transboundary context
- work out on this basis a suitable zoning according to UNESCO's Statutory Framework for Biosphere Reserves. It should enable a fully functional biosphere territory unit in each country yet ensuring a consistent zoning system beyond state boundaries (e.g. taking into consideration transboundary watershed issues for the protection of abiotic resources or transboundary home ranges of a viable snow leopard population etc.)
- implement a focal point for each biosphere territory unit in each country
- set up a (lean) transboundary coordination structure to ensure that joint projects can be put in place, yet respecting and ensuring the countries' full independence. This could be e.g. a permanent secretariat jointly financed by the member countries or it could be implemented by a rotating system where the focal points take the overall coordination function in turns.

- set up a co-ordinating steering group representing various administrations and sciences as well as the authorities in charge of the protected areas, representatives of local communities, interested and affected groups, including youth, and of the private sector and (regional) non-governmental organisations. The task is to put the goals into action by transboundary joint projects and apply for the necessary financial and human resources both nationally and internationally.

- use the world network of biosphere reserves consisting presently of 18 transboundary biosphere reserves or territories for the exchange of experience and as models, in order to tailor an appropriate structure for a future transboundary biosphere territory in the Altai.

More practical recommendations on transboundary biosphere territory coordination can be found at UNESCO (2000) and downloaded at: http://unesdoc.unesco.org/images/0012/001236/123605m.pdf

Last but not the least: Although transboundary processes take a lot of time and effort, they will be worth it and in the long run provide a better future for nature and new opportunities for the people.

Contact: doris.pokorny@brrhoenbayern.de
©photos: Vogel and BRR

REFERENCES

1. Grebe R. et al. (1995): Biosphärenreservat Rhön. Rahmenkonzept für Schutz Pflege und Entwicklung. Neumann-Verlag.
2. UNESCO (1996): Biosphere reserves. The Seville Strategy and the Statutory Framework of the World Network. UNESCO Paris.
3. Pokorny D., Whitelaw G. (2000): "Sustainability through Transdisciplinarity? The Biosphere Reserve Concept as Opportunity and Challenge". In: Transdisciplinarity: Joint problem solving among science, technology and society. International Transdisciplinarity Conference 2000. Proceedings (Workbook 1 Dialogue sessions and idea market) pp. 425-430.
4. UNESCO (2000 ed.): Recommendations For the Establishment and Functioning of Transboundary Biosphere Reserves in: Seville+5. Proceedings on the International meeting of experts, Pamplona, 25-27 October 2000; pp. 55-58.

THE LINKS BETWEEN POVERTY AND ENVIRONMENT: THE RATIONALE FOR ENVIRONMENTALLY SUSTAINABLE RESOURCE USE, WITH APPLICATION TO LAND MANAGEMENT IN THE ALTAI REGION

D. H. SMITH
Economist, Division of Policy Development and Law, United Nations Environment Programme, Nairobi, Kenya

Abstract: This paper outlines the links between the environmentally sustainable use of land and water resources and sustainable economic development, including poverty reduction. In many parts of the world the degradation of land and water resources is worsening while the social and economic conditions of people are not being improved. In some parts of the world, both poverty and environmental degradation are increasing. One reason for this is the perception held by some that the sustainable management of the environment and economic development are competing priorities. While there are specific cases where measures to end environmentally unsustainable actions will restrict human use of environmental resources in the short term, over the long term environmentally unsustainable land and water use will reduce the social and economic benefits to humans provided by the environment. Thus, the environment should not be treated as a competitor but as a core component of the natural resource base of human social and economic development. The paper provides specific examples relevant to the Altai region to illustrate the nature of the link between environment and development, including poverty reduction. It also outlines aspects relevant to regional security and provides elements for environmentally and economically sustainable land management.

Keywords: Economic development; environment; land management; poverty; sustainable development, transboundary management; water management.

1. INTRODUCTION – LAND DEGRADATION

Land, along with water, is a fundamental requirement for human survival and socio-economic development. While levels of dependency on land resources for socio-economic development vary in different parts of the world, in regions like the Altai, land remains a key economic resource. As the recent draft report on the Altai biosphere proposal highlighted, the Altai region's economy is heavily dependent on land-based resources - agriculture is the dominant economic activity, with, for example 76% of the population employed in the agricultural sector of the Altai Republic. (ECO 2004).

Unfortunately, land degradation is a serious problem in parts of the region and also globally. As a result, the economic and social well-being of hundreds of millions of people is being directly or indirectly threatened by environmentally unsustainable land use, which is resulting in major physical damage and diminished economic benefits.

The extent of land degradation globally is illustrated by the following: (UNEP 1999; 2002):

- It has been estimated that 23% of all usable land has been affected by degradation to a degree sufficient to reduce its productivity.
- In the early 1980s, about 910 million hectares of land were classified as 'moderately degraded', with greatly reduced agricultural productivity.
- A total of 296 million hectares of soils were 'strongly degraded' and 9 million hectares 'extremely degraded'.
- Around 3,600 hectares, or 70% of the world's drylands (excluding hyper-arid deserts), are degraded.
- Some 25-30 million hectares of the world's 255 million hectares of irrigated land were severely degraded due to the accumulation of salts.
- An additional 80 million hectares were reported to be affected by salinization and waterlogging.
- In the 1980s it was estimated that about 10 million hectares of irrigated lands were being abandoned annually.
- About 2,000 million hectares of soil, equivalent to 15% of the Earth's land area, has been degraded through human activities.
- Worldwide about 1.2 million square kilometres of land have been converted to cropland in the last 30 years, with such habitat loss a principal factor affecting 83% of threatened mammals and 85 % of threatened birds.

Soil erosion is a major factor in land degradation, and has severe effects on soil functions – such as the soil's ability to act as a buffer and filter for pollutants, its role in the hydrological and nitrogen cycle, and its ability to provide habitat and support biodiversity. The main causes of soil degradation include overgrazing (35%), deforestation (30%), agricultural activities (including irrigation) (27%), overexploitation of vegetation (7%), and industrial activities. (1%) (UNEP, 2002).

It is a fact that environmentally unsustainable land use over time reduces land productivity and therefore the social and economic benefits produced by such land. Thus, over time, environmentally unsustainable land use can increase poverty.

Additionally, if land use in one country has negative impacts on neighbouring countries, increased political tension can result. For example, illegal forest burning in Indonesia caused serious air pollution in areas of Malaysia and in Singapore, raising political tensions. Deforestation in one area of the Altai could have significant downstream impacts on another area.

Table 1: Extent and Causes of Land Degradation

Degradation extent	Cause
580 million ha	Deforestation — vast reserves of forests have been degraded by large-scale logging and clearance for farm and urban use. More than 220 million ha of tropical forests were destroyed during 1975–90, mainly for food production.
680 million ha	Overgrazing — about 20 per cent of the world's pasture and rangelands have been damaged. Recent losses have been most severe in Africa and Asia.
137 million ha	Fuelwood consumption — about 1 730 million m^3 of fuelwood are harvested annually from forests and plantations. Woodfuel is the primary source of energy in many developing regions.
550 million ha	Agricultural mismanagement — water erosion causes soil losses estimated at 25 000 million tonnes annually. Soil salinization and water logging affect about 40 million ha of land globally.
19.5 million ha	Industry and urbanization — urban growth, road construction, mining and industry are major factors in land degradation in different regions. Valuable agricultural land is often lost.

Source: UNEP 2002

In recognition of the seriousness of the land degradation problem, the nations of the world have agreed to take action to halt land degradation –

through adopting Agenda 21, the Johannesburg Plan of Implementation at the World Summit on Sustainable Development, the UN Convention to Combat Desertification and the Convention on Biological Diversity. The World Summit on Sustainable Development (WSSD), held in 2002, highlighted the natural resource base of sustainable development. Thus, in relation to land and water, to achieve sustainable economic development, including poverty reduction, it is necessary to manage our land and water resources sustainably. It is stressed that this is not primarily for the sake of the environment; rather it is to secure sustainable economic and social development for people, especially poor people.

Yet, land degradation problems are worsening in many parts of the world. There are a number of reasons for this, including a lack of capacity, inappropriate land policy and management techniques, insufficient financial resources and a lack of political commitment. A failure to apply integrated land and water management is a key problem.

Another reason why insufficient action is being taken to reduce land degradation is that the perception still exists in too many cases that the issue is one of development versus environmental protection. That is, that reducing land degradation comes at the expense of economic development and that protecting the environment will hamper economic development. While in specific cases, there will be instances where human activities will have to be restricted to achieve environmentally sustainable land or water management, in general terms over time, unless environmentally sustainable land and water use are achieved, people will suffer economic and social costs. That is, over time, environmentally sustainable land use is an essential factor in achieving sustainable economic and social well-being.

2. THE COSTS OF ENVIRONMENTALLY UNSUSTAINABLE LAND USE

The table below outlines some of the negative economic and social impacts of environmentally unsustainable land use. It is not exhaustive, but highlights that reduced production, increased input costs and health costs are real problems – all of which reduce net incomes. Thus, over time, attempts to reduce poverty will be inhibited by environmentally unsustainable land use.

Table 2: Costs of environmentally unsustainable land use

Environmentally unsustainable action	Result	Economic impacts
1. Overgrazing	• Reduced grass growth • Soil degradation • Erosion	• Decreased numbers of livestock • Decreased productivity • Reduced incomes
2. Water pollution from pesticide & fertiliser use and/or from farm livestock effluent	• Water use restricted, depending on degree of pollution. • Increased incidence of water borne diseases	• Reduced farm production due to restricted availability of safe water • Increased water treatment costs • Increased cost of obtaining safe water • Increased incidence of ill-health and medical treatment costs. • Decreased incomes
3. Depletion of water resources through excess withdrawals from surface & groundwater sources	• Decreased water availability	• Decreased production as less water is available. (Fewer livestock, less crops) • Increased costs of obtaining water • Decreased net incomes.
4. Inefficient irrigation	• Soil salinization • Other soil degradation • Water waste	• Larger water bills • Decreased soil productivity, production & incomes. • Ultimately, complete cessation of livestock or crop production.
5. Deforestation – i.e. failure to manage forests for sustainable yields & complete cutting of areas of forest	• Erosion causing increased sedimentation downstream, reduced water retention due to increased run-off, soil degradation, less timber. • Increased vulnerability to flooding	• Decreased productivity downstream from sedimentation • Decreased rainfall as deforestation impacts on localised climate patterns. • Decreased productivity on site over time if erosion & soil degradation occurs • Decrease in timber resources. • Decreased incomes & increased costs.

In the Central Asian region, the case of the Aral Sea basin is the best known example of how environmentally unsustainable land and water use can have serious economic consequences. The environmentally unsustainable use of land and freshwater resources for agricultural production in the Aral Sea basin has decreased lake water levels and quality severely. As a result, the fishing industry has collapsed. In addition, the inefficient use of irrigation water in this semi-arid region has led to salinization and a subsequent decrease in agricultural production. Serious human health problems have arisen with wind blown dust contaminated by agricultural chemical residues. (UNEP, 1993; UNDP, 1996).

Parts of the Altai region suffer from overgrazing and deforestation. Water pollution and excess water withdrawals are also significant problems in parts of the region, and are also related to land degradation.

Two specific examples from countries in the Altai Region reinforce the points made above (International Water Management Institute):

Xian County, Fuyang River Basin, North China Plain: Formally dependent on surface water resources, increasing water demand and upstream developments have led to an increasing dependence on groundwater in this area. However, excess withdrawals have made the water table fall substantially. In order to obtain sufficient water, tubewell depths have increased from 25 metres in the 1980s to 200 metres plus in the 1990s. The total cost of a 200 m tubewell is about US$3,900, nearly four times the cost of a 25 m tubewell. As water supplies have declined, grain yields have declined by 25%.

Jambul, Kazakhstan: Located in the Aral Sea basin, this area has suffered from increasing salinization due to poor irrigation practices. Salinization is reducing farm productivity and contaminating drinking water supplies. Grain yields are declining as a result of salinization, from 4 tonnes per hectare to less than 2, in one case study.

In summary, unsustainable land use imposes costs through reduced production and health problems. While there is much focus on the costs of environmental protection, there is not enough focus on the costs of using environmental resources such as land and water unsustainably. This is partly because these costs are often borne by those who have the least say - the poor. Over time, poverty can be increased by unsustainable land and water use. However, this link between poverty and the environment is not sufficiently understood.

A key to understanding the contribution of environmentally sustainable land and water use to economic development, including poverty reduction, is the concept of ecosystem services.

3. ECOSYSTEM SERVICES: THE BASIS FOR THE CONTRIBUTION OF THE ENVIRONMENT TOWARDS ECONOMIC DEVELOPMENT AND POVERTY REDUCTION

The environment generates social and economic benefits for humans through ecosystem services which are defined as the conditions and processes through which natural ecosystems contribute to human economic and social well-being. Ecosystem services provide benefits for humans through:

- Provisioning: - natural resources used for economic activities
    e.g. Food, fuel, plant and animal products, energy, fiber, non-living resources and water

- Regulating - life supporting functions for humans
    e.g. purification of air and water, mitigation of floods and droughts, decomposition of wastes, generation and renewal of soil

- Enriching - cultural and religious services
    e.g. spiritual components, aesthetic values, social relations, education and scientific values

Relating the impacts of land degradation listed in the table to the ecosystem services listed above indicates how land degradation, over time, reduces the ability of land to provide provisioning, regulating and enriching services. Overgrazing that results in reduced grass cover, erosion and soil degradation reduces production and incomes, reduces the ability of land to purify itself from pollution, increases the risk of flooding and inhibits the renewal and regeneration of soil. Land degradation also impacts on aesthetic values – for example, tourists do not wish to spend money to view eroded and deforested areas.

It is also important to note the interdependency between the components of ecosystem services. For example, over-harvesting, overuse, misuse or excessive conversion of ecosystems into human or artificial systems damages regulating services, which in turn reduces the flow of the provisioning service provided by ecosystems.

All people depend on services provided by ecological systems. Yet, the poor are more heavily dependent on these services than the rich, since the rich can buy food, or clean water or build appropriate shelters to isolate themselves from economic and social problems caused by land degradation, or other environmentally unsustainable behaviour.

Unfortunately, the link between environmental sustainability and poverty reduction is not sufficiently understood or integrated into economic developing planning. Thus, poverty reduction strategies fail to reflect that ecosystem services are vital contributors to human social and economic well-being. More specifically in the context of land use, poverty reduction strategies fail to include policy, management and capacity building actions to achieve environmentally sustainable land use in a manner that reduces poverty.

Over time, in order to reduce poverty on a sustainable basis, it is necessary to use resources such as land and water in an environmentally sustainable manner. A key to achieving environmentally sustainable land and water use is to fully integrate the objective of the achievement of environmental sustainability into poverty reduction strategies, plus national and sectoral development strategies and plans. In terms of specific management approaches, integrated land-water management is the most important tool. This management tool is referred to as Integrated Water Resources Management (IWRM), and despite its name, it is integrated land and water management. It is also integrated economic-social-environmental management.

In recognition of the importance of integrating environmental sustainability into economic development and poverty reduction strategies, the United Nations Environment Programme has a poverty and environment project. This project aims at increasing the capacity of developing countries to mainstream environment and ecosystems in their development strategies. Such mainstreaming will ensure that poverty reduction is not undermined by unsustainable use of environmental resources.

The project will assist countries to identify how environmental sustainability and poverty reduction are linked in their country. That is, how specific environmental resources such as land contribute to the economy and how environmentally unsustainable use of such resources will make it more difficult to reduce poverty in the long term. Then the projects will demonstrate how such environment-poverty linkages can be integrated into poverty reduction strategies and national development plans in order to ensure development and poverty reduction strategies are sustainable. For example, the project will demonstrate how countries can ensure that the policies, management tools, legal frameworks and institutions in their national development and poverty reduction strategies are consistent with sustainable land and water use in order to ensure that economic growth and poverty reduction is not undermined by environmentally unsustainable behaviour.

The United Nations Development Programme (UNDP) has a similar and larger programme, called the Poverty-Environment Initiative, and UNEP and UNDP are acting in close collaboration, with a view to developing a global partnership on poverty and environment.

4. ENVIRONMENT FOR DEVELOPMENT

While the perception that the issue is one of environment versus development, the stress at the World Summit on Sustainable Development (WSSD) on the natural resource base of sustainable development, plus the discussion above, highlight that it is more accurate to refer to how sustainable use of the environment and economic development, over time, reinforce each other – particularly in countries heavily dependent on natural resources such as land and water. That is, it is more realistic to refer to environment for development.

While some may argue that there is no choice but to use the environment in an unsustainable manner – for example, to continue to overgraze because there is no alternative – these arguments are rarely based on an assessment of the costs and benefits of such environmentally unsustainable actions over time. What, for example, are the true costs of overgrazing - in terms of lost productivity in the long term? Or the true costs in terms of the cost of restoring degraded land? (Even if restoration is possible, it is likely to be more expensive than sustainable management would have been in the first place and to take many years to achieve). Often environmentally unsustainable activities merely postpone the day when alternative economic activities have to be found. In other words, sustainable management now will save money later.

In summary, it is a mistake to assume that environmental sustainability inhibits economic development over time. Rather, the environment is an essential pillar of economic development and thus environment sustainable resource use should be a key objective of economic planning.

5. CONTRIBUTION OF ENVIRONMENTALLY SUSTAINABLE LAND USE TOWARDS REGIONAL SECURITY IN THE ALTAI REGION

If there are transboundary impacts from land degradation, then this could lead to increased political tension. While in the case of transboundary waters, impacts are more immediate and direct when compared with land, environmentally unsustainable land use can have transboundary impacts also. There may be negative impacts on the environment in one country that increases risks and costs and decreases economic benefits in other countries.

For example, deforestation in the upper part of a water basin can cause sedimentation, increased risk of flooding and unstable or reduced mean water flows in downstream countries. Deforestation can also lead to reduced rainfall in adjacent regions as climate patterns can be altered by deforestation. As another example, overgrazing in one country may lead to pressure for livestock to use pasture in a neighbouring country. (In some border areas of Kenya and Uganda, cross-border armed clashes over grazing occur, particularly when overgrazing and drought combine to put great pressure on land resources).

Environmentally sustainable land use over time maintains the quality and quantity of the benefits generated from specific land resources over time. Where there are transboundary impacts, this reduces the potential for increasing competition over a diminishing resource. Therefore the potential for disputes is reduced when environmentally sustainable management is practiced.

The application of transboundary management can also reduce the potential for disputes. For example, if there is great pressure on pasture in one country, but across a nearby border, pasture lands are under-utilised, managing the relevant pasturelands as an integrated, transboundary whole and permitting cross-border grazing could well reduce environmentally unsustainable land use and enhance cross border relationships.

A number of factors –political, geographical, economic, social and environmental – combine to determine the state of regional security and it is necessary to fully include the environmental dimension in regional security considerations. Given the transboundary nature of land and water environmental inter-linkages in the Altai region, it is important that a transboundary approach is taken to manage the region, to maximise the benefits generated by the region as a whole and to share them equitably to, *inter alia*, reduce the potential for regional disputes.

In summary, environmentally sustainable land use managed on a transboundary basis is consistent with reducing the potential for regional insecurity. Below key elements for environmentally sustainable management are set out.

6. ACHIEVING ENVIRONMENTALLY SUSTAINABLE LAND USE: SOME KEY ELEMENTS, INCLUDING TRANSBOUNDARY ASPECTS

Arguably, the most important element for achieving environmentally sustainable land use is real commitment by the relevant decisions makers at all levels that environmentally sustainable land use is a core objective of land use policy. With such commitment, it is far more likely that the necessary policies and management plans will be developed and implemented.

Where such commitment is not present, highlighting that environmentally unsustainable land use will hinder economic development and increase poverty over time can be important in achieving the necessary commitment.

Full integration of environmental sustainability into poverty reduction strategies, national development plans plus sectoral policy and management plans.
As indicated above, the rationale for this is that economic development and poverty reduction are enhanced by environmentally sustainable land use and that using land resources in an environmentally unsustainable manner reduces the economic and social benefits that land generates.

Application of integrated management approaches based on environmentally sustainability as a core goal.
- Ecosystem management

Ecosystem based management ensures that the relevant ecosystem inter-relationships – plus their economic and social implications - are taken into account when managing the environmental issues under consideration. A key aim of ecosystem management is to ensure that the social and economic benefits generated by ecosystems are not degraded to the extent that human economic development is undermined.
- Integrated land and water management including integrated environmental-economic-social management

Integrated water resources management (IWRM) is integrated land and water management, and is the internationally accepted best practice management process for land and water management. Land and water are closely linked in ecological, economic and social terms, thus IWRM incorporates integrated environmental, economic and social management, and is also consistent with sustainable development, which has three elements - social, economic and environmental.

- Multi-objective planning.

A key tool for integrated, inter-sectoral approaches is multi-objective planning that bases management decisions on an integrated assessment of environmental, economic and social factors. The elements of multi-objective planning are:

    i   Cost-benefit analysis from the national perspective
    ii   Cost-benefit analysis from the project or regional perspective
    iii   Environmental impact analysis
    iv   Social impact analysis (usually non-monetised.)

Cost-benefit analysis should include economic, social and environmental impacts. If it does not, then it will not include all the relevant impacts. Some impacts cannot be costed, but they should at least be identified.

UNEP has a great deal of experience in transboundary management issues, and the steps below indicate a sound basis for successful transboundary management, incorporating the elements listed above. The importance of different steps will vary according to circumstances, and steps may occur at the same time or in a slightly different sequence. However, commitment from all parties is an essential first step. From the perspective of the international community, a Ministerial level commitment by all countries is a clear signal that makes support realistic, including in terms of fund raising. It is important to emphasise that transboundary management does not mean that governments lose sovereignty. Rather it means that governments work together to manage a transboundary resource.

6.1 Commitment from all states to address all issues on an equitable basis

As indicated above, without this commitment, successful transboundary management to the greatest mutual benefit is very difficult. An example of how such commitment can be stated is when relevant Ministers release a declaration stating that their respective countries will work together on a specific issue to achieve a certain goal, outlining key elements and principles for the way forward.

6.2 Adoption of legal frameworks for integrated management

Legal frameworks for integrated management should be based on broader UN Conventions and agreements, if applicable, and do not need to be complicated documents covering many pages. The purpose is to set out principles, objectives and negotiation mechanisms to address disagreements. More specific legal agreements can follow specific programme recommendations.

6.3 Establishment of institutional mechanisms with clear mandate & funding

Successful and cost-effective transboundary programme design and implementation requires establishment of institutional mechanisms that are dedicated to the transboundary issue under the consideration. It is essential that any transboundary institutions have a clear mandate and the staff and financial resources necessary for their task. Dedicated institutional mechanisms can also assist in raising finance from donors, who prefer to have a legally established mechanism to which they can provide financial support. It is stressed that such transboundary institutions do not mean any one country loses sovereignty. Typically, transboundary institutions make re-commendations to a body made up of representatives of each country. This could be a ministerial level council with decisions taken by consensus.

6.4 Development of a comprehensive transboundary management programme and projects

To realise the potential economic, environmental and social benefits of a transboundary environmental asset, management - in an environmentally sustainable manner - requires an operational programme and projects designed and implemented on a transboundary basis. Experience shows that this should include both an assessment (or diagnostic) study and a strategic action programme including subsidiary projects for the transboundary asset under consideration, (For example, these projects could cover capacity building, economic development, reforestation, pastureland management, water quantity and quality, biodiversity preservation etc). These projects should be prioritised and implemented in an integrated manner.

- Assessment (diagnostic) study
  - Identify transboundary environmental, economic & social inter-linkages.
  - Identify environmental, economic & social issues and their underlying causes.
  - Identify opportunities for economic development based on sustainable use of ecological resources
  - Identify priorities for environmentally sustainable development and protection

- Strategic action programme
  - A comprehensive programme consisting of projects and processes to address the issues and achieve the priorities identified in the assessment study. This could include reform to national policy and laws.

6.5 Benefit sharing frameworks

Central principles of transboundary management include benefit maximisation and benefit sharing. That is, seeking to maximise the benefits generated by the transboundary asset as a whole and then sharing those benefits in a mutually agreed, equitable manner. (Note – equitable sharing does not mean sharing resources or their benefits equally, it means sharing them in a fair manner, reflecting different resource endowments, development and protection efforts combined with transboundary inter-linkages and impacts.).

A core difference between national and transboundary management is that benefits and costs in one country can be changed by actions by another country. Related to this, to maximise benefits from a transboundary asset as a whole - for example a river or biosphere reserve – could mean taking action in one country that generates more benefits in a different country. (For example, planting trees in a border area that has few immediate benefits in the planting country, but improves water flow and quality in the neighbouring country, which gains economically and in turn increases the overall benefits generated by the transboundary asset). The first country may not agree to this unless there is a mechanism by which it will share in the benefits generated in the next country or in the costs incurred in planting the trees. Similarly, action by one government could create costs in another country – for example, polluting a river near a border with the pollution flowing across the border to cause illness and water treatment costs.

Thus, while a central aim of transboundary management is to maximise benefits, it is important that benefit-sharing mechanisms are developed so that benefits are shared in an equitable manner. For example, if a country dams a river and benefits from generating electricity, but imposes costs and environmental damage in a downstream country, it could compensate the downstream country financially or with electricity.

The recommended approach is to develop a plan to maximise overall environmental, social and economic benefits from the transboundary asset, and then negotiate benefit sharing, using a previously agreed benefit sharing framework. Then implementation of the plan is more likely to be supported by all parties.

6.6 Policy and law harmonisation

Management of a transboundary asset, such as a river, contiguous pasturelands, or a biosphere reserve, will partly depend on the policy and laws in each state; therefore it is important that these policies and laws are in harmony. That does not mean that identical laws and policies are required. It does mean that, at least, any inconsistencies that hamper joint management of a transboundary asset to maximise and share benefits in an environmentally sustainable manner need to be removed. For example, if an upstream country has pollution regulations that permit significant pollution, which then impacts on a downstream country, it is recommended that these pollution regulations be reformed to reduce pollution.

In closing this section it is important to stress that transboundary management is a voluntary exercise freely entered into by the relevant countries. The role of the United Nations is, upon the request of governments, to assist.

7. CONCLUSION

Land degradation is a serious problem in parts of the world, including in parts of the Altai Region. Environmentally unsustainable land use is a threat to economic development – including poverty reduction – and the environment. Over time, land degradation - through, for example, overgrazing resulting in soil erosion - reduces incomes, increases costs and worsens poverty. This is a reflection of the fact that land and water are vital components of the natural resource base of sustainable development. Ecosystems generate social and economic benefits for humans and these benefits are, over time, decreased and even destroyed by environmentally unsustainable activities such as overgrazing, inefficient irrigation and deforestation. Unsustainable land use can also increase political tensions if there are transboundary impacts. Thus, to help achieve economic development and reduce poverty and to reduce the potential for tension between states, land resources should be managed in an environmentally sustainable manner.

REFERENCES

1. International Water Management Institute (undated) Comprehensive Assessment of Water Management in Agriculture: Implications of Land and Water Degradation on Food Security. Colombo.
2. ECO Consulting Group, 2004. Draft Feasibility Study for a Transboundary Biosphere Reserve in the Altai Mountains. Bonn.
3. United Nations 1993. Agenda 21. United Nations. New York.
4. United Nations Environment Programme 1993. The Aral Sea - Diagnostic Study for the Development of an Action Plan for the Conservation of the Aral Sea. Technical Report, Freshwater Unit, Nairobi.
5. United Nations Development Programme. 1995. The Aral In Crisis. Report presented at Tashkent, Uzbekistan.
6. United Nations Environment Programme 1999, Global Environmental Outlook 2000 UNEP/Earthscan, Nairobi/London.
7. United Nations Environment Programme 2002, Global Environmental Outlook 3 UNEP/Earthscan, Nairobi/London.
8. United Nations Environment Programme/International Institute for Sustainable Development, 2004. Exploring the Links Human Well-being, Poverty and Ecosystem Services. Nairobi, Kenya.

CONDITIONS AND TRENDS IN NATURAL SYSTEMS OF THE ALTAI-SAYAN ECOREGION

A. MANDYCH
Institute of Geography, Russian Academy of Sciences, Moscvow, Russian Federation

Abstract: Results of preliminary assessment of forests in the Russian part of the ecoregion, pasture lands in the Western Mongolia and Katun river watershed in the North of Altai Mountains are outlined in the paper. High North – South contrasts in natural, socio-economic and other conditions across the natural, administrative and state boundaries of the region have been revealed in the process of the research. They could be regarded as some particular phenomenon usually existing on large interregional or intercontinental frontiers. Those interregional contrasts would cause effects on the future health of natural systems in the ecoregion.

Keywords: Altai, Sayan, Western Mongolia, Katun' River, ecosystems assessment.

1. INTRODUCTION

For the last five years, WWF Russia together with WWF Mongolia carries out an ecoregional program on long-term biodiversity conservation in the Altai-Sayan ecoregion. Since 2001, WWF in co-operation with UNDP and some other international organizations initiated projects for ecoregion biodiversity conservation in the Altai-Sayan, which is the main value of the ecosystems and their services. The projects are as follows:

- Ensuring long-term Conservation of Biodiversity in Altai-Sayan ecoregion, four-year WWF Russia projects (1998-2002), which was to a great extent focused on biodiversity inventory and assessment in the ecoregion. One of its results was the first version of the ecoregion GIS providing a comprehensive basis for the further works;
- The PDF B phase of UNDP/GEF project "Biodiversity Conservation in the Altai-Sayan Mountain ecoregion" initiated by WWF Russia;
- UNDP/GEF project "Conservation and Sustainable Use of Biodiversity in the Altai-Sayan ecoregion in Mongolia" initiated by WWF Mongolia.

- Project of WWF Russia, conceived for the use of the potential of protected areas in the southern territory of the Altai Republic for the social and economic development of the region.

The implementation of the Projects has revealed that the state of biodiversity in the ecoregion and its future depend not only on an impact of local population and economy on natural ecosystems but also on a number of social, economic and political factors, which are often originated geographically far abroad (e.g. due globalization processes). It became obvious now, that the biodiversity conservation in ASER cannot be approached outside appropriate concepts or strategies of sustainable development for its population, economy and nature as parts of a holistic unity, integrated in the regional ecosystem. The situation seems even more intricate if someone will take into account that the ecoregion belongs to four states, which follow their own development strategies and treat nature in accordance with their economic, resource and other potentialities.

Realistic concepts of ASER sustainable development cannot be elaborated without careful, sometimes very fine analysis of the interaction between natural ecosystems and society at different scales. Such goals were set up in the Millennium Ecosystem Assessment Program, which WWF Russia and WWF Mongolia decided to join to in 2002. As a result the project: "Millennium Ecosystem Assessment: Altai-Sayan ecoregion" was established in 2003. Work on the project commenced in August 2003. It was carried out by efforts of Russian and Mongolian WWF. Since the beginning of the project the preliminary review and assessment of key selected ecosystems namely forest and grazing ecosystems were completed. The basin of the Katun River, which drains the central part of the Altai Mountains (as an inland freshwater ecosystems and water services), and ecologically oriented behavior of rural householders in the ecoregion were assessed, too.

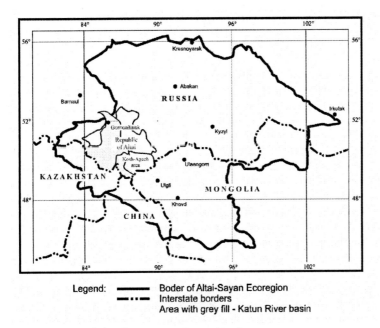

Fig. 1: Location of Altai-Sayan ecoregion and study areas

2. ALTAI-SAYAN ECOREGION: GENERAL CHARACTERISTICS

2.1 Geographic scope

The Altai-Sayan ecoregion (total area of 1 065 thousand km2) is situated in the intermediate zone between the boreal forest belt of the Northern hemisphere, mountain ranges of Altai and Sayan and semi-deserts and deserts of China and Mongolia. Its area belongs to the territory of Russia (62%), Mongolia (29%), Kazakhstan (5%) and China (4%) (Figure 1). Great latitudinal and longitudinal extents of the ecoregion, significant variance in the height and landscapes, and its location in the centre of Asia on the border between sub-continental divides of the basins of the Arctic Ocean and closed drainage depressions of Central Asia, cause considerable spatial and temporal gradients in climate and ecological conditions. This peculiar combination of conditions and the complex geological structure of the Region underpinned the origin of a rich diversity of natural ecosystems and their ways of functioning caused a great variety of ecosystem services provided by the biodiversity of the ecoregion at local, regional, national and sub-global levels.

2.2 Biosphere and biodiversity

The Altai-Sayan is one of the key WWF Global 200 ecoregions. Its biological, landscape, historical and cultural diversity is considered very unique. The ecoregion is one of the most distinctive examples of a conglomeration of the world's diverse habitats. Its ecosystems are known to be among the richest in the North Eurasia in terms of biodiversity. Moreover, the region represents the most complete sequence of altitude vegetation zones in Siberia. The area contains geographically distinct biomes, consisting of high mountain taiga, mountain tundra, a mix of forests, desert and semi-deserts, steppes and wetlands that share a large number of species, dynamics and environmental conditions (Kupriyanov, 2003; Shvarts and Shestakov, 2000).

Forest ecosystems occupy half of the entire territory of the ecoregion; steppe and highlands occupy 24 and 22%, correspondingly, whereas semi-desert and desert make up 4% (Table 1).

Table 1: Landscapes of the Altai-Sayan ecoregion

Name	Area, thou. km^2	Area, %
Forest	535	50
Steppe	249	24
Highlands, including	234	22
Mountain tundra	175	16
Alpine and sub alpine meadows	59	6
Desert	47	4

This ecoregion gives life to two of the world's ten largest rivers: the Ob' and the Yenisei, with a total watershed of over 5.5 million km^2. The ecoregion also contributes to the drainage of Lake Baikal, while in Mongolia, the Altai Mountains drain into the closed water system of the Great Lake Basin. Water quality and runoff of the largest Siberian river basins over an area equal to Europe in size depend on the future preservation of ecological processes in the Altai-Sayan ecoregion. The "health" of the Altai-Sayan ecoregion is crucial to the health of freshwater ecosystems that stretch far outside the immediate area.

Forests in the Altai-Sayan ecoregion are of crucial global importance. They contain the world's largest unbroken stretches of Siberian Pine forests (some trees are up to 700 years old) of the highest quality and dark coniferous taiga. This taiga consists mainly of endemic Siberian fir forests that have many relict and endemic vascular plants and mosses. Siberian larch and birch represent the remaining forest areas. Siberian forests occupy more than 535 thousand km^2. Approximately 26% of the forests in the ecoregion are located in the protected areas of IUCN I-IV categories.

As a natural dividing line between the Siberian boreal forests and the Central Asian deserts, the ecoregion hosts a unique species diversity of mammals, birds, vascular plants and mosses.

Vertical climatic variations and the isolation of various intermountain depressions determine the richness of the Altai-Sayan flora (3726 species). More than 10% of plant species are endemics and 76 species are listed in the Red data Book (Kupriyanov, 2003). Altai-Sayan ecoregion is defined by the WWF/IUCN Centres of Plant Diversity project as one of the world's centres of plant diversity (The Secretariat…, 2001).

There are 691 species of vertebrate animals in the region, 49 of them are listed as rare and endangered (Kupriyanov, 2003). The Altai-Sayan ecoregion forms the world's northernmost habitat of the Snow Leopard, locally called Irbis (*Uncia uncia*) and the Argali (*Ovis ammon ammon*), the world's largest species of wild sheep. Other mammals include Mongolian saiga antelope, Siberian maral, Asiatic ibex, Manul, Musk deer, Siberian mountain goat, Reindeer, Sable, Marmot and others. The Altai-Sayan ecoregion is home to more than 400 bird species, including several rare ones, such as the Altai ular, Cenerous vulture, Golden eagle, Dalmatian pelican, Great bustard, and Demoiselle crane, among others. The ecoregion is also an important habitat for endemic fish species, such as the Altai ide, Mongolian grayling, and Siberian catfish.

2.3 Ethnical and cultural diversity

For many centuries, the region has been at the crossroads of European and Asian civilizations, and thus is home to great historical treasures. The ancient history of the region is so unique that many historians and archaeologists call it *"the cradle of civilization"*. The ancient historic monuments are integrated into the natural landscape in such a way that it forms a harmonious and inseparable unity. Thousands of petroglyphs, cave paintings, antique burial mounds, menhirs, steles, tumuli and other ancient monuments are found in the area, some even as ancient as the Egyptian pyramids. One of the oldest sites, Malaya Syya in Khakasia, dates back to 35,000 BC.

More than 20 different indigenous ethnic groups that have lived together for centuries inhabit the Altai-Sayan ecoregion. Various ancient cultures (Scythian, Turkic, Ugro-Finnish, Iranian, Chinese and others) "merged" together in the region. A variety of languages of the Slavic, Turkic, and Mongolian families are spoken today. Different nationalities including Russians, Mongolians, Chinese, Kazakhs, Uighurs, Uzbeks, Altainians, Tuvinians, Buryats, Shores, Khakasians, Teleuts, Soyots and others live in the region making it really multicultural.

The lifestyle and traditional occupations have much in common through the region and are based on the traditions of nomadic cattle-raising tribes. Nowadays, old traditions connected with pagan customs and shamanism, national holidays and traditional art (e.g. the unique musical art of throat singing) are being revived.

The population consists of approximately 1.5 million people, with the majority still living in rural areas. Most of the territory still remains scarcely populated. The average density is among the lowest in Eurasia. In general, people in the Altai-Sayan ecoregion rely heavily upon local natural resources for their livelihood. Most of the inhabitants of the Mongolian part of the region are herdsmen whose economic activities are based on livestock and pasture management. People in the Russian part are engaged mainly in farming, cattle breeding, and mining, and many people earn a living by hunting and using variety of forest resources. Most of the Altai-Sayan region remains among the least economically developed regions of all four countries of the ecoregion. Poverty, widespread unemployment and a lack of alternative economic activities are serious socio-economic problems that could have a negative impact on both natural resources and the biodiversity of the ecoregion.

Exploitation and strain on natural resources are increasing rapidly in the ecoregion. Protection of Altai-Sayan biodiversity depends in many ways on the ability of the local communities to preserve traditional land use patterns. Such patterns do not only ensure the sustainable use of nature, but also ensure that particular natural sites, towards which local people hold a sacred attitude and special value, will remain respected and undisturbed.

3. DRIVERS OF CHANGE

3.1 International and political

From the viewpoint of administrative division, the Altai-Sayan ecoregion is a complex formation situated on the borders between four states: Kazakhstan, China, Mongolia and Russia. The Russian part of the ecoregion belongs to 8 large administrative territories of the Russian Federation.

Within the previous two decades, problems of effective use of resources and the environmental potential of ecosystems experienced radical changes in the geopolitical space. Until the late 1980's, motives for social and economic development within the larger part of the ecoregion were dominated by ideological doctrines characteristic for the development type based on extensive exploitation of natural resources and neglect to nature protection constraints.

The disintegration of the Soviet Union resulted in the appearance of new states, Kazakhstan and Russian Federation, and states including the ASER territory clearly demonstrated the tendency to establish priorities of their development, starting from their own interests. Over recent years, states of the ecoregion have faced a number of problems including consequences of unpractical industrial development, underdeveloped economic infrastructure and increasing deterioration of environment. It becomes evident that a number of acute problems can only be resolved on the base of efficient use of resources and services of natural ecosystems, and conservation of rich natural biodiversity. Given the division of the ecoregion by national borders, international cooperation is clearly the main prerequisite to overcome handicaps existing on the way to sustainable social, economic and environmentally favourable development of the Altai-Sayan ecoregion.

Regional administrations of the countries included in the ecoregion increasingly understand the necessity of joint action targeted on biodiversity conservation and efficient use of ecosystem resources and services. This understanding had an effect on a number of international and inter-regional treaties. The most important of these include the following documents.

In 1998, representatives of Kazakhstan, China, Mongolia and Russia met in Urumqi (China) to agree on the organization of a trans-boundary strict nature reserve and launch joint programs to conserve biodiversity. These intentions make provision for development of land tenure systems that consider ecological requirements and traditions of local people, generation of environmentally friendly power systems and transportation infrastructure, development of tourism and trans-boundary communications of local population, collaboration in the field of protection of natural matters of cultural, historic and religious heritage.

In 1998, the Republics of Tyva, Khakassia and Altai in the Russian Federation signed an agreement on the protection of the environment. All parties expressed their commitments to develop a package of urgent measures to conserve biological and landscape diversity within the territories of the participants of the Agreement, to develop national strategies for conservation of the snow leopard and Altai argali sheep as key components of ecosystems in the ecoregion and to create an ecological framework of strictly protected natural territories.

In 1999, an international forum held at the initiative of the WWF in Belokurikha (Russia) adopted the Altai-Sayan Initiative for the Next Millennium, which was signed by leaders of the governments of the Republic of Altai (Russia), the Republic of Tyva (Russia), the Republic of Khakassia (Russia), governors of Krasnoyarsk Region (Russia), the Aimags of Bayan-Ulgii, Khovd, Khuvsugul and Uvs, and also by the directors of the Mongolian and Russian WWF offices. Kemerovo Region and Eastern Kazakhstan Region joined the treaty later. Participants of the Initiative recognize the existence of a tight correlation between protection of natural processes and economic development. They stated that conservation of natural processes should be the main goal of regional development.

In March 2003, organizations representing state governments of Altai Region (Russia), Bayan-Ulgii Aimag (Mongolia), Eastern Kazakhstan Region (Kazakhstan), the Republic of Altai (Russia), Xingjian Uygur Autonomous Region (China) and Khovd Aimag (Mongolia) resolved to establish an International Steering Board called "Altai, Our Common Home" in agreement with a resolution of the International Forum of the same name. The main aim of the Board is to unite efforts of state bodies, businessmen, scientists and representatives of public organizations to create optimum conditions for the development of all parts of Altai Region.

The main challenge faced by the population of the Altai-Sayan ecoregion, irrespective of states and nations, is the necessity to realize the idea of long-term sustainable development. It is necessary to schedule the framework and conditions that would allow natural ecosystems (including biological diversity) and the community represented by local people to form mutually supplementary and supporting parts of the ecosystem of the ASER as a whole developing unit. An important step towards this goal would be integrated assessment of the natural environment of the ecoregion. The assessment will help to shape the current condition and trends in the change of ecosystems, determine their true social and biosphere value, recognize key driving forces causing unfavourable changes and, what is most important, to develop recommendations for decision makers of all levels on due practical measures to achieve sustainable and environmentally friendly social and economic development of the Region.

3.2 Social and economic

The population of the Altai-Sayan ecoregion is 5.5 million. Territories of different countries are significantly diverse in population density (Table 2). This probably acts as a driving force of spatial differentiation of human impact on ecosystems within the region.

Multinational composition of the population is characteristic of the ecoregion. Representatives of 40 nationalities inhabit the Russian part of the Region, whereas the Mongolian part is inhabited by 2 nationalities, the Chinese part by 13, and Kazakh part by 11. The Altai-Sayan ecoregion is home to many indigenous peoples, totalling 1.5 million persons (Climatic Passport, 2001) united in approximately 20 ethnic groups. The origin of their ethnic and cultural peculiarities is rooted in the early ages. Harmonious coexistence with nature, and feeling a part of it, are clearly expressed characteristics of these people.

Table 2: Population of ASER

Region	Population		Population density, persons per km^2
	Thousand persons	%	
Kazakhstan	373	6.8	6.2
China	590	10.7	13.7
Mongolia	556	10.1	1.2
Russia	3987	72.4	2.4

From the early 1990's, the transition from a centralized to a market economy started in Kazakhstan, Mongolia and Russia, causing a dramatic economic decline, which has not yet been overcome. The main consequence of this decline was a significant decrease in living standards of the population, which in turn intensified previously existing environmental problems and awoke new threats to biodiversity conservation. Although details of this process differ in Kazakhstan, Mongolia and Russia, its consequences for people and nature were very much the same.

A profound decrease and deterioration of former agricultural enterprises in Kazakhstan moved local people to subsistence farming conditions. Cattle breeding, agricultural work and exploitation of natural ecosystem biological resources became major income sources for most of the people. The level of income of the population decreased to critically low rates and amounts to just a few US dollars a month.

Livestock privatisation, increase in herder household numbers and the increase in goats within the herd structure in Mongolia became primary driving forces causing rapid pasture decay over large territories. Significant loss of cattle due to particularly severe winters and summer droughts considerably aggravated the economic situation over the last three years. As a result, around 40% of the Mongolian part of the ecoregion is below the official poverty line (WWF 2002). Budgets of all somons situated in this area are subsidized by the state (Enkhtsetseg et al., 2002).

Industrial and agricultural activities of state enterprises have almost completely stopped in the Russian part of the ecoregion, resulting in a virtually zero economy level. Thus, for instance, the Republic of Tuva is currently subsidized from the federal budget at a rate of 96%, and the Republic of Altai by 90%. In this situation, the population's income almost solely depends on subsistence farming and exploitation of any available resources of natural ecosystems.

Other driving forces also contributed to a stronger human impact on ecosystems of the ecoregion. These forces include a low technological level of industries working in the Region, dominating primary processing of natural resources, low effectiveness of resource and power utilization and a lack of ecological considerations. Moreover, the Region is deprived of opportunity to financially support conservation activities. State control over exploitation of natural resources is weak.

The low level of social and economic human well-being in the ecoregion promoted an increase of human impact on natural ecosystems. This impact is expressed in forest reduction, pasture overgrazing, mass poaching, and excessively intensive exploitation of ecosystem resources by local people, redundant recreation, etc.

Direct human influence on ecosystems provoked action of secondary, yet powerful forces causing ecosystem transformation. These include forest fires, habitat loss, fragmentation of landscapes, pollution of environment, and violation of regimes in strictly protected natural territories. All this occurred along with the weakening of the system of state management in economy and the environment under new economic conditions.

3.3 Climate change

The present-day climate of the Altai-Sayan ecoregion is characterized by significant spatial contrasts in the magnitude and rate of changes (Kharlamova, 2000, 2002; Revyakin and Kharlamova, 2003).

Meteorological observation in the Altai-Sayan ecoregion indicates progressive warming. Kharlamova (2000) has analyzed air temperature variations at the Barnaul meteorological station located northwest of the region and ascertained the clear trend of increasing the mean annual air temperature – 2.8°C for the past half century (Figure 2). The increase of air temperature differs by seasons. It is 3.4°C from November to March and much lower in the warm period – 2.3°C.

The analysis of the mean annual, seasonal and monthly air temperatures of the Altai Mountains indicates that the ecoregion is in the area of their rising, which particularly strengthened since the 1970's. Intensive increase of the mean annual temperature is observed in all altitudinal belts. However, the change of temperature differs in meteorological stations. The mean air temperature trends in altitudinal belts of the Altai Mountains by seasons for 60 years (1935-1994), 50 years (1945-1994), 40 years (1955-1994), 30 years (1965-1994) are shown in Table 3 (Paromov, 1999).

As a whole, the mountain part of the Altai is characterized by a deceleration of the increase of the mean annual and seasonal air temperature with height. If in the high-mountain and middle-mountain belts (outside intermountain depressions) the rise of mean annual temperature was 0.23°C per decade for 60 year and it was only due to increase of winter temperature (0.27°C per decade), in the foothills the temperature growth occurred in all seasons, excluding autumn (Table 3).

Faster warming is observed in the intermountain depressions and the lower Katun basin, in the foothills, where the rate of temperature increase comprises 0.40°C and 0.42°C per decade, respectively. The contrast between temperatures in the valleys and surrounding mountain slopes gets lower, indicating climate softening.

Table 3: Characteristic values of the trend (°C per 10 years) for the areas of the Upper Ob basin classified by this indicator

Period (years)	Season	Areas of close values of the trend parameter			
		High mountain	Intermountain depressions	Low mountains	Lower Katun
60	Year	0.23	0.40	0.27	0.42
	Winter	0.27	0.55	0.33	0.51
	Spring	no	0.16	no	0.24
	Summer	no	0.10	no	0.14
	Autumn	no	no	-0.12	no
50	Year	0.27	0.45	0.32	0.54
	Winter	0.40	0.71	0.52	0.73
	Spring	no	no	no	no
	Summer	no	0.11	no	0.18
	Autumn	no	no	-0.14	no

Despite air temperature, annual precipitation change is not so clearly pronounced. The curve of annual precipitation of the Barnaul meteorological station (Figure 3) shows that the increase is within normal annual fluctuations for the past decades.

The examination of seasonal precipitation over the same period of observation has no single meaning as temperature of the surface air layer. Winter precipitation decrease is seen at meteorological stations of foothills, low altitudinal belt and intermountain depressions. There is not an appreciable tendency in precipitation change of this season in the middle and high mountain belts. On the whole, the total precipitation in spring and autumn is stable at all altitudes of the Northern Altai, excluding high mountains where it increases in spring and, during recent years, in autumn. Summer precipitation has decreased in all altitudinal belts, but the change of sign of tendency occurred during the past 30 years.

However, this increase did not evolve into a steady trend. The annual precipitation in the foothills, low mountains and intermountain depressions decreases, and is stable at middle altitudes. It increases, mainly due to growth of spring precipitation, on watersheds of high mountains, which slopes are exposed to moisture-laden western air masses.

Climate warming is also observed throughout West Mongolia while the rate is different in its various parts. The general estimation of increasing the mean annual air temperature in Mongolia (Batima, 2002) corresponds to the data of the Barnaul meteorological station (Figure 4).

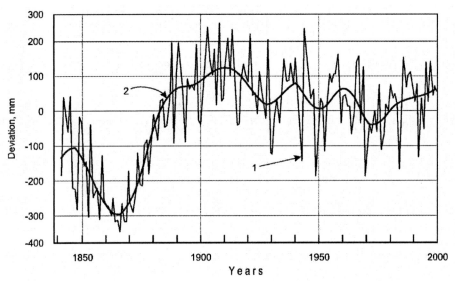

Fig. 3: Deviation of annual precipitation from long-term average at Barnaul meteorological station: 1 - true deviations, 2 - deviations curve smoothed by low frequency filter (Revyakin and Kharlamova, 2003)

Forest Ecosystems

Analysis of condition and trends of forest ecosystems in ASER has been carried out at three spatial scales: for the whole Russian part of the ecoregion, for the Republic Altai and its Kosh-Agach area (Figure 1). Hence the forests

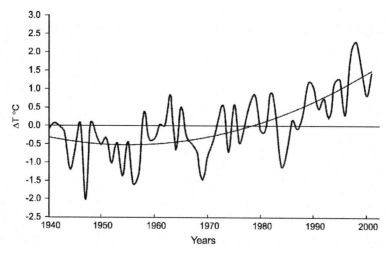

Fig. 4: Standard deviation of normalized air temperature in 1940-2001 (Batima, 2002)

were characterized at regional, sub regional and local scales. The assessment of forest ecosystems of the Russian Altai-Sayan ecoregion in general (national scale) is approximate.

4. FOREST ECOSYSTEMS

4.1 Ecosystem outline

Forests of the Altai-Sayan ecoregion constitute certain integrity according to geobotanic, floristic and silvicultural attributes. They are classified as a single complex of boreal South-Siberian-Mongolian forests (Polikarpov et al., 1986). Forests in the region mostly exist in the form of mountain altitudinal belts. If the topography of the region is plain, there would be zonal vegetation types such as steppe and forest-steppe. Only a small part in the northern ecoregion lies at a latitude of taiga.

The mountain topography of the most part of the region, considerable latitudinal and longitudinal extension, diverse lithologic and edaphic factors predetermine very high spatial inhomogeneity of forest habitats in the Russian ecoregion. As a result the forest cover has a very complex structure. It is

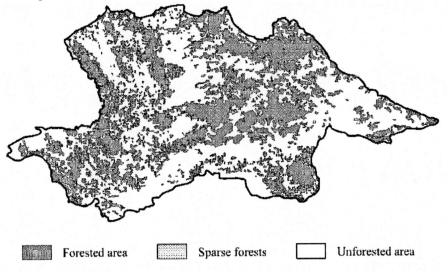

Fig. 5: Forest cover of the Russian part of ASER

characterized by the clear hierarchic structure of forest biogeocenoses and their complexes sketchily shown in Table 4.

Objects of ecosystem analysis can be forest types considered as integrated forest communities and conditions, which regularly recur within certain areas with relatively homogeneous zonal conditions of climate and soil (Smagin et al., 1980). Forest types originate hierarchically dependent landscapes of different size. There is a zonal-provincial complex of forest types (ZPC) and altitudinal complex of forest types (AC).

ZPC is a set of forest types classified by similar climatic and soil conditions of a natural zone for plain topography. The size of such spatial objects is enough for mapping at 1: 10 000 000 – 1:20 000 000 scales.

AC prevails in the Altai-Sayan ecoregion. It is a set of forest types classified by similar zonal-provincial and altitudinal conditions of climate and soil of an altitudinal belt for mountain topography. Such spatial objects can be mapped at 1:100 000 – 1:1 000 000 scales. Every AC is different from upper and lower ones in heat and moisture supply and the specific composition of lower layers of forest stand.

Such classification units as class or formation of forest types, which combine all forest types with one tree species, are also used to visualize spatial distribution of forest types and their components for more suitable analysis. Figure 1 is a reduced copy of the map of forest type formations of the Altai-Sayan ecoregion, which are classified by dominant tree species.

Simple and well-known characteristics of forest cover and ecosystems were used for the preliminary assessment of the state and trends of forest ecosystems described in the report. They consist of the attributes, as follows:
- Size of forest cover, tree species, burnt-up and cut areas; it takes the form of the percentage of the total area or hectares or square kilometres.
- Timber resource per area; it takes the form of m^3/ha
- Rate of changing the timber resource per unit of time; it takes the form of m^3/ha a year
- Forest stand productivity index. It is represented in marks.
- Forest regeneration; it takes the form of qualitative wordy scores: poor, fair, good, etc.

The Republic of Altai of 92.9 thousand km^2 occupies the highest part of the Central Altai and the northern foothills. It is entirely located in the Ob River basin. The southern border runs from west to east along the ranges of Kholzun, Listvyaga and partially Katunskiy, Yuzhno-Altaiskiy and Saylyugemskiy. The ranges exceed 2000 m, reaching in some areas 3500 m. The highest point of the Altai Mountains is Mount Belukha (4506 m) in the Katunskiy Range.

The eastern border of the Republic of Altai coincides with the divide between the Ob and Yenisei basins. It runs from north to south along the ranges of Poskay, partially Abakanskiy, Shapshalskiy and Chikhachev to the joint with the eastern offshoots of the Saylyugemskiy Range. The ranges reach 3000-4000 m.

The western border coincides with the border of the Altai-Sayan ecoregion, the northern one runs along the crest of the Niyskaya Griva Range.

Table 4: Hierarchic levels of the forest cover in the Altai-Sayan ecoregion

Categories	Scale of occurrence or mapping	Characteristic
Types of forest, biogeocenosis or ecosystem	1:10 ths – 1:100 ths	Object of ecosystem analysis of the lowest rank
Groups of forest types	1:10 ths – 1:100 ths	Genetically close forest types with one dominant tree species and with similar composition of low vegetation layers and soils of the same type or subtype
Ecogenetic or landscape series of forest types	1:10 ths – 1:100 ths	Set of groups of forest types with the same row of soil formation
Zonal-provincial complex of forest types (ZPC)	1:10 million – 1:20 million	Combination of forest types with similar climate and soil within a part of a province of a natural zone
Altitudinal complexes of forest types (AC)	1:100 ths – 1:1 million	Combination of forest types of an altitudinal belt with similar zonal-provisional and altitudinal features of climate and soil
Formations of forest types (ZPC or AC classes)	1:1 million – 1:20 million	Set of all forest types with one tree species

The Republic of Altai is a mountain country, consisting of a set of ranges going radially from the highest part of the Central Altai. Narrow river valleys or vast intermountain depressions divide the ranges.

The Kosh-Agach area is located in the southwestern part of the Republic of Altai and borders Mongolia, China and Kazakhstan (Figure 1). The forest area totals 308.6 thousand ha, including 186.6 thousand ha of the forest-covered area (Gunya et al., 2002). All the forests belong to Group 1. The dominant species is larch. All the forests grow under extreme conditions at altitudes over 1200-1400 m above the sea level. The timber reserve totals 28.4 million m^3, the reserve of mature and over mature forests is 15.2 million m^3.

The share of forests is lowest in Mountain Altai – 7.5%. However the value of forests of the Kosh-Agach area is very high, mainly due to their water-control and erosion-control functions.

4.2 Condition and trends

As last century so at present the main factors of forest ecosystems dynamics in ASER are natural regeneration and successional processes, fire, climate fluctuation, cutting operations and grazing. The analysis of Forest Inventory data shows that no radical transformations of forests have been observed at scales of the entire ecoregion and Altai Republic in 1970-2000.

Forest fire is one of the leading and steadily acting drivers of forest ecosystem change. According to the contemporary standards, the rate of forest fire in the ecoregion is fair (Furyaev, 2002). Annually it affects nearly 0.1% of the forested area of the ecoregion. The total number of forest fire in the Altai-Sayan ecoregion in 1997-2001 was 1630 and the area of burnt-up forests was about 500 km^2. However, the potential of fire is different in various parts of the ecoregion owing to natural and climatic conditions.

The very high rate of forest fire is characteristic for the Republic of Tyva where the annual number of fire ranged from 12 to 24 per 1 million ha in 1997-2001 and the burnt-up area totalled about 350 km^2.

The comparatively low rate of forest fire in the Republic of Altai is caused by prevailing coniferous forests of low inflammability and high precipitation in the warm period. However, about 250 forest fires occurred in an area of about 80 km^2 for 1997-2001.

During the past years forest ecosystem change caused by climate warming is discussed in some publications. It is argued that the upper timberline in the Chuya depression (Kosh-Agach area) has raised much toward the alpine and mountain-tundra belts (Modina et al., 2002; Ovchinnikov et al., 2002). It is noted that the upper timberline in the Central Altai has risen since the middle of the 19th century (Adamenko, 1985).

The mean annual air temperature increases in the Altai part of the ecoregion during the past decade (Kharlamova, 2000, Revyakin and Kharlamova, 2003). It is characterized by considerable spatial irregularity of warming and the increase of the mean annual air temperature caused by dominating increase of winter temperature and nearly non-changing summer temperature. Hence there is a doubt that the response of forest ecosystems to increasing mean annual air temperature can occur under the lower increase of mean temperature of the vegetation period. One of the factors behind the fluctuation of the upper timberline may be natural biogenic dynamics (Polyakov and Semechkin, 2001; Ovchinnikov et al., 2002). There are no publications about explorations made by methods that are specially designed for revealing changes of the upper timberline in the Altai-Sayan ecoregion induced by climate warming. Moreover, forest transformation near the upper timberline can be induced not only by climatic factors and can occur not only as the extension of trees upward. Tree species can simultaneously migrate at opposite directions or stay in the same place (Leak and Graber, 1974). It is a fact that the change of the upper timberline upward is observed only in several places, its magnitude is low and we cannot say about considerable redistribution of altitudinal forest belts.

Economic activities have little influence on forest ecosystems at a scale of the Altai-Sayan ecoregion or Republic of Altai. The main human-induced factors affecting forest ecosystems are logging and fire. The logging area has been much reduced for the past decade. It totally accounts for less than 1% of the forest-covered area of the Altai-Sayan ecoregion and less than 0.1% of the Republic of Altai. Logging is not practiced, at least legally, in the Kosh-Agach area.

Change in the age-structure of forests is one of the evident transformations of the forest ecosystems in ASER. Within ASER there are reliable data on forest age-structure for the Altai Republic. They were collected for five intervals between 1966 and 2002. Analysis of the data revealed the following specific feature of the age-structure dynamics in the region:

Coniferous stands are dominated in the regional forests (occupy near 83% of the total forested area).

From 1966 up to 2002 the area of middle-age coniferous forests increased from 17.0 to 33.6%.

During that time the area of deciduous forests remained practically constant.

During 1966-2002 the age composition of forest ecosystems of Altai Republic was changed as follows: young forests increased from 2.6 to 6.2%, middle-age forests – from 18.4 to 31.7%, and area of mature/over mature stands decreased from 64.7 to 43.5% (Figure 6).

Changes in age-structure of forest ecosystems in Altai Republic are caused by specifics of the history of deforestation processes in the region. The intensive economic development of the region since early 1900es was attended by significant increase of cutting operations and human induced forest fires. All this has resulted in the change of the forest age-structure towards its rejuvenation.

The important consequence of the forest age-structure changes in Altai Republic is almost doubling of carbon sequestration potential. The matter is that sequestration of atmospheric carbon (CO_2) by forests depends on the successional stage of forest ecosystems. The rate of carbon accumulation is greatest in middle-age stands (Kolchugina and Vinson, 1995). At regional scale, the carbon sequestration potential depends on the area of forests, in which carbon accumulation is still not compensated by carbon emission due to respiratory process. To a first approximation change in the intensity of sequestration in a region could be evaluated through an analysis of age-structure of its forest cover. It is necessary to conduct additional research to obtain more correct estimates of the regional rate of carbon accumulation by forest ecosystems.

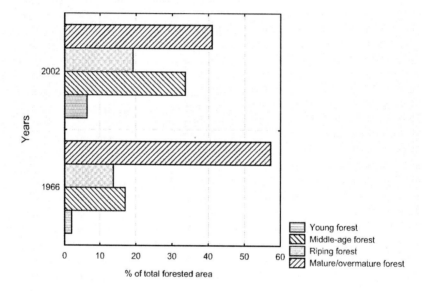

Fig. 6: *Forest age-structure of the Altai Republic, 1966-2002 (Alimov et al., 1989; Forest Inventory, 2003)*

To a certain extent, the conclusions above may be extrapolated over the rest territory of ASER, where the various mountain forest ecosystems are dominated with almost the same history of economic development and natural disturbances.

The eight most important forest ecosystem services are represented in table 5. They include timber and wood products as well as a number of not timber goods and services of forests in ASER. During previous decades, timber and firewood were taken up as the most valuable forest resources at national and regional level. Others, not timber products of forest ecosystems, are considered as less valuable or as resources of minor importance, which are mainly useful for local people and economy. Such a view has reflected the ways of natural resources use dominated in those times. They were focused on quick receiving of a profit with ignoring probable future adverse consequences of such a practice for people and economy.

At present, it is recognized that in ASER the value of other, not timber forest products, such as biodiversity, conditions for recreation, medicinal plants, etc., can be not only comparable but even exceed the timber and firewood values in long-range outlook. However, the relative values of different products and services of forest ecosystems with respect of their use by human are not determined yet. In a few numbers of publications, the values of not timber forest products are considered from the standpoint of their consumer and technological properties (Pozdnyakov, 1973). Therefore, the determination of justified economic estimates of various goods and services of forest ecosystems is an urgent task. Such estimates could be a basis for development of the long term strategy of sustainable use of forest ecosystems in the ecoregion. At that, the extents to which one forest product/service could be substitute by another might be determined.

The analysis of Table 5 reveals some peculiarities of the estimations of ecosystem resources and services.

4.3 Plausible future

The existing different and often not very correct data on the forest ecosystem behavior in various locations under outer impacts indicate that their future depends on fire, human activities, climate change and pests. Now, fire is a leading driver by an effect on ecosystems. Its role will increase in the nearest future owing to human activities and climate change.

The contemporary wood harvest in the region is not a factor of irreversible transformation of the forest cover in the nearest decades. Even if it is considerably increased, if logging distribution is specially designed, the forest cover of ASER will keep its resource and environment-forming potential.

One of the likely and appreciable effects of climate change on forest ecosystems will be not the change of the upper timberline but the increase of the fire rate. The climatic factor may also spur pests.

The above judgments are based, as shown above, on diverse, unsystematic and specially unanalyzed information. Scenarios for forest ecosystems in ASER for nearest 30-50 years are targets for the following stage of the study.

4.4 Summary

The present-day state of the main resource of forest ecosystems, timber and firewood, in the Altai-Sayan ecoregion as a whole is estimated good. Even though the score is reduced to fair or even poor in some localities that are under human or economic impact or where logging is practiced.

In the visible future, forest ecosystems of the region will continue providing people and economy with demanded resources and carry on environment-forming functions. However it will be possible only in case of keeping on the state control of forest resources use at least at the contemporary level.

The leading driver of changing forest ecosystems in the ecoregion in the nearest decades is fire. Its impact will increase owing to human being, economy, climate change and pests.

The existing data and outputs of exploration are not enough to reply to the question about an impact of climate change on forest ecosystems and the consequences. It is urgent to conduct a special study concerning an impact of climate change on forest ecosystems of the ecoregion.

Table 5: Forest ecosystems' services in the Russia part of Altai-Sayan Ecoregion: current state and trends of change

	Kosh-Agach area		Republic of Altai		Russian Altai-Sayan
Goods / services	Importance for local population	State and trends	Importance for local population	State and trends	State and trends
Timber and firewood	Low	Poor ↘	High	Normal →	Good →
Meat and fur of wild animals	Low	Fair →	Medium	Fair ↘	Fair →
Biodiversity	Medium	Fair ↗	High	Normal →	Normal ↘
Medicinal plants	High	Poor ↘	High	Fair ↘	Fair ↘
Habitat of harvested animals and plants	Medium	Fair ↘	High	Fair ↘	Poor ↘
Territories of indigenous people living	High	Normal ↗	High	Normal ↗	Fair ↘
Chain of hydrologic cycle; flood control	Medium	Fair ↘	High	Normal →	Normal →
Cultural, historical, aesthetic, religious values	Medium	Fair ↘	Medium	Fair ↘	Normal →

Legend for trend: → - no changes; ↘ - worsening, decreasing; ↗ - improving, increasing.

5. GRAZING ECOSYSTEMS

5.1 Ecosystem outline

The Mongolian part of the Altai-Sayan Ecosystem (hereinafter referred to as Western Mongolia) is 304 thousand km^2, or about 20% of the country. It is administratively divided into 87 soums of six western aimags (Table 6).

The population of the region is about 537 thousand people with the mean density of 1.3 persons per km^2. The inhabitants are irregularly distributed. It is 2 persons per km^2 in Bayan-Ulgii Aimag while it is 0.5 persons per km^2 in the Gobi-Altai Aimag.

Western Mongolia is characterized by the diversity of natural landscapes. Alpine meadows, forest-steppe and desert-steppe, which form the grazing resources, cover about 59% of the area. The other part is occupied by barren deserts, rocks, waters and nival and glacial landscapes.

Steppe landscapes prevail in western Mongolia, accounting for more than 46% of the area. They occupy planes, depression bottoms, terraces of big river valleys, as well as mountain slopes up to 2300 m above the sea level and up to 2550 m on favorable expositions of slopes (Chistyakov and Seliverstov, 1999).

Steppe ecosystems in the region are classified into two main groups. The first one presents communities of perennial xerophylous turf grass resistant to low temperature in winter. Such steppe landscapes occur in the northern part of the region.

Table 6: Population and administrative units (Statistical Yearbook 2002)

Aimags	Number of soums	Area in ASER, km^2	Percentage of total area	Population in 1999, thousand persons
Bayan Ulgii	12	45 861	98.8	100.0
Khovd	16	75 547	73.8	94.5
Uvs	19	63 765	99.4	98.4
Zavkhan	14	28 234	56.0	104.0
Gobi-Altai	14	47 118	63.5	74.1
Khuvsgul	12	43 491	66.8	65.6
Total	87	303 836		536.6

Desert-steppe landscapes prevail in the southern part. They exist under low precipitation, very high heat supply in a warm period and usually under unfavorable soil and lithological conditions, such as sand, salt ground, etc.

The term "grazing ecosystems" is used in the paper to accentuate one of the most important features of the ecosystems of Western Mongolia, i.e. permanent use by wild and domestic animals for grazing. Actually, the category includes quite different ecosystems, such as Alpine meadows and tundra, forest-steppe, steppe and semi desert. The floristic composition of these ecosystems, their productivity, feed value and resistance to different impacts range much. The population of the region and its main activities are mostly found in steppe landscapes.

5.2 Conditions and trends

The overview of the present-day conditions and trends of changing main ecosystem services in West Mongolia is shown in Table 7. Eight basic ecosystem services of four ecosystem types were assessed.

Animal husbandry is a basic economy of the region. It makes up about 70% of GDP, being the basis of the social and economic development of rural areas and well-being. Hence the feed resources provided by natural ecosystems play a key role in the people well-being[5].

Table 7: Grazing ecosystems' services in the Mongolian Altai-Sayan ecoregion: current state and trend of change

Goods / services	Type of natural landscape			
	Alpine meadow and tundra	Forest-steppe	Steppe	Desert-steppe
Fodder	Good ↘	Fair →	Poor ↘	Poor ↘
Firewood and timber	Not assessed	Fair →	Poor ↘	Poor ↘
Meat and fur of wild animal	Poor ↘	Poor ↘	Poor ↘	Poor ↘
Biodiversity	Fair ↘	Fair ↘	Poor ↘	Poor ↘
Habitat of harvested animals and plants	Fair ↘	Fair ↘	Fair ↘	Fair ↘
Area of nomadic style of living	Fair ↘	Good ↘	Fair ↘	Fair ↘
Cultural, historical, aesthetic, religious values	Not assessed	Fair →	Good →	Good →

Legend for trend: → - no changes; ↘ - worsening, decreasing; ↗ - improving, increasing.

According to Table 7, a supply of feed from steppe ecosystems, which are potential for the economic development of West Mongolia, is inadequate. A supply of wood, meet and fur of wild animals, and biodiversity has the same

---

[5] Nomadic herding practiced in West Mongolia for millennia has a significant effect on the structure and floristic composition of grazing ecosystems. However, the effect was similar to wild herbivorous animals in intact landscapes. Hence nomadic herding in West Mongolia was more or less harmonized with the internal interactions and processes of these ecosystems. That is why ecosystems under grazing balanced with their natural carrying capacity are recognized as nearly intact. But we should note that on the way to such state of regional ecosystems, wild herbivorous animals were extinguished or at least their number was reduced (Kozlov, 1948).

value. The state of medicinal plants and habitats of wild animals and plants is little better. This indicates a poor state of grazing ecosystems.

The contemporary changes of the ecosystem services show continuing degradation.

The transition of Mongolia to the market economy started in the beginning of the 1990's triggered changes. The most important ones are shown in Figure 7.

The primary factor controlling the magnitude of changes in grazing ecosystems for the last 13 years is abolishment of herding collectives and privatization of the most part of the state livestock. It has caused a series of sequential aftermaths that have a great impact on grazing ecosystems, environment of the region and human well-being. The most important ones are as follows:

1) After the abolishment of herding collectives, herders themselves became the ones to take decisions on all issues of husbandry – from grazing to marketing. The number of herding farms and livestock has increased for a short period (Table 8). Herders immediately face the problem of selling finished products. The earlier facilities were not intended for the considerable increase of animal production. Under new conditions, many functions of the state became ineffective or stopped. Hence herders started roaming closer to cities and big inhabited areas where they could sell their products. As a result, a great number of livestock is concentrated within limited areas around big settlements, pastures are overgrazed and even some of them are irretrievably destroyed[6].

Table 8: Total livestock population trend, in million heads (Ecoregion Conservation 2002)

Year	1930	1960	1995	1990	1999	2000	2001
Million heads	23.6	23.0	25.9	28.6	33.6	30.2	26.1

2) The privatization does not touch the pasture irrigation system, which was established in the pre-perestroika years and which mostly consists of wells. Most of them are ownerless and destroyed. The number of active wells

---

[6] For millennia, Mongolia has developed a specific traditional approach to the use of grazing areas. It includes the formation of a particular species ratio composition of livestock consisting of a combination of camels, horses, cattle, goats and sheep. The specific percentage of each species in the livestock leads to the most uniform grazing of the whole range of grazing species, which differ much in nutrition value. Such practice makes the efficiency of the use of pasture areas higher, normally preventing them from overgrazing.

decreased by 40% till 2000 compared with the late 1980's. It reinforced an impact on accessible grazing lands, overgrazing and partial degradation.

3) Since the early 1990's, the price of goat wool (cashmere) considerably increased. It became a trigger for the increasing goat number from 30% to 60%, i.e. twice from 1990 to 1999. This, in turn, resulted in more intensive overgrazing.

4) The ownership of grazing lands is not legally regulated yet. Most of them are still public lands. Overgrazing in many areas and lack of grazing resources are causes of conflicts between herders. This finally reinforces an impact of nomadic herding on grazing ecosystems.

5) The uneven development of herding for the past decade has increased the magnitude of irrational use of grazing lands. It partially depends on the involvement of new, young herders who are not experienced in nomadic herding. It is also caused by weakening the state management of grazing lands and poor control of their proper use. So, the control of pests of grazing lands such as white mouse is nearly stopped. The growth of the number of this pest considerably decreased feed resources. As the earlier described cases, it makes an impact on grazing lands higher.

Climate warming will result in more intensive aridization of Western Mongolia. This will have an unfavorable impact on the productivity of grazing ecosystems and, by the opinion of some Mongolian scientists, it already has. Thus, climate change has decreased the productivity of grazing lands by 20% to 30% for the past 40 years (Batima, 2002). The changes correspond to increasing air temperature but the effects are transformed by the landscape structure where they occur. It is seen in the case of considerable decrease of the productivity of grazing lands in the Gobi-Altai (Erdenemandal) while north (Ulgii) the change is moderate (Figure 8). Climate warming will also increase biomass produced by grazing ecosystems in high mountains that is already found by experimental field observation (Batima, 2002).

5.3 Likely future

Based on the assessment of the contemporary state of grazing ecosystems and their trends in the region, the most important analyses of their likely future were chosen. They are, as follow.

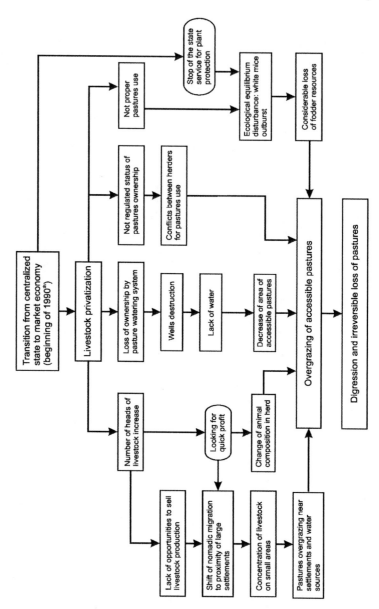

Fig. 7: Some processes of grazing ecosystems change in the Western Mongolia during last decade

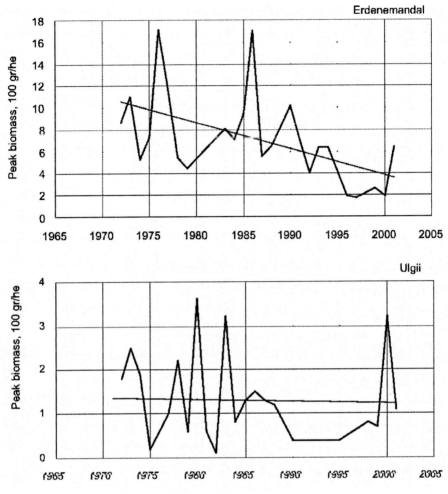

Fig. 8: Trends of peak biomass (Batima, 2002)

1) The contemporary trend of the socio-economic development of the region. The scenario is aimed at searching instruments to re-direct the process to sustainable development when the human use of grazing ecosystems is in compliance with their actual productivity rather than determining dates and details of the collapse of the "pasture – human population and livestock" system.

2) Future grazing ecosystems under the rational use of their resources. The scenario is aimed at analyzing means (social, economic, technological, legal, etc.) for long-term self-sustaining functioning of grazing ecosystems. The reality of achieving the scenario's goals is assessed as well.

3) Future grazing ecosystems of West Mongolia under considerable climate warming. The scenario is aimed at describing the spatial heterogeneity of climate warming effects for grazing ecosystems and determining possible measures to resist adverse effects.

5.4 Responses

Based on the assessment of the state and trends of grazing ecosystems, it is possible to define priorities in assessing the social response to the present-day situation and future development. It is obvious that the transition to the market economy caused the loss of some functions of the state, which were important at any economic systems. They are the development of elements of a social strategy related to the use of natural resources and sustainable nature management. Decision making on the use of grazing ecosystems at the present-day conditions and in future was passed from the national to the local level. It is clear that decision making at this level cannot adequately take into account many factors of regional and national levels.

The detailed analysis of the social response to the use of grazing ecosystem services is made in 2004 and 2005.

5.5 Summary

Overgrazing and over harvesting of natural resources in West Mongolia is a result of the impoverishment of country population who has very few facilities for surviving and who depends much on natural resources.

The transition of Mongolia to the market economy caused some adverse effects. One of them related to an impact on grazing lands is caused by the cessation of seasonal grazing, which has been practiced by Mongolian nomads for millennia. Seasonal herding provided a more equal impact on all grazing lands of the region. Abandoning the traditional herding practice, long distance from sales markets and social services, hindered access to water sources at many grazing lands were accompanied by the great increase of livestock concentrated in small areas around administrative centers and near water sources. It caused considerable overgrazing or even destruction of grazing lands in some places.

6. KATUN' RIVER BASIN[7]

6.1 Introduction

The assessment of the state and trends of the hydrological regime and water resources of the Katun Basin purposes several aims. First, since the Katun drains glacial watersheds of the South and Central Altai, it is possible to describe effects of climate warming observed in the Altai on the hydrological system of the river basin. Second, the population and economy of the region have an growing impact on the water resources and hydrology of the Katun River. It is pronounced in direct water withdrawal from rivers and other sources, water pollution and conditions change of forming water resources. The good state of the Katun basin, its ecosystem is a prerequisite for developing economy and raising well-being of the residents. Third, the Katun basin is one of two parts of the headwaters of the vast Ob River system. Hence the conditions for forming the water resources of the region, hydrological regime and water quality are of great importance for the state of downstream of the Ob, people dwelling on its banks.

6.2 Ecosystems outline

The Katun River springs from glaciers of the Altai highest mountain – Mount Belukha (Figure 9). The watershed is 60, 900 km^2, the length – 688 km. The major tributaries of the Katun are the Koksa, Argut, Chuya, Ursul, Chemal, Sema, Isha and Kamenka Rivers (Semenov, 1969).

The biggest tributary of the Katun is the Chuya with the watershed of 11 200 km^2, the river is 320 km long, it springs from glaciers of the Saylyugemskiy Range, crosses the Chuya and Kuray intermountain depressions and flows into the Katun River in the middle-mountain zone. The Argut (232 km long) and many other tributaries of the Katun River, with the watershed altitudes averaging from 2200 to 2500 m, spring from glaciers of the Central and South Altai. The greater part of their runoff is formed in the glacial and nival zones (Table 9).

---

[7] Co-author Semenov V.A.

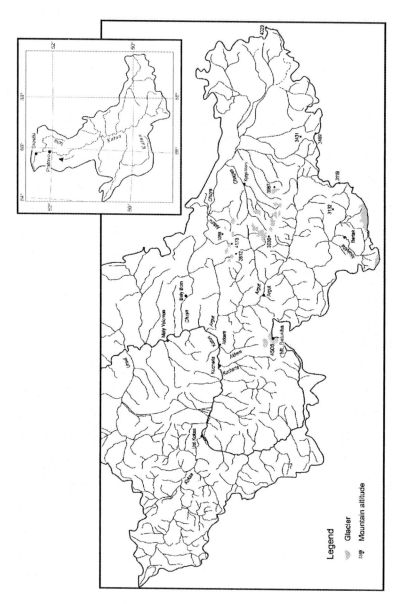

Fig. 9: Katun River basin

The share of ice melted water in runoff amounts to 40-60% for rivers, which drain glaciers and which watersheds' average altitude exceeds 3000 m. It decreases to 10-15% at sub mountain reaches of the rivers. However, melted water from snowfields and seasonal snow cover constitute to 30-50% of the Katun River flow in the high mountain part of its basin. Precipitation contributes to runoff less than 20% and the ground water – 30%. The share of precipitation increased in river flow in the 1990's and the beginning of the present century that has influenced on composition of annual runoff. That was especially pronounced in high-mountains.

Table 9: Annual runoff components of the rivers with glaciers in the watersheds

River-point	Water shed, km²	Water-shed average height, m	Annual runoff components, %			
			groundwater	snow	rain	glacier
Aktru – Alpinist camp	33.4	3100	10	15	15	60
Akkem – Akkem hydrometeostation	78,9	2900	9	35	13	43
Argut (Akalakha) – Bertek hydrometeostation	602	2600	12	40	20	28
Argut – Argut	7070	2500	21	43	20	16
Katun – at mountain Belukha foothills	56.0	2700	10	27	14	49
Katun – Maly Yaloman	3680	2280	25	40	22	13
Katun – Srostki	58 400	1770	28	44	20	8
Kucherla – Kucherla	627	2300	22	30	21	27
Chagan – Kyzyl-Many	372	2800	11	32	18	39
Chuya – Bely Bom	10 900	2340	33	31	18	18

Since thermokarst processes occur in the widespread permafrost zone, the share of the ground component in the river runoff will change.

The main spatial features of precipitation and maximum snow reserves in the basin are their increase with elevation and decrease at the same elevations from periphery towards inner areas.

The total water loss for evaporation from the river basin reduces from 500 mm and more in the lower basin at an altitude of 400–600 m to 150–200 mm in high-mountain intercontinental areas and intermountain depressions at an altitude of more than 2000 m that is explained by different heat regimes of the basin high zones and lack of precipitation in high-mountain intermountain depressions.

The greatest mean annual runoff is formed in the high-mountain zone of the Katun Range in the southern Katun basin, reaching 800 to 1000 mm. The worse runoff conditions are found in intermountain depressions, e.g. it is 50 mm in the Chuya depression (Komlev, 1966; Semenov, 1969).

The total mean annual runoff of the Katun basin is 19.5 km^3. About half of the volume is formed in the upper basin before the Chuya mouth.

6.3 Conditions and trends

6.3.1 Glaciers

There were 400 glaciers, totaling 626 km^2 in the Katun basin at the beginning of 1981 (Revyakin, 1981).

The Argut River gets the most amount of water in the Katun basin. The Chuya and Upper Katun, upper the Argut mouth, provides approximately the same volume of water – some more than a quarter of 1 cubic kilometer (Galakhov, 1999). The calculation of the glacial runoff of the rivers of the Central Altai shows that it is 10% to 15% lower in years with favorable conditions for glaciers and 25% to 30% higher in years with unfavorable conditions. The ice resources of the Central Altai are 150 km^3, including 27% in glaciers, 69% in permafrost, 4% in seasonal snow cover, 0.4% in frazils (Galakhov, 1999).

The observation of the Altai glaciers fluctuation has been carried out since 1835, in the past decades also by researchers of the Tomsk University. It indicates the recession of more than fifty glaciers within the Altai mountain system.

Glaciers decreased between 8 and 40% for the observation period. While decreasing, the largest glaciers come apart. Some glaciers disappear completely.

Small glaciers are more stable. That is because of the higher location and location in shadowy kars on the northern slopes of mountain rangers.

The total number of glaciers increased by 25% due to the disintegration of large glaciers for the century and a half (Narozhny, 2001). The glacial area in the Katun basin decreased by 25% for the period. A third of this decrease occurred after 1952. Although glaciers recessed in general during the past decades, its intensity increased 1.5 times for low valley glaciers and reduced approximately the same value for high mountain glaciers.

In the alimentation areas, the surface of glaciers considerably declined, which predetermines their further recession within the next 10–15 years.

6.3.2 Hydrology

The analysis of the hydrological observation for trend, with the verification of the reliability with several methods proposed by the World Meteorological Organization (Semenov et al., 1994), shows that the runoff of the Katun and Chuya rivers was characterized by a negative trend from the 1930's to the middle 1970's (Table 10) while the runoff of the tributaries with the high share of ice feed (the Akkem) did not change significantly.

Table 10: Parameters of intensity (b) and statistical significance (γ) of changing flood runoff by periods

River - point	Watershed, km²	Observation period	Whole observation period		1941-1975		1951-1975	
			b, mm	γ, %	b, mm	γ, %	b, mm	γ, %
Katun – Srostki	58 400	1936-1977	-1.483	99	-1.430	82	-1.525	54
Chuya – Bely Bom	10 900	1932-1974	-0.095	93	-0.044	69	-0.277	66
Akkem – Akkem	79	1952-1975	10.113	59			10.113	59

The lower and middle Katun (at Tyungur and Srostki settlements) flood and mean annual runoff is characterized by insignificant negative change for 1951-1980. The same tendency had remained for the following period up to 2000, but with changing the negative trend to stability. The stability of the mean annual runoff is also characteristic of the middle-mountain zone (the Koksa) while the runoff of the low-mountain zone is characterized by a negative tendency (Semenov, 1997).

The annual runoff of low and high mountains with a considerable part of ice and snow feed is characterized by smoothing seasonal differences. The rivers of intermountain depressions show the stable annual distribution of runoff in all moisture phases (Narozhny, 2001; Paromov, 1999; Skotselyas, 1975).

The tendency of a decreasing difference between seasonal runoff in high and low mountains, which is revealed by the analysis of annual seasonal runoff, is confirmed by a significant increase of an coefficient of runoff self regulation during the course of an year in the rivers of those zones (Narozhny et al., 1998) and is a response of runoff to climate change, seasonal precipitation first of all. The Katun runoff remains without change owing to the combination of the opposite tendencies in the runoff of its tributaries.

Table 11: Runoff change of the Katun and Koksa Rivers from 1950 to 2000

River - point	Watershed, km²	Mean runoff for the period, l/sec. km²		Runoff norm before 1970, q_0, l/sec. km²	Ratio	
		1981-2000, q	1951-1980, q_{cp}		q/q_{cp}	q/q_0
Katun – Srostki	58 400	18.6	19.4	19.6	0.96	0.95
Koksa – Ust-Koksa	5600	13.3	14.6	14.6	0.91	0.91

6.3.3 Human impact

Human-induced impacts on the Katun hydrological regime and water resources of the basin are diverse. Impacts are induced in the process of direct water use or are indirectly pronounced, e.g. under the change of the hydrological cycle caused by logging. The main economic activities in the river basin that have an impact on the condition of natural waters are as follows:

- Communal water supply, agriculture and industry
- Waste discharge into water bodies
- Water use for power generation
- Irrigation
- Forestry
- Water transport
- Recreation

The water consumption in the Katun basin accounts for an insignificant percentage of the mean annual runoff (about 0.18% in 1992). Hence an impact of water withdrawal on the hydrological regime is extremely low. It is notable that water use considerably decreased for the past decade that is caused by the deep economic crisis (Table 12).

Human-induced impacts on natural waters are most pronounced through pollution. It is caused by waste discharge into water bodies, fertilizer and chemicals discharge from arable lands, traffic, poor treatment of used water, etc. In general, the river water quality in the basin is close to natural. The main characteristics of the water chemical composition at the lower Katun (near Platovo village) in 2002 were, as follows: salt content 88.2-186.9 mg/dm³, pH -7.2-7.5, COD – 3.2-6.9 mgO/dm³, BOD_5 – 0.31-1.99 mgO/dm³, organic matter content – 1.18-2.64 mg/dm³, water hardness – less than 3.35 mg-equ/dm³. The contend of dissolved oxygen in the Katun was at least 9.0 mgO_2/dm³ in the year that indicated, together with high positive Eh values, a good oxygen regime of the river.

However the human-induced decrease of the river water quality is found in the densely populated and economically developed foothill areas and intermountain depressions (Table 13). The content of some pollutants exceeds maximum permissible concentrations (MPC). Table 6 shows the content of some pollutants that exceeds MPC in the Katun water.

Table 12: Water withdrawal from surface water bodies and aquifers in 1992-2002, mln cub.m

Year	1992	1993	1994	1995	1996	1997	1998	1999	2000	2001	2002
From surface waters	22,2	14,6	13,4	12,2	10,5	8,15	7,52	2,31	4,55	1,51	1,81
From aquifers	12,6	10,4	10,9	12,7	7,67	11,2	8,26	7,34	7,35	7,02	7,16
Total	34,8	24,9	24,3	25,0	18,1	19,4	15,8	9,65	11,9	8,53	9,05

The existing data of the surface and ground water pollution in the Katun basin show that the present-day human-induced impact on the natural waters does not deteriorate them. The chemical composition of most rivers is determined by natural factors. However, the further development of the economy and water supply in the Katun basin may cause the quick increase of the river water pollution if appropriate measures to prevent from water pollution and to treat communal, agricultural and industrial wastes are not taken.

The hydropower resources of the Katun are estimated 55 billion kWh (Ryabchikov, 1978). Their use began in 1934 when a low-powered hydropower plant on the Chemal River was built. It still operates at present. The construction of a hydropower plant on the Chuya near Aktash was stopped due to design mistakes. In the late 1990's the Katun Power Plant near Yeland village was designed but it was not completed. Now the small Altai Power Plant on the Katun 235 km upper the confluence of the Katun and Biya (near Yeland village) is designing. The main features of the reservoir are 10 times smaller than those of the Katun Power Plant that was projected earlier. The area of the reservoir is 12.1 km^2 and the volume is about 0.2 km^3.

An agricultural impact on the water resources of the Katun basin is comparatively low. It includes the use of river water for irrigation of intermountain depressions within the Chuya, Koksa and Ursul watersheds. The irrigated lands totaled about 8 thousand hectares in the early 1990's (Bannikova, 2001). Water used for irrigation was 22.4 million m^3 or about 0.11% of the Katun mean annual runoff in 1992. This volume reduced to 0.76 million m^3 till 2001 due to economic decline.

The forest cover that played an important role in protecting and regulating the river runoff in the Katun basin was significantly reduced at the first half of the 20th century; forest in the Chuya Basin was completely destroyed, which intensified desertification (Yaskov, 1997). In the middle of the 20th century, water bodies suffered from the destruction of scrub and drainage of wetlands during melioration. Nowadays, the main threat to forest ecosystems is fire and pests with which the combat was stopped during the economic decline. Selection felling is basically practiced.

Table 13: Content of some pollutants in the Katun water near Platovo Village in 2000 -2002

Pollutants	Range of content, mg/dm^3			MPC, mg/dm^3	Exceeding MPC in 2002
	2000	2001	2002		
NH_4^+	0.01-0.50	0.39-1.01	0.02 - 0.56	0.39	Up to 1.4 times
NO_2^-	0.01-0.05	0.02 - 0.07	0.01-0.06	0.02	Up to 3 times
NO_3^-	0.09-1.03	0.17-0.43	1.04-10.92	9.1	Up to 1.2 times
Petroleum products	0.02 - 0.32	0.01-0.41	0.01-0.30	0.05	Up to 6 times
Phenols	0.001-0.009	0.001-0.007	0.002-0.019	0.001	Up to 19 times
SAA	0.02 - 0.33	0.01-0.25	0.05 - 0.44	0.5	–
CLA	0.00-1.50	0.00 - 0.99	0.01-0.93	3.9	–
BOD_5	0.10-1.06	0.39-2.59	0.15-1.00	3.0	–
COD	1.10-5.10	3.50 - 7.70	2.51-3.35	15.0	–
Hg^{2+} µg/dm^3	0.02 - 0.08	0.005 - 0.07	0.01-0.09	0.01	Up to 9 times

The lower Katun up to approximately 30 km is navigable. The building of the modern navigable channel began in 1978 and requires river-training works (Berkovich et al., 2000). Till 1984 twelve straightening constructions were built on the river section 4 to 22 km from the mouth and the major bank strengthening reached 10 km. While reconstructing the natural riverbed, it was also deepened. In the early 1980's, intensive extraction of sand and gravel began in the lower Katun. This made the level of the Katun lower and changed the floodplain water supply.

The inundation of the lower floodplain (0 to 2 m) in the 1980-1990's decreased 50 days compared with the previous period, middle floodplain (2 to 4 m) – 20 to 30 days, and high floodplain (4 m and above) – 3 to 6 days. Now the zone of flooding is limited by the course floodplain of 0.5 to 1 km wide. Hence the high floodplain was xerophytized and that was accompanied by drying out floodplain lakes and streams, degrading wetlands and meadow. The effects of changing flooding duration are supplemented by overgrazing of floodplain meadows and clearing forest and scrub for arable lands. Cut forests do not regenerate in the floodplain owing to climate conditions.

Recreation, treatment and tourism in the high and low mountains of the Katun basin in the soviet time were mostly organized and so accountable. The annual tourist flow to the Alati was 40 to 50 thousand people at that time (Arefyeva and Chudova, 1994). In 1990, about 40 thousand tourists visited Mountain Altai, then there was a decline of organized tourism. In the late 1990's it was changed by the increase of "wild" tours organized by a great number of companies in Siberian cities and independent autotourists. That is why tourist account is difficult. Unofficial estimates show that the number of tourists visiting Mountain Altai in 2000 was 224 000, in 2001 – 402 780, in 2002 – 582 400, for 9 months of 2003 – 413 000.

The account of amateur tourists, autotourists in particular, is very complicated. Most of them go to the bank of the Katun and its tributaries and in forest to collect nuts of Siberian pine, berries and mushrooms. Despite organized tourists who pass over established routes, amateur tourists may do maximum damage to tree roots, shrubs and herbage. They are also local threats to water and near-water ecosystems.

6.4 Plausible future

It is quite possible that climate warming will cause the further increase of annual air temperature in Mountain Altai while effects on seasonal temperature and precipitation are different. Many climatologists forecast precipitation increasing, which had to occur. Some Russian and Chinese experts substantiate its insignificant change and even decreasing in some seasons. The calculation made by the methods of Sherstyukov and Isayev (1999) shows that air temperature is increasing in summer and decreasing in winter and spring, precipitation decreasing in winter and increasing in summer in the region up to 2010. The annual precipitation will increase, however evaporation loss will increase, too.

The change of runoff and water resources in the Katun basin caused by climate warming is indefinite because of different respond of various parts of the basin system to climate warming. This issue is an objective of further study.

A human-induced impact on the Katun basin is still temperate. However, the increase of population and economy will likely cause surface water pollution.

6.5 Summary

The river network of the Katun basin drains the northern macro slope of the Altai high mountains where about 400 glaciers of more than 600 km^2 are located. Considerable climate warming is observed in the region during the past decades which results in glacier recessing and shrinking. Hence, the Katun basin may be used for monitoring of the respond of a large river system and its ecosystem of a sub global scale to climate change.

The runoff change of the Katun tributaries for the past 4-5 decades had opposite trends. The compensation of the opposite runoff trends as the response to climate change was the reason that there was no trend in the Katun runoff from 1951 to 2000. However, the trend is found to smooth differences between the seasonal runoff in the high and low mountains of the Katun basin. It is explained by the different responses of the rivers with runoff forming in different elevation zones to seasonal changes of air precipitation.

The water of the Katun basin is used for communal water supply, agriculture, industry, irrigation, power generation, water transport and recreation. Direct water withdrawal for economic needs is insignificant and has no noticeable impact on water bodies.

The population and economy has an impact on the water quality of the rivers and lakes. Although the general water quality remains close to natural, water pollution may be very high in some localities. The further economic development in the Katun basin may cause the quick increase of river water pollution. Appropriate measures to treat communal, agricultural and industrial wastes should be taken to prevent water from pollution.

Katun River has large hydropower resources that are almost not used now. They should be used only through small hydropower plants whose construction and maintenance will not cause adverse environmental effects. So it will largely improve the power supply of the Republic of Altai and a part of electric power may be supplied to neighboring regions.

The most valuable resource of the Katun basin is forest, which is of high water protection value and provides high biological diversity of the region. Nowadays, the main threat to forest ecosystems is forest fire and pest outburst. The role of these factors considerably increased for the past decade owing to poor fire and pest control. Drastic measures are required to neutralize these threats.

Great and diverse facilities for recreation are a feature of the Katun basin. There are facilities for alpinism, walking tourism, rafting, climatic healing, and balneological treatment. The number of visitors to Mountain Altai is quickly growing. In 1990 about 40, 000 tourists visited the region, in 2002 the number of tourists reached 582, 400 people, for 9 months of 2003 – 413, 000. The number of summer cottages, inns, hotels, motels and so on is quickly growing on rivers and lakes. The unregulated development of recreation infrastructure may be a serious factor of polluting natural waters of the river basin.

7. CONCLUSION

The main result of the initial assessment stage in the Altai-Sayan ecoregion is a preliminary overview of the state of key ecosystems which provide ecological services for local population and economy (forests, grazing ecosystems, water). Besides ecosystem assessment, the initial evaluation of the population attitude towards ecosystem importance and role in the society life was conducted. This overview allows identifying necessary next steps for the more comprehensive assessment.

The key features of the ecoregion identified as the main for the further assessment and understanding of the ecosystem services are as follows:
- Assessment of the state and trends of forest and grazing ecosystems for the past 10-20 years showed the very high level of the North – South contrasts of natural, social, economic and cultural characteristics through the natural and administrative boundaries. These contrasts are outside just "transboundary" effects and could be regarded as sub global

phenomenon. These interregional contrasts would cause effects which will determine the future status of the ecosystems in the ecoregion and thus need full assessment;
- Substantial part of the population in the ecoregion survives on cattle breeding using ecosystem services provided by grassland (pasture) ecosystems. Current state of grazing ecosystems in the ecoregion is at the threshold of destruction (especially in most easy access pastures) due to overexploitation during last years as a result of collapse of the centralized economic management system and elimination of the state support to the livestock producers;
- Forest ecosystems suffer from the same driving forces as grazing lands. Economic and subsequent social crisis enormously increased direct uncontrolled use (often illegal) of all type of forests resources especially those easy to derive with extensive methods (e.g. non timber biodiversity resources). However, despite these negative trends forest ecosystems of the ecoregion in general are still in good condition. At this stage, overexploitation has a limited scale close to and around settlements. Forests in the ecoregion are much more stable ecosystems (rather) than grasslands. At the same time growing timber demand and intensification of economic activities in some parts of the ecoregion pose additional threats to forests which could lead to negative biodiversity dynamic tendencies in the coming 5-10 years. Forest ecosystems could become the main source of services for local communities in the coming years;
- Meteorological stations in the Basin of Katun River show an increase in temperature and trends of increasing melting of glaciers and snow. Different parts of the Katun River basin show differentiated reactions on climate changes (temperature and precipitations). Despite differentiation of response in parts of the basin, the river basin in general as a system is still quite stable and its general flow characteristics are maintained without changes;
- There are no proves or current research which could confirm changes in the ecosystems caused by climate change in the ecoregion;
- Transformations of political and economic systems in the countries of the ecoregion took place from the top and are clearer at national and regional levels rather than at local. New local governance structures are still under formation. Economic activities at local level were dramatically affected by political changes at national level. At the same time, local resources use patterns still keep on practising traditions of the previous political period. Differences in the social-economic and political situation and developments in the countries of the ecoregion determine unequal and sometimes controversial processes in the similar ecosystems;

- There are evidences of changes in attitude of local rural population (in the Russian part of the ecoregion covered by sociological study) towards an increase of awareness about environmental issues and interest in sustainable use of biological resources;

In 2003, the initial stage of the assessment revealed the key issues to be investigated and assessed. Those issues include:

1. Assessment of the natural water resources as a key element of the ecosystems in the ecoregion playing an important role for the human wellbeing in the entire ecoregion and as a service for neighbouring (downstream) regions. This assessment will include: i) water resources quantity and quality, ii) relationships with functioning of key selected ecosystems, iii) access to water resources in the different parts of the ecoregion, iv) response on climate change, v) use in economic activities (water supply, hydro energy, water transport, recreation, medicinal use etc.). Special emphasis will be done for the assessment of the role of wetlands in the ecosystem structure and functioning of the ecoregional biodiversity. Scale: ecoregional, national, basin and local.

2. Assessment of the current status and trends of biodiversity (at indicator species and ecosystem levels). Key ecosystems include: nival-glacial systems, alpine tundra and meadows, forest, forest-steppe, steppe, desert-steppe, wetlands, agricultural ecosystem. Assessment should identify major drivers of changes and current mechanisms of changes in ecosystem goods and services. Special focus should be done on the assessment of the carrying capacity (sustainability thresholds) of the ecosystems of different types and their potential to provide services and goods under different levels of anthropogenic pressure. Scale: ecoregional, national and sub regional (for biodiversity status and trends) and local level (for carrying capacity and services production under pressure).

3. Investigation of social, political, institutional and legal driving forces determining responses of local communities to ecosystem changes in the ecoregion. Investigation should be focused on assessment of adequacy of the current community organization (institutional, legislative, legal, psychological aspects) as related to the character and scale of contemporary ecosystem changes. Scale: local.

4. Investigation of specific features of human well-being related to specific ecosystems (based on traditional knowledge and land use systems) for different ethnic groups and indigenous communities in the ecoregion. Assessment of the potential of using the existing traditional knowledge in the ecoregion for the organization of sustainable use in the ecoregion. Scale: sub regional, local.

5. Development and analysis of scenarios. Current situation review revealed the following potential developments to be investigated:
conditions and potential volumes of services which grazing ecosystems will provide to maintain traditional and/or current cattle breeding patterns a swell as

changes in the grazing system needed to ensure a sustainable and comprehensive use of available pasture resources;
conditions for the long-term balance between various resources and services and different practices of their use in forest ecosystems. Forest ecosystems under various development scenarios (intensive forestry oriented / industrial and mining development / biodiversity use oriented);
key types of the ecosystems under climate change.
Scale: ecoregional, national, sub-regional and local.

6. Assessment of the role of current governance systems and their trends in the different parts of the ecoregion for potential changes of the resources use patterns and incorporation of ecosystem services into economic policies. This assessment will include:
Analysis of the institutional, legal and social basis for the population's reaction on the ecosystem changes. Scale: local, sub-regional, regional, national;
identification and analysis of socio-economic driving forces and conditions, causing ecosystem changes and determining appropriate responses of local communities. Scale: local, sub-regional, regional, national.

Acknowledgements

I wish to thank Drs. Alexander S. Shestakov and Chimed-Ochir Bazarsad for discussions various details of the study and for contribution of data and materials for it. Many thanks Drs. A. Sokolov and S. Surazakova for their essential materials as well. This study was initiated and sponsored by WWF Russia Program Office through Grant No. A240/Ru109201/GLP.

REFERENCES

1. Adamenko, M.F., 1985, Reconstruction of Thermal Regime Dynamics for Summer Months and Gglaciation in the Altai Highlands Region in XIV-XX Centuries. Theses of dissertation, Novosibirsk, 16 p. (in Russian)
2. Areyev, V.E., Chudov A.V., 1994, Tourism in the Altai. Altai State University, Barnaul, 119 p. (in Russian)
3. Bannikova, O.I., 2001, Assessment of Natural Resources and Environmental State of Intermountain Depressions. Dissertation thesis, Kaluga, 25 p. (in Russian)
4. Batima, P., 2002, Living with Climate Change. 1st draft of unpublished Synthesis Report within framework of the ongoing project "Potential Impacts of Climate Change and Vulnerability and Adaptation Assessment for Grassland Ecosystem and Livestock Sector in Mongolia" financed by GEF, implemented by Global System for Analysis, Research and Training (START), Third World Academy of Sciences (TWAS), UNEP, 91 pp.
5. Berkovich, K.M., Chalov R.S., Chernov A.V., 2000, Science of River Channels Ecology. GEOS, Moscow, 332 p. (in Russian)
6. Biological Resources and Natural Conditions of Mongolia. Joint Russian-Mongolian Interdisciplinary Biological Expedition. 1995. Vol. XXXIX, Ecosystems of Mongolia: Distribution and Current State. Moscow, "Nauka", 224 p. (in Russian)
7. Chistyakov, K.V. and Yu.P. Seliverstov, 1999, Regional Ecology of Little Altered Landscapes. North-West of the Inner Asia. Sankt-Petersberg University Publishing House, 264 p. (in Russian)
8. Climatic Passport of Altai-Sayan Ecoregion, 2001, Issue 1. Moscow, WWF, 26 p.

9. Enkhtsetseg, B. et al., 2002, Socio-economic Situation in the Altai-Sayan (Mongolia).
10. *Ensuring Long-term Conservation of Biodiversity in Altai-Sayan* ecoregion, 1999, Proposed tasks, results and activity for the project. WWF Russian Program Office, Document with the summary after International Forum "Altai-Sayan – XXI Century", December 1999, (unpublished), 62 p. (in Russian)
11. Furyaev, V.V., 2002, Biodiversity Conservation by Measures to Decrease of Forest Fire Threats in Altai-Sayan Ecoregion. Krasnoyarsk, (unpublished), 34 pp. (in Russian)
12. Galakhov, V.P., Mukhametov R.P., 1999, Glaciers of the Altai. Nauka, Novosibirsk, 135 p. (in Russian)
13. Gunya, A.N., Drozdov, A.V., and G.S. Samoilova, 2002, Landscapes and land-use in Kosh-Agach area of the Altai Republic. Proceedings of the Academy of Sciences, Geographical series, **5**:83-90. (in Russian)
14. Kharlamova, N.F., 2000, Dynamics of thermal regime in the continental regions of Russia over last 160 years. Proceedings of the Altai University, 3(17):56-58. (in Russian)
15. Komlev, A.M., Titova Yu.V., 1966, Runoff Formation in the Katun Basin (Altai Mountain). Nauka, Novosibirsk, 155 p. (in Russian)
16. Kordonsky, S.G., 2003, Activity of Tyungur Village community in tourist business. Personal communication.
17. Plant Communities of Tyva. Novosibirsk, 1982, A.V. Kuminova, ed., Nauka, Novosibirsk, 204 p. (in Russian)
18. *Biological Diversity of Altai-Sayan Ecoregion*, 2003, A.N. Kupriyanov, ed., CREOO "Irbis", Kemerovo, 156.
19. Leak, W.B., and R.E. Graber, 1974, A method for detecting migration of forest vegetation. Ecology, 55(6): 1425-1427.
20. Modina, T.D., Drachev, S.S. and M.G. Sukhova, 2002, On the global climatic change on the Mountain Altai territory. Proceedings of regional scientific and practical conference, Tomsk, p. 161. (in Russian)
21. Narozhny, Yu. K., Paromov V.V., Shantykova L.N., 1998, Features of the annual runoff of the Altai rivers. Transactions of Glaciological Studies, **84**:34-40. (in Russian)
22. Narozhny, Yu. K., Okishev P.A., 1999, Dynamics of the Altai glaciers at the regression phase of the minor ice age. Transactions of Glaciological Studies, **87**: 119-123. (in Russian)
23. Narozhny Yu. K., 2001: Resource Assessment and Tendencies of Glaciers in the Aktru Basin (Altai) for the Past Century and a Half. Transactions of Glaciological Studies, 90, pp. 117-125 (in Russian)
24. Nemolyaeva L., 2001, Ensuring Long-Term Conservation of the Altai-Sayan ecoregion, WWF Project: Achievements and Lessons Learned. WWF Russia, 17 p.
25. Ovchinnikov, D.V., and E.A. Vaganov, 1999, Dendrochronologic characteristics of Siberian larch (Latix sibirica L.) at the upper forest limit in the Altai Highlands. Siberian Ecolog. J., **2**: 145-152. (in Russian)
26. Ovchinnikov, D.V., Panyushkina, I.P., and M.F. Adamenko, 2002, millennial tree-ring chronology of larch stands in the Altai Highlands and its utilization for reconstruction of summer temperature. Geography and natural resources, **1**: 102-108. (in Russian)
27. Paromov, V.V., 1999, Water Resources of the Upper Ob: Contemporary Assessment and Tendencies. Dissertation thesis, Institute of Geography of Siberia and the Far East, Irkutsk, 24 p. (in Russian)
28. Polikarpov, N.P., N.M. Chebakova, D.I. Nazimova, 1986, Climate and Mountain Forest in the Southern Siberia. Nauka, Novosibirsk, 226 p. (in Russian)
29. Polyakov, V.I., and I.V. Semechkin, 2001, Biogenic dynamics of cedar forests in the West Sayan Mountains. Siberian Ecolog. J., **6** : 667-673. (in Russian)
30. Revyakin, V.S., 1981, Natural Ices of the Altai-Sayan Mountain Region. Gidrometeoizdat, Leningrad, 268 p. (in Russian)

31. Revyakin, V.S. and N.F. Kharlamova, 2003, Climate change in the Inner Asia in XIX-XX centuries. Proceedings of International Symposium "Climate and Environment Change in the Central Asia", Ulan-Ude, 2003, pp. 57-63. (in Russian)
32. Ryabchikov, A.M., Gvozdetsky N.A., Konisshev V.N., Salisshev K.A., Saushkin Yu.G. (edd.), 1978: Altayskiy Kray. Atlas, vol. 1. GUGK, Moscow, Barnaul, 222 p. (in Russian)
33. Resources of Surface Waters of the USSR, 1969, Vol. 15-1: Mountain Altai and Upper Irtysh. V.A. Semenov, ed., Gidrometeoizdat, Leningrad, 318 p. (in Russian)
34. Semenov, V.A., Alekseeva A.K., Degtyarenko T.I., 1994, River runoff change in Russia and adjacent areas in the 20th century. Meteorology and Hydrology, 2: 76-83. (in Russian)
35. Semenov, V.A., Klimova O.V., 1997: Regional distribution, human-induced and climatic change of the river runoff in Mountain Altai, in: Natural Resources of Mountain Altai. Gorno-Altaysk, pp. 138-141 (in Russian)
36. Sherstyukov, B.G., Isaev A.A., 1999, Method of multiple cyclicity for analyzing time series and long-term forecasts by the example of a heating season in Moscow, Meteorology and Hydrology, 8: 24-28. (in Russian)
37. Shvarts, E. A., Shestakov, A.S., 2000, Protected areas and sustainable development: a chance for biodiversity or a new self-deception, in: *Nature Conservation: Bridges between West and East. Publikation zur Tagung "Forum fur Wissen" Swiss Fed.*, Inst. for Forest, Snow and Landscape Research, Birmensdorf, pp. 23-29.
38. Shvarts, E. A., Shestakov, A.S., 2002, Protected areas: contribution to sustainable development of Russia, in: Transition to sustainable development: global, regional and local levels. Foreign experience and Russia's issues. (Series: Sustainable development: problems and prospects. Issue 1.), KMK Publisher, Moscow, pp. 287-297. (in Russian)
39. Shvarts, E. A., 2003. WWF Russia project "Ensuring Long-term Conservation and Sustainable Use of Biodiversity in Altai-Sayan ecoregion". Manuscript, 8 p. (in Russian)
40. Shvarts, E.A., 2003: Ecological and Geographical Problems of Biodiversity Conservation in Russia. Dissertation thesis, 49 p. (in Russian)
41. Skotselyas, I.I., 1975: Calculation of Annual Runoff Distribution for Unstudied Rivers of Mountain Altai. Transactions of the Kazakh Research Hydrometeorological Institute, 48, 36-59 pp. (in Russian)
42. Smagin, V.N., S.A. Il'inskaya, D.I. Nazimova, et al. 1980. Forest Types in the Southern Siberia. Novosibirsk, "Nauka" – Siberian Branch, 336 p. (in Russian)
43. The Secretariat of the Convention on Biological Diversity. 2001. Global Biodiversity Outlook. Secretariat of the Convention on Biological Diversity, Montreal, 282 p.
44. WWF, 2002. Altai-Sayan Ecoregion Conservation Action Plan. Final Draft. WWF.
45. Yaskov, M.I., 1997: Desertification of the Chuya Steppe – State and Prospects. In: Natural Resources of Mountain Altai. Gorno-Altaysk, pp. 164-167 (in Russian)

TRANSBOUNDARY BIOSPHERE TERRITORY "ALTAI": EXPERT EVALUATION FOR THE ESTABLISHMENT

YU. VINOKUROV, B. KRASNOYAROVA, S. SURAZAKOVA,
RUSSIA
Institute of Water and Environmental Problems (SB RAS), Barnaul, Russia

Abstract: The article describes the territories of Altai Mountains (by example of the Republic of Altai): population structure, national-ethnic composition, natural-climatic conditions, and social-economic level of development, state structure and economic/nature protection activity. An analysis of the establishment and development of the Transboundary Biosphere Territory "Altai" is given. A consideration of advantages and limitation, problems and perspectives of organization is discussed.

Keywords: Transboundary Biosphere Territory, Altai Mountains, international cooperation, preservation of a biodiversity, Republic of Altai, regional development, remote areas, objects of World Heritage "Altai – the Golden Mountains", environment, steady development.

The idea of establishing a Transboundary Biosphere Territory "Altai" (TBT "Altai") has been discussed by various power structures and scientists from different countries for a decade already since this territory is distinguished by the unique natural complexes and numerous cultural-archeological monuments. Here, the interests of biodiversity conservation and sustainable development of at least 4 countries, i.e. Kazakhstan, China, Mongolia and Russia, are concentrated. The contiguous territories of the countries mentioned above are similar in population structure, national-ethnic composition, natural-climatic conditions and the social-economic level of development but differ greatly in peoples' world outlook, state structure and economic/nature protection activity. Each country sticks to its own legislation, carries out its own economic and environmental policy that often doesn't coincide with interests of neighboring countries and even the world community ones. Moreover, all these transboundary territories are remote areas with undeveloped economy and a low level of social service.

Many problems of social-economic and natural-environmental character which occurred in the transboundary regions with high ecological status are solved by developed countries by means of the biosphere reserves establishment. First, the establishment and operation of biosphere reserves are regulated by international resolutions adopted at the UNESCO conference in Seville in 1995 and 5 years later the resolutions were confirmed in Pamplone. Second, the status of a biosphere reserve along with high biosphere importance of the territory envisages the availability of the zone for a development or biosphere site to carry out economic activity but with mandatory observance of ecological imperative. Thus, the realization of the local population's interests along with the conservation of unique natural complexes within the boundaries of the reserve is provided; in other words, a unity of man and nature is reached.

At present, the Feasibility Study for the Establishment of a Transboundary Biosphere Territory in the Altai Mountains has been accomplished. This work was made in two stages. First, national working groups of experts were formed. These groups performed the Feasibility Study for the Establishment of TBT "Altai" for each country-participant. Then, the project feasibility study was prepared based on the generalization and analysis of national reports.

The paper presents the main points of the National Report of Russia that appeared due to discussions among German and Russian (Novosibirsk, Barnaul, Gorno-Altaisk) experts and consultants from St.Petersburg, Moscow, Irkutsk, Krasnoyarsk. First of all, they analyzed the situation in the border region of Russia – the Altai Republic. They examined the social, demographic and economic development; national-ethnic peculiarities; and cultural-historic traditions. They also placed stress on the high biospheric status of the territory, which has five objects of the world's natural heritage, and formulated the main subjects of their research. The report gives much attention to the protection and regulation of the use of natural resources, the institutional system, public organizations, training of specialists and ecological education. Finally the experts, authors and participants of this research estimated the possible creation of the Transboundary Biosphere Territory, determined the basic trends and concrete objects of development, and substantiated the necessary expansion of protected natural territories and change in the status of existing ones. The creation of the natural reservation, like any other interference in nature, brings both positive and negative changes to the environment and social life of the local population. The main tasks of the experts are: to estimate all the pros and cons of the creation of TBT 'Altai'; to minimize the negative sides of the project; and to substantially increase the positive ones in the following stages of the project. The creation of the Transboundary Biosphere Territory would not only increase the competitive positions of the territory in the market of ecosystem services, but also contribute to the improvement of the standard of living in the region as a whole. In the report, special attention was given to the realization of measures on sustainable development of the Transboundary Biosphere Territory.

1. REPUBLIC OF ALTAI: THE SITUATION ANALYSIS

The Russian part of the Transboundary Biosphere Territory is represented mostly by the Republic of Altai situated in the heart of the Asian continent, in the south of West Siberia, to be exact. It covers the territory of 92.9 th.km^2 including farming lands -19%, forests - 47%, aquatic areas - 0.9% and others - 33.9%.

1.1 Natural importance of altai

The Altai Mountains are situated in the center of Eurasia at the boundary of two natural zones of the north hemisphere, i.e. humid boreal and arid deserted-steppe. The mountain system occupies the most elevated part of the North-Asian continental watershed - the largest tectonic-morphological Asian structure dividing river runoff and water resources by the basin of the Arctic Ocean and internal-drainage basins of internal Asia. The Altai is distinguished by complicated geological and relief structures, various natural complexes, diverse vegetation and wildlife.

Altai (that is situated between vast steppe territories of West Siberia and Kazakhstan in the north and in the west, desert-steppe Mongolian plateaus in the south, significant forest massifs of West Sayany in the east) is the powerful center of biodiversity and biological species expansion for the great part of Siberia and contiguous regions of Kazakhstan, China, Mongolia, Tuva, Khakassia. The Altai Mountains are the important part of one of 200 world ecoregions, namely the Altai-Sayan, which is of great interest due to its biodiversity conservation problems. Note, that a lot of studies and international projects were dedicated to this region.

Special protected natural territories of the Republic of Altai are represented by an extensive net including the objects of different environmental status. Let us consider only the ecological or nature protective status of Altai Republic. Here, the area of SPNT occupies more than 22% of the territory. The Altaisky State Reserve and Katunsky Biosphere Reserve (which are the oldest beyond the Urals) cover more than 10 th. km^2 that makes up 11.1% of the whole territory of the Republic or about 50% of the protected territories. The rest of the natural protected area is represented by the following three reserves of republican importance: Kosh-Agachsky, Sumul'tinsky and Shavlinsky as well as by Turochaksky zoological reserve of regional importance where activities on the beavers' conservation are conducted. Besides, there are two natural parks: "Belukha" and "Katun". The first one is situated in Ust-Koksinsky region, and the Belukha Mountain, the highest peak of Asian Russia and the Object of World Heritage, is found here. The territory of the second one embraces the right bank of Katun River within Chemal and Maima regions. All in all there are 42 natural monuments.

Besides, special protected natural territories include the "Ukok" zone of rest which is unique in Russia. At present, it is a reserve of regional importance with a limited period of validity till 2004. At the same time, the "Ukok" zone of rest along with Altaisky and Katunsky reserves, Belukha Mountain and Lake Teletskoye is one of the sites of the Object of World Heritage "Altai – the Golden Mountains".

Because of the low nature conservation status of the majority of special protected natural territories and the lack of governmental support (legal, manpower, financial), the current net of SPNT is not able to function in a proper way. Here, poaching as well as disturbance of the unique Altai flora and fauna habitats is observed.

According to the experts' assessment, the proposed net of SPNT is impracticable because of the current practice of maintenance of the reserves' functioning and lack of real financing even for existing reserves.

Along with the current conservation types, such new ones as natural-economic and ethnic-cultural parks have been recognized. Their establishment is aimed at the conservation of not only natural but ethnic-cultural complexes and the sites of traditional nature management.

Among the successfully functioning natural-economic parks are the "Chui Oozy" and "Uch Enmek". The experts have proposed to establish another 7 parks including Severokatunsky and Seminsky natural parks and five ethnic-cultural ones.

1.2 National-ethnic and cultural-historic values and traditions

Altai is a distinguished global phenomenon from the point of view of its unique ethnic- cultural diversity and spiritual basis for root population and other people. This unique nature is reasoned by some peculiarities of its historical-cultural development.

Altai is the ancient heart of civilization with a 1-1.5 mln. year old history.

Altai is rich in historic monuments of all eras. In recent years, a lot of discoveries have been made on the Ukok plateau.

Altai is the origin motherland of all Turk peoples, proven by numerous archeological findings. This Region is the heart of ethno genesis and intensive intercultural interaction.

Altai always was a region of difficult access. Various tribes and peoples took refuge in the mountains. It is conceived, that several root ethnic groups of Eurasia (e.g. Turk, Mongol, Slavic, Ugro-Finnic ones) consider Altai as their historical original motherland.

The historic memory of Altai is really unfathomable: cultural layers of migrating people were not "washed away" (like it was in lowlands and valleys), v.v. they were "stored". Hence we can find here sustainability of ethno cultural traditions that keep world outlook of centuries-old history [5,6].

In Altai, such religious-cultural traditions as Orthodox, Buddhism, Islam and Pagan came together and formed intricate synthetic forms. The Russian population is mostly orthodox, the Kazakhs are Moslems but Altai people practice different religions. Still some of the latter practiced Shamanism, a traditional religion of Siberia, exists in the far East and the North. Shamanism appeared in ancient times as a form of public perception when people personified the nature and its forces. Other Altai people are Burkhanists (Tibet branch of Buddhism). In the 19th century, many Altai people adopted Christianity. Such a place of concentration of religion confessions occurred here because Altai was the crossroad of different civilizations.

For Altai people Altai is a sacred, inspired, idolized place inhabited by spirits of mountains, rivers, fire, taiga, forests, animals, etc. Altai is described by root people as majestic, grand, blessing and sacred. Not only the beauty of nature, but also the vivifying energy of Altai is of great importance. The Altai language has preserved its ancient features and epic works that are still widely spread.

1.3 Social economic conditions and differentiation of development

Against the background of high ecological and ethnic-cultural status, the level of social –economic development of Republic of Altai and living standard of its population are, on the contrary, very low.

According to the figures on social-economic development, the Republic of Altai enters into the group of subjects of the Russian Federation with low a level of development. It occupies the 63rd place of 89. The Republic of Altai is characterized by a low relationship between the average income per capita and the mean living wage per capita, share of population with income lower than the living wage, retail turnover volume and paid services.

The Republic of Altai is an independent subject of the Russian Federation. Its territory makes up 92.9 th. km^2 and the population - 205.2 th people, 3/4 of population is rural and live in 10 administrative regions. The whole urban population (53.3 th people) live in Gorno-Altaisk city, the capital of the Republic of Altai.

The economy of the Republic of Altai is mainly agrarian. Gross regional product (GRP) increased 3.5 times since 1998 and in 2001 it made up 5654.0 mln. Rubles (at a current price). GRP per head also increased 3.4 times since 1998 and in 2001 it constituted 27.61 th. Rubles. In 2001, the production of goods made up 2196.5 mln. Rubles or 38.9% of GRP, the servicing made up 2803.6 mln. Rubles (40.7% of GRP), goods taxes constituted 20.4% of GRP. The largest contribution into goods production is made by agriculture, industry, and civil engineering; into the goods production - trade and commercial activity in sale of goods and services, insurance and nonmarket services. Their share in service production constituted 22.8; 12.6; 46.1%, correspondingly.

Key industries are the production of building materials, food industry and non-ferrous metallurgy.

Leading branches of agriculture are cattle-, sheep-, and goat-breeding, yak, maral and dappled deer farming. Fruit growing is developed. Among traditional branches are beekeeping and hunting. In agriculture, 80% falls on cattle breeding and 20% on plant growing.

The regions of the Republic differ in natural-climatic conditions and area as well as in the distance from the administrative center and transport junctions. It causes a high differentiation of the regions in employment rate and production pattern in agriculture; pasture and plough load, percentage of forest land and potential for the development of merchantable wood. Among 11 subjects of regional development, the largest one is Gorno-Altaisk town. More than a quarter of the Republic's population lives here, which in 2001 made up 53.1 th. people. Gorno-Altaisk town shows the highest development of social infrastructure, employment rate in economy and at the same time the unemployment rate and liquidity of local budget due to native income. We sought to give the official description of each of the 10 rural administrative regions of Republic of Altai. Maiminsky and Kosh-Agachsky regions are the most polar ones. Maiminsky region takes the first place in 9 items of 22 criteria of social-economic development, and it occupies the last place in area per head. In contrast, Kosh-Agachsky region takes the last 6 places and only two of the first ones - on the area of agricultural land and sheep stock per head. This region is characterized by a very severe climate, remoteness from the center, the highest unemployment rate and a very low level of industrial production per head as well as by low volume of retail turnover. As for Maiminsky region it is situated in low mountains and is distinguished by a mild climate and advantageous position. It is the place of thoroughfares' intersection; the distance to the nearest railway station (Bijsk town) is about 100 km. The region is noteworthy for the highest density of population and the density of settling. Along with Gorno-Altaisk town it acts as the infrastructure center occupying a prominent place in the volume of retain turnover, paid services and other infrastructure indices. The share of able-bodied citizens and economic employment mainly in state and services organizations are large.

Among economically developed regions are Chemalsky and Ust-Koksinsky. Chemalsky region has optimal natural-climatic conditions, and the level of recreation development is high. However, this kind of activity makes a small contribution into the region's budget and social economic development. The region ranks 6-8th on the majority of social and infrastructure indicators and the 9th on the share of income in local budget. The employment is rather high (3-4 places in the employment rates in economy and population age structure). At the same time it is valid to say that the first pay of the able-bodied citizens is out of public activity. It is due to the concealed income from tourist business or food production at the private farm that increases the living standard of the local population but doesn't replenish the city budget.

Ust-Koksinsky region refers to the highly developed agrarian ones. It is the first in agriculture ratio in the gross regional product and the area of arable land per head. Shebalinsky, Ust-Kansky and Ongudaisky regions are agrarian ones as well. In Ust-Koksinsky and Shebalinsky regions agriculture plays a significant role and is of trade importance. Ust-Kansky and Ongudaisky regions specialize in cattle-breeding.

Turochaksky, Choisky and Ongudaisky regions relate to the forest territories where forest yield per head makes up from 5.8 th. m^3 (Choisky region) to 11.7 th. m^3 (Turochaksky region). However, the level of merchantable wood processing has lowered significantly, and nowadays wood processing doesn't determine the regions' economy. For instance, Turochaksky region despite of heavy stocks of wood, placer gold and the unique Lake Teletskoye takes 8-9 places in almost all social-economic indicators and it occupies the last place in horse and goat heads. Choisky region rates first in output of industrial products and it occupies the last place in population and small cattle heads. Ust-Kansky region has the highest employment in economy and the highest output of agricultural products and livestock per head. At the same time the region demonstrates the lowest indices and poor development of social sphere. Ongudaisky region can be referred to the highly developed agrarian ones. It comes third in the area of agricultural lands and cattle density and rates first only in accommodation supply per capita. And finally, Ulagansky region ranks last in 6 items of economic and social development, for instance, output of agricultural products per head and accommodation supply. The region is high only in the list of the area and goat stock per capita.

The analysis carried out points to the low level of social-economic development of the regions in Republic of Altai as well as to the unbalance with natural potential. Extra seasonal exploitation of natural resources mainly for recreational use takes place.

Infrastructure of the Altai Republic territory is very poor. Geographical peculiarities of the Republic are responsible for the development of two modes of transport: motor transport (90% of all kinds of transportation) and aircraft. Railways and navigable waterways are not available. The nearest railway station can be found in Bijsk town that is 100 km far from Gorno-Altaisk town. There are three airports (Gorno-Altaisk town, Ust-Koksa settlement and Kosh-Agach settlement) and several helipads in the region. At present all airports do not function because of the emergency state of landing strips and unprofitableness of small aviation. Therefore transportation of passengers and goods is performed by motor transport.

International relations are maintained through "Chuisky pass", a single highway of federal importance connecting the Republic with Mongolia and Altai Krai. At present the construction of a Tashtagol-Turochak highway that connects the Republic with Kemerovo region has been completed. The construction of Talda-Karagai (Ust-Koksinsky region)-Ridder connecting the regions of Republic of Altai with East Kazakhstan is planned. The reconstruction of a motorway between the Republic and Tyva is conducted.

Other projects of road-building including motor communication between Russia and SUAR (China) through Kanas pass are under consideration.

The current motorway system is one of the undeveloped ones in West Siberia.

Power supply in the Republic is accomplished from "Altaienergo" system by two power lines – "Bijsk-Gorno-Altaisk" double-circuit line and "Bijsk-Smolenskoye-Cherga" single-circuit one. The total rated discharge capacity makes up 75 MW, while power demand during winter exceeds 100 MW. Most of administrative regions in the Republic get the power through single-circuit lines and don't have redundancy.

In many back settlements power supply is accomplished by low-power diesel power stations functioning for 5-6 hours a day.

Heating is provided owing to small boiler houses using the coal imported from Kuzbass and stove heating. Annual import of coal makes up 300 th. tons.

About 3 th. tons of gas (liquefied) is delivered by motor transport only for population needs.

The proven coal supply in the Chagan-Uzunsky coal-field (Kosh-Agachsky region) and the Pyzhinsky coal-field (Turochaksky region) as well as the unique natural and hydro power energy resources are of limited use. There is some experience in construction and operation of small (Chemal HES), mini- (Kairu HES, Ulagansky region, Chulyshmanskaya valley has been put into operation; the construction of Dzhzatorskaya HES on Tyunya river, Kosh-Agachsky region has been started), and micro-HES (Turochaksky and Ulagansky regions).

The current market infrastructure is in the making. Bank and credit institutions operate. Most of them are the branches of central banks of the country and other regions. Service-market is slightly developed.

Communication service is hardly developed. Telephone communi-cation is not available in more than 25% of settlements, and almost half of the settlements do not have local telecasting.

Social sphere and housing and communal services are backward. About 30% of schools require thorough repairs and 15% are in emergency situation. Running water is not available in 80%, central heating – in 83%, sewer system – in 87% of schools. Only 10% of schools are equipped with modern services and utilities.

A similar situation is typical for objects of health protection and culture.

Even in pre-reform time the Republic occupied last places in population provision by the objects of social sphere. By now the situation has become worse. A recent earthquake has aggravated the situation. In the last decade the construction of first-class accommodation lowered by a factor of 3.2, schools – by a factor of 19. Most of local hospitals are placed in the non-type buildings that are more than 50 years old, the lack of hot water and sewer system is observed in more than half of buildings. According to the data from the Committee on Statistics the rate of wear of capital assets of social sphere constitutes 30-40%. However, this is an optimistic index owing to the inflationary component in the assessment of depreciated and replacement cost of capital assets. The observed depreciation makes up 50-60%.

Sharp territorial differentiation between the population living standard and the development of the social sphere is still a problem of great concern. The analysis of living standard factors in the regions shows that the mean monthly income per employee varies by a factor of 5, trade turnover per capita - by a factor of 5.7, hospitals – 2.7, doctors – 2.5.

1.4 Economic and ecological policy of nature management: conflicts and solutions

Economic policy in the Republic of Altai was made through the principles of extensive use of natural resources and in doing so, the involvement of natural resources varied in different territories of the Republic and periods of economic development.

At present (post-reform period), agricultural lands have reduced by 16% and agricultural holding – by 6%. Their transformation caused by a decrease in croplands and hay meadows is observed. Total cattle load per unit of area has decreased as well. However, these facts are not positive even with relation to biodiversity conservation since in some cases the liquidation of large-scale agricultural factories has disturbed the seasonal pasture rotation, and the land load within settlements has increased. Pasture degradation is observed in all settlements with the population of more than 200 people and it is disastrous in settlements where population is 600 and more people. The depression makes up about 90% of the grass vital capacity. Besides, the reduction of croplands (growing of forage crops under crop rotation) causes the increase of cattle load on natural forage land.

Agricultural lands are alienated mainly to recreation, road-building and mining organizations.

Agriculture is still a key branch in the republic's economy and traditionally receives financial and economic support. It is a result of not only economic policy of the region but of the social as well. The greatest part of the region's population lives in rural areas and is engaged in self-sufficiency. The marketable value of agriculture is low due to low productivity and high production price as well as to the lack of modern marketing.

The processing of agricultural raw materials is performed at 9 large-scale and a numerous small-scale enterprises. The index of the actual volume of the food industry continues to fall.

In road-building the reconstruction of the federal highway "Chuisky pass" has been completed. The motorways connecting the Republic of Altai with Kemerovo oblast and Khakassia have been commissioned. The construction of the advanced highway to the Republic of Tyva has been started. The design work on "Assessment of construction investments for Novosibirsk-Kosh-Agach-Kanas pass" highway has been carried out. The improvement of other home roads has been done. On the one hand it requires the land alienation; on the other it increases the investment and recreation attraction of the region that finds the reflection in economic indicators. Over the past few years, the investment attraction of the Republic has increased and tourism has become a dynamically developing sector of the national economy. To maintain these sides of economy, the government makes a series of efforts including the construction of new recreation enterprises. The legislative basis allowing a combination of interests of objects of tourism, local population and municipal formations has been worked out.

In forest management the economic recession is observed. Not less than 10% of the wood-cutting area has been developed. The quality of commercial wood processing is low. The ineffective activity in forestry reduces the investment resources of this branch. The lack of funds for forest management, sanitary, an improvement of felling and forest protection is the case. As a result, the centers of Siberian silkworm spread and forest fires are observed in the forests including the ones of the first protection class.

Mining industry is represented mainly by prospecting worksmen's cooperative associations for stream-gold mining. Mines "Vesely" and "Priisk Altaisky" are functioning, the Kalgutinsky tungstenmolibdenic mine has resumed its operation, and the development of building materials is under consideration. The major deterrent for an intensive development of mineral-raw material resources is the lack of power. At present the construction of Katun HPS is the subject of wide speculation, and the appropriate decision has been made by the government. Construction and further exploitation of large HPS allows creating new qualified jobs, to improve power supply to population and economics of the Republic. It could be an additional stimulus for providing a sustainable social and economic development as well as for cooperation among the Bolshoi Altai countries within the framework of the "Altai" TBT project (thorough processing of mineral-raw and agricultural resources).

Despite of available rich natural potential, the Republic of Altai is still the economically backward region. The ecological status of the Republic proclaimed in the early 1990th assumes the realization of the policy of environment safety and minimal depletion of natural resources. However, such a policy can't reach a practical implementation because of some economic, social, national-ethnic, environmental, legislative and standard-legal reasons.

Economic reasons:

- The lack of own power bases leads to insufficient energy supply and the impossibility of a rational complex use as renewable (forest) as non-renewable (mineral-raw) and generally used (stone, sand, clay, etc.) resources;
- The lack of professional people;
- The lack of reprocessors (for the last 60 years none of wood processing enterprises has been constructed)
- The lack of road transport infrastructure (roads are available only among settlements, the lack of railway transport).

Social reasons

- Disposition of local population, the lack of skills for intensive land tenure;

- Conservation of territories of traditional nature management (hayfields, pastures, incidental forest exploitation, distant pasture cattle-breeding, etc.);
- The lack of compensation mechanism (direct and indirect) in the event of alienation of sites of traditional nature management.

National-ethnic reasons:

- National-ethnic composition of the local population;
- Solicitous attitude of local population to natural resources and individual natural complexes;
- Conservation of the cult of nature, idolization of individual natural objects and phenomena.

Ecological reasons:

- The uniqueness of landscape complexes combining natural resources of different function, for instance the Pyzhinskoye coal deposit in the place of cedar forest with wedging out of mineralized water);
- Extensive anthropogenic load under natural complexes exploitation for recreation purposes;
- Contraction of biodiversity natural habitat and ecological corridors.

Regulatory-legal reasons:

- Imperfection of legislative and regulatory-legal base in nature management and production sharing that doesn't reflect and support the interests of the subject of RF and doesn't induce to the development of processing branches and industrial processing of natural resources (forest, generally found, water);
- Grant of authority to local administration for natural resources control and regulation taking into account the interests of local population.

Due to the current economic traditions, natural and historical peculiarities, the common lands are used as a forest stock, pastures, hunting grounds and recreation stock. Since under these circumstances conflicts can be provoked it is necessary to delimit departmental, public and interpersonal interests.

Possible conflicts result from the restriction of natural resources recreational use by local population, incidental forest exploitation, forest utilization for private use at the farms (grant on lease, enclosure of riversides, etc.).

2. ESTABLISHMENT OF "ALTAI" BIOSPHERE TERRITORY AS A MODEL FOR SUSTAINABLE DEVELOPMENT

The problems of transboundary cooperation for biodiversity conservation and socio-economic development of four countries (Kazakhstan, China, Mongolia and Russia) have been discussed for more than 10 years. The issues on boundary cooperation were repeatedly discussed at intergovernmental meetings and scientific-practical conferences, workshops and symposiums. The Altai Declaration was signed on the initiative of the Republic of Altai at the Conference on discussion of perspectives for the development of Central Asia (CoDoCa) held on September 13-18, 1998 in Urumchi (China). Later the bi- and multilateral agreements, protocols, notes on boundary cooperation among Altai countries in economy, trade, science, education, culture, biodiversity conservation and environment protection were signed. These include the Spiritual-Ecological Charter of Altai-Sayan region ("Our Common Home Altai" International Conference, July 3-7, 2000, Aya settlement, Altai Territory, Russia); note of mutual understanding of Transboundary Biosphere Territory establishment signed by the official agents from Germany, Kazakhstan, Mongolia and Russia on March 12-16, 2002 in Novosibirsk, Russia; Altai statement (International Conference within the framework of the Year of Mountains, September 24-27, 2002, Gorno-Altaisk town, Republic of Altai, Russia), etc.

Various aspects of the possible establishment of "Altai" TBT including territorial boundaries, conservation of biodiversity and cultural heritage, socio-economic development of boundary regions, etc. were discussed at international meetings in different times and countries. The main thing that joins all the participants of the meetings is their wish for conservation of natural integrity, national-ethnic and cultural originality as well as for the mutual cooperation among Altai countries.

The establishment of the TBT Altai as any other natural–economic project realization has both positive and negative consequences. The thorough analysis and feasibility study makes it possible to solve the issue of possibility and necessity of the TBT establishment. The expert assessment revealed positive and negative aspects of the TBT establishment in respect to biodiversity conservation and sustainable social-economic development of the projected territory and the Republic on the whole.

2.1 Alternative options of economic development of republic of altai

The Republic of Altai and its regions were repeatedly the objects of strategic planning. Various options as the models of nature conservation and nature management were proposed. The examples of an integrated solution for problems of economically and ecologically sound development are available.

Nature protection projects - the development of these projects was due to international and regional reasons. It should be mentioned that the Altai-Sayan region is one of 200 centers of biodiversity conservation. Besides, 5 objects situated in Republic of Altai are put in the List of Objects of World Natural Heritage. The ecological-economic zone (region) has been established and is functioning for 10 years in the Republic.

- Concept of noosphere–biosphere approach when the whole territory of the Republic (86%) was proposed to consider as the Object of World Heritage. The authors prepared and submitted to UNESCO the documents to put Katunsky and Altaisky reserves, Belukha Mountain, zone of rest "Ukok" and Lake Teletskoye in the List of World Natural Heritage. However, this way of development is not very promising from a position of local population life support;
- WWF project on "Long-term biodiversity conservation in the Altai-Sayan ecoregion" (1997), has been developed, its implementation is aimed mainly at conservation of key species and territories;
- UNDP Capacity 21 project "Development and realization of local strategies on sustainable development in Republic of Altai, Russia" (2001). Relying on the pilot study of the activity of the local population and administrative bodies Kosh-Agachsky, Ongudaisky and Chemal'sky regions were taken as the model ones. Special strategies were developed for these regions. Small-scale projects including the support of "Choi Oozy" and "Uch Enmek" natural-economic parks and the development of Gorno-Altaisk botanical garden have been implemented;
- UNDP-GEF project "Conservation of biological diversity in the Altai-Sayan ecoregion" (Russia, Kazakhstan, China, Mongolia, 2001). The application for the development of large-scale project on the Russian territory of Altai-Sayan region was submitted and approved by GEF in January 2004.

2.2 Projects of social-economic development

- "Ecological-economic zone of Altai"-initiative of the Altai Republic government on establishment of the offshore zone" [12](1996). The ecological way for economy development in Republic of Altai has been accepted, businessmen have got the "tax recess". Unfortunately, many firms registered in the free zone operated outside the Republic that resulted in small benefit to the Republic and population;
- "Eurasian continental bridge"- the Russian-Chinese initiative to construct transport pass through the Kanas passage (1996-2001). Stimulation of cooperation between two countries was thoroughly discussed at the national level and by community. The Russian side has prepared the "Assessment of investments to the construction of Kosh-Agach - Kanas

pass highway". At present the highway is functioning up to Lake Kanas, and about 60 km of it is to be constructed;
- "Eurasian continental bridge" - Mongolian variant of social-economic development of transboundary territory (2001). The reconstruction of highways in Mongolia connecting the Altai okrug in China, Khunsantsui - Dayan-Nur settlement in Mongolia and check point in Tashanta settlement (Russia) - Tsagan-Nur (Mongolia) is proposed;
- "Power engineering model"- industrial development due to construction of Katunskaya HPS and gas pipeline" (1986-2001).

2.3 Models of sustainable development

- Concept of sustainable social-economic development of Republic of Altai;
- Concept of transboundary biosphere reserve "Altai" (near boundary Altai).

The projects mentioned above differ greatly depending on goals and priorities proposed by the authors and developers of these strategies. Among them are so called "traditional" models. One of such traditional models is typical for Russia on the whole, but not acceptable for Altai. It is based on the development of regional natural resources. Another traditional model keeps the priorities of the Soviet period specific for Altai, i.e. the development of the agrarian sector of economy and processing of agricultural and forestry products. Both ways of development are dead-end ones for the Republic of Altai because of the local population outlook (first option) and low compatibility (second option). Moreover, both options hardly consider the ecological status of the territory.

Thus a number of projects are based on national-ethnic peculiarities of nature management carried out by the local population as well as on its world outlook. It is so called "aboriginal" model or Republic. In our opinion, the disadvantage of this model is in authors' rejection of globalization processes that occur even in remote Altai. Undoubtedly, 1) few people would like to "come back to nature" and 2) such an approach can't provide high living standard for the population and may lead to poaching and destructive nature management.

A number of attempts to combine ideology on biodiversity conservation and sustainable development were made but not implemented because of the lack of financial, material, institutional, intellectual (mainly skilled managers) resources.

Why is the idea of TBT establishment attractive? It makes possible to learn how to combine interests of nature (biodiversity conservation and the environment on the whole) with the interests of local population [on the one hand, traditional way of life, on the other hand- creation of up-to-date infrastructure (transport, information-communication, market, etc.)], implementation of new activities and technologies. In this case there is a chance to preserve the most valuable natural complexes.

2.4 Options of biosphere territory establishment

In the course of expert assessment of TBT "Altai" establishment three options of its boundaries were considered.

- "Bolshoy Altai" includes the whole territory of the Republic of Altai and a small part of the Altai Krai (the Tigiretsky reserve). Positive aspects of this proposal: TBT comprises all objects of World Natural Heritage and the Tigiretsky state reserve situated in Altai Krai at the border with Kazakhstan. Wide opportunities (territorial) to preserve biodiversity, to have different prospects for social-economics development of national territories and transboundary cooperation exist. However, in this case an extremely vast territory with numerous national and economic objects of development will be formed. As a result TBT area may reach 300 th.km^2. Such a territory will be poorly controlled and coordinated decisions concerning different subjects (conservation objects and development subjects of 4 countries) will be hardly made.
- "Near boundary Altai" – the territory is limited by the Katunsky biosphere reserve, its "biosphere polygon", "Ukok Plateau" area with northern spurs of South Altai and Tabyn-Bogdo-Ola as well as Dzhazator river valley that is about 10% of the Republic territory. In this case we have 3 objects of World Heritage to be preserved but economic development and cooperation of countries-participants are restricted.
- "Middle Altai" occupies 1/4 of the Republic territory. It is a near boundary territory restricted in the north by the Ursul river basin (except for Teletskoye lake basin). The Russian territory has distinctly oriented chains of ridges, i.e. South- Chyusky, North- Chuisky, Katunsky, Tirektinsky, Aigulaksky, Kuraisky, Chikhachevsky and dividing them such depressions as Chuyskoy, Kurayskoy, Uymonskoy, Abayskoy, Bertekskoy, Samakhinskoy, Tarhatinskoy and some other hollows.

According to the experts assessment, the latter variant seems to be the most acceptable after some alterations made. It is reasonable to include into "Altai" TBT the Russian territory situated southward the Katun and Chuya rivers watershed up to the Katun mouth.

It is expected to establish the Biosphere Territory mainly within the boundaries of two municipal formations: Kosh–Agachsky and Ust-Koksinsky administrative regions. For this purpose the following activities should be undertaken:

a.) For biodiversity conservation

- To extend Katunsky biosphere reserve through the construction of Yungur claster site (Kosh-Agachsky region);
- To increase the material and technical basis of the current Katunsky reserve and "Belukha" natural park, Ust-Koksinsky region;
- To legalize the status of the "Ukok" zone of rest as the natural park "Ukok zone of rest", to carry out its administration registration and landscape planning;
- To establish the Board of special protected natural territories in Kosh-Agachsky region exercising control over Kosh-Agachsky, Sailugemsky and Shavlinsky state reserves.

b.) For sustainable development of the territories mentioned above

- To establish the following three biosphere sites with different traditions and land use:
- Dzhazatorsky (Kosh-Agachsky region) – traditional land use by the Kazakhs - distant pasture cattle-breeding; processing of milk, meat, wool, fluff and skin of cattle; stocking up and processing of wood; trades;
- Kokorinsky (Kosh-Agachsky region) – traditional land use by Altai and Telengit people; distant pasture cattle-breeding; processing of sheep breeding products; trades;
- Uimonsky (Ust-Koksinsky region) – Old Russian land use – agriculture, cattle-breeding, beekeeping, trades.
- To establish enterprises – points of increase:
- To construct hydro-wind-heliopower units in the Kosh-Agachsky region; factory for processing of sheep- and goat-breeding products (meat, wool, fluff and skin), camel-breeding in Kosh-Agach settlement;
- To create the international company for the development of Asgat silver-ore deposit, Kosh-Agachsky region;
- To construct the experimental maral-breeding farm (storage and preserving of horns, complex use of sideline products), Ust-Koksinsky region;
- To construct the factory for vacuum-impulse drying of animal and phytogenous raw material, Ust-Koksinsky region.

The establishment of transboundary biosphere territory is a longstanding and complicated process that along with political decision of the countries, regional and municipal authorities requires significant investments as for its construction as for operation. It is evident that the funds from the national budget are not sufficient. The attraction of other donors, such as international organizations and governments of foreign countries, World, European and Asian Banks for Development, charitable organizations and foundations is necessary. WWF, GEF, Government of Germany and the Netherlands can be considered as the potential sponsors.

A series of actions mentioned in this project are included in the Plan on biodiversity conservation in Altai-Sayan ecoregion [5] submitted to GEF for consideration. Some of them are specific and the rest are presented for the first time since the project assumes the search of not only the ways for biodiversity conservation but for sustainable development of the local population as well.

***

Scientific, political, organizational and economic bases for "Altai" TBT establishment were discussed in Russia and other interested countries. They were approved and supported by public organizations and state authorities at different levels.

It was during the last two years when the issues of boundary cooperation and the establishment of "Altai" TBT were repeatedly discussed at the meetings of Coordination Council "Our Common Home Altai" with the participation of the heads of legislative authority from Altai regions (Kazakhstan, China, Mongolia, Russia), scientists, public organizations (September, 2002, Belokhurikha town; March, 2003, Barnaul; August, 2003, Gorno-Altaisk town, Russia); as well as at the international symposiums (March, 2002, Novosibirsk, Russia; July, 2002, Bayan-Ulgiy, Mongolia; March, 2003, Brussels, Belgium; August, 2003, Barnaul, Russia; September, 2003, Novosibirsk, Russia; November, 2003, Ulanbaatar, Mongolia, July and August, 2004, Urumchi, China, etc.). Experts gave reports on the necessity, possibility and prospects for the establishment of "Altai" TBT giving considerable attention to positive and negative sides of the project. The decision to carry out the expert assessment of "Altai" TBT establishment was made. The materials presented are the result of the assessment in the Russian part of the Transboundary Territory. The results were approved by world public, the representatives of national and regional managerial bodies. At present practical moves for its implementation should be taken.

CHAPTER 5

LANDSCAPE PLANNING AS AN INTEGRATIVE TOOL FOR SUSTAINABLE DEVELOPMENT

LANDSCAPE PLANNING AS A TOOL FOR SUSTAINABLE DEVELOPMENT OF THE TERRITORY

German Methodology and Experience

D. GRUEHN
Austrian Research Centers – systems research GmbH, Vienna, Austria

Abstract: Landscape planning has been established in Germany as a legal instrument in the early 1970ies. Because of its federal structure, 16 different systems of landscape planning have been developed in Germany. In spite of that, there does exist generally accepted tasks of landscape planning according to article 13 German Federal Nature Protection Act. What are the main tasks? Landscape planning has to support the realization of nature conservation goals and principles, to draw requirements and measures of nature conservation and landscape management, and to give reasons and arguments for these requirements and measures. In the last years the question concerning the effectiveness of planning instruments became more and more important in Germany. Thus, at Berlin University of Technology several research projects concerning this topic have been carried out. By means of statistical methods significant effects of landscape plans have been proved. The most important factors for succeeding of landscape planning have been isolated, too.

Keywords: Landscape Planning; German Federal Nature Protection Act, planning system; effectiveness; sample survey; statistical methods.

1. PLANNING SYSTEM IN GERMANY

The planning system in Germany is characterized by a multi-dimensional diversification of planning instruments according to their tasks, scale, spatial extension, responsible authorities, commitment effects, participation and some other federal peculiarities. The German planning system includes not only landscape planning, but also regional and physical planning, agricultural planning, forestry planning and others.

Concerning the legislation in the matter of nature conservation and landscape management it should be considered that the German parliament has only competences to issue framework legislation in landscape planning, regional planning and some other topics. That means federal states in Germany are allowed to create peculiar legal binding details, for example to limit landscape planning on open land as it has been done in North Rhine-Westphalia since the 1980s.

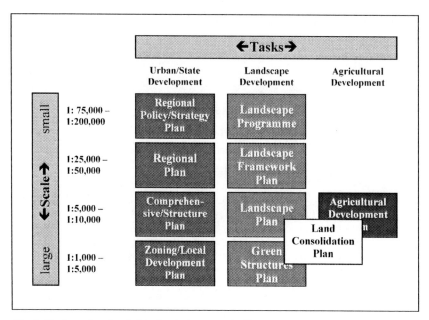

Fig. 1: Tasks and scale of planning instruments in Germany

The different planning instruments in Germany according to their tasks and scale are presented in figure 1. There are three main tasks, urban or state development as well as landscape development and agricultural development. As it is shown each planning instrument has a specific scale. Whereas the green structures plan is an instrument for landscape development on a large scale, the small scaled regional policy plan is used for state development. The coincidence of landscape planning and urban/state development instruments is obvious. Each planning instrument has its counterpart on the same scale; nevertheless there are exceptions in some federal states. Overall, landscape planning has four different planning instruments, landscape programme, landscape framework plan, landscape plan and green structures plan.

Finally, the land consolidation plan is something special. It can be used for both, landscape and agricultural development, and it is regularly worked out in smaller scales, that means in greater detail.

The planning instruments also differ according to their commitment effects and the participation procedure (fig. 2). Contrary to all expectations participation is not an obligatory element of planning processes in Germany. If the commitment effects are soft or even missing, participation procedures are not necessarily needed. A good example for this fact is the landscape

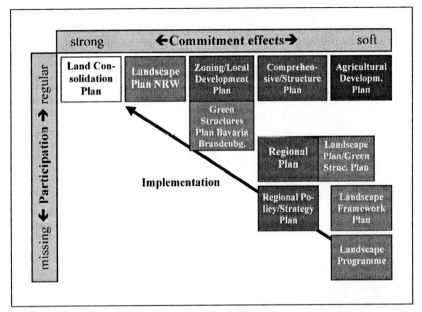

Fig. 2: Commitment effects and participation of planning instruments in Germany

programme. Nevertheless, there does exist planning instruments with only soft commitment effects and a regular participation processes (e.g. agricultural development plan). On the other hand strong commitment effects demand for a regular participation, because of their potential impact on proprietary rights. Examples are the land consolidation plan or the landscape plan in North Rhine-Westphalia. These instruments are suitable for an implementation of the proposed measures at a high degree. In any other cases the implementation possibilities are considerably worse.

2. TASKS, INSTRUMENTS, CONTENTS AND METHODS OF LANDSCAPE PLANNING IN GERMANY ACCORDING TO FEDERAL NATURE PROTECTION ACT

In article 13 German Federal Nature Protection Act the tasks of landscape planning have been defined. The main tasks are
- to support the realization of nature conservation goals and principles,
- the drawing of requirements and measures of nature conservation and landscape management and
- to give reasons and arguments for requirements and measures of nature conservation and landscape management.

Especially because of the aspect mentioned before there can be no doubt that landscape planning has something to do with communication. On the one hand landscape planning has to inform, on the other hand it has to persuade. If landscape planning can not persuade, it will not succeed.

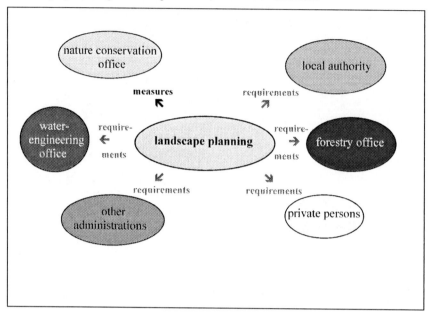

Fig. 3: Landscape planning's addressees

Figure 3 shows landscape planning's different addressees. All these addressees are requested to contribute to nature conservation and landscape management goals within their own legal competences. Due to juridical reasons landscape planning in Germany is only permitted to propose concrete measures against nature conservation offices. In any other cases landscape planning should recommend so called requirements (of nature conservation and landscape management).

The contents of landscape planning have been fixed in article 14 German Federal Nature Protection Act.

Contents are therefore
- a statement on present and expected conditions of nature and landscape,
- to put goals and principles of nature conservation and landscape management into concrete terms and
- an evaluation of present and expected conditions of nature and landscape due to goals and principles of nature conservation and landscape management.

That leads to the question what these goals are in detail? According to article 1 German Federal Nature Protection Act the goals of nature conservation and landscape management are defined as sustainable ensuring of

Landscape Planning as a Tool for Sustainable Development

1. landscape balance e.g. natural systems,
2. regenerative capacity and sustainable utilization of natural resources,
3. animal and plant kingdom and species habitats and
4. variety, uniqueness, beauty and recreation value of nature and landscape.

Putting the goals and principles of nature conservation and landscape management into concrete terms they can be described as sustainable ensuring of following landscape functions (Gruehn and Kenneweg, 1998, Krönert, Steinhardt and Volk, 2001):

- habitat (-function),
- visual quality / recreation (-function),
- landscape history (-function),
- erosion protection (-function),
- biotic production (-function),
- groundwater protection (-function),
- groundwater recharge (-function),
- water retention (-function),
- water reservoir (-function),
- restoration of (surface) waters (-function),
- bioclimatological function,
- air regeneration (-function) and
- noise attenuation (-function).

After this it is obvious that landscape planning in Germany is not only an instrument for the protection of rare species, it also serves as security for sustainable land use and human culture.

The landscape plan Nennhausen, Brandenburg, is a good example for the above mentioned methodological approach within the legal standard procedure (GfU, 1999). It was worked out in a scale: 1:10,000. 13 landscape functions were being considered. Special GIS-supported methods were used for the evaluation of landscape functions. A high value of one or several landscape functions is a sign of an area for protection measures or requirements. Areas for development measures can be deduced from areas with a low value of one or several landscape functions.

The integrated concept contains quality objectives for all landscape functions. The landscape plan includes measures and requirements of nature conservation and landscape management for different addressees as well as recommendations for the implementation of these goals. A participation procedure has been carried out within the process of drawing up a comprehensive plan for Nennhausen community.

3. EFFECTIVENESS OF LANDSCAPE PLANNING IN GERMANY

Due to the discussion on deregulation in the last years in Germany the question concerning the effectiveness of planning instruments became more and more important. The main problem in the past was the lack of available valid data on planning instruments effectiveness.

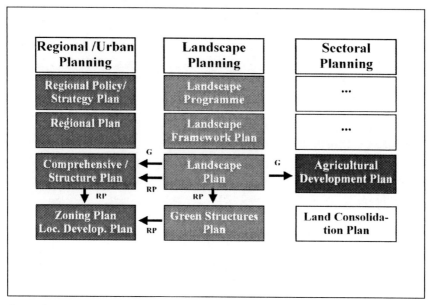

Fig. 4: Recent research concerning the effects of landscape planning in Germany

Thus, at Berlin University of Technology several research projects concerning this topic have been carried out (Gruehn, 1998; Gruehn and Kenneweg, 1998; Gruehn and Kenneweg, 2002a; Gruehn and Kenneweg, 2002b). By means of statistical methods significant effects of landscape plans have been proved within several sample surveys.

The subject of these research projects has been the effect of landscape planning instruments towards urban planning as well as agricultural planning in Germany (G) or in Rhineland-Palatinate (RP) as it is shown in figure 4. Further, Wende (2000) has made an investigation concerning the effects of landscape framework plans on environmental impact studies. In the following, the most important results according to Gruehn and Kenneweg (1998) as well as Gruehn (1998) are being presented (fig. 5–8).

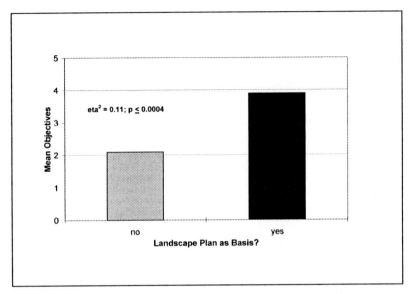

Fig. 5: Consideration of nature conservation objectives in comprehensive plans in Germany in dependence on landscape plans as basis (n = 414)

Since landscape planning has been introduced into German planning system, both, planners and politicians have been full of expectation that nature conservation interests would be considered to a higher degree as it used to be. On the other hand there has been a vital skepticism against the effects of landscape planning. Actually a significant effect of landscape plans on comprehensive plans has been detected (fig. 5). In comprehensive plans with a landscape plan as basis, nature conservation and landscape management objectives have been considered two times higher than in comprehensive plans without landscape plan. Thus, the landscape plan is a fundamental information basis for the comprehensive plan. The eta^2 value (0.11) indicates that 11 % of the variation of the dependent variable (mean consideration of nature conservation and landscape management objectives in comprehensive plans) can be explained by the factor variable (landscape plan as basis). Because of the range of eta^2 (from 0 to 1) the conclusion is admissible that there must be other important factors.

Which factors could also play a vital role in this matter? From a theoretical point of view it could be the quality of landscape planning. In this context a high quality means a complete processing of legal goals and requirements. For example, a limitation of contents on visual landscape assessment aspects would be insufficient.

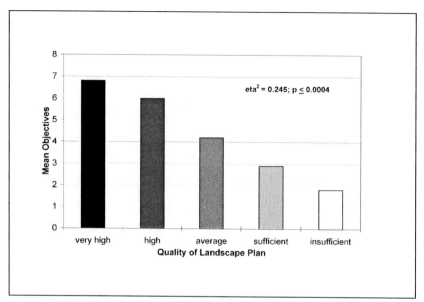

Fig. 6: Consideration of nature conservation objectives in comprehensive plans in Germany in dependence on landscape plans' quality (n = 164)

So, hypothesis concerning the effect of landscape plans' quality on the consideration of nature conservation and landscape management objectives in comprehensive planning was tested, too. The significant results recommend accepting the above mentioned hypothesis (fig. 6). The higher the landscape plans' quality, the more nature conservation interests have been considered within urban planning. A high quality of a landscape plan also means high persuading effects. If reasons and arguments of landscape planning are strong, landscape planning can persuade its addressees. If it cannot persuade, it cannot succeed (fig. 6). Further, the fact is mentionable, that comprehensive plans with a landscape plan of very high quality as basis do consider about 7 (of totally 13) nature conservation objectives on an average. The eta^2 value (0.245) indicates a rather strong effect.

Figure 7 presents the results of several analyses of variance (ANOVA). Thus, the effect of different variables on the consideration of nature conservation and landscape management objectives in comprehensive plans is comparable. The most important factors are:
- the different practical experiences in the 16 federal states of Germany (eta^2 value = 0.34),
- the quality of landscape plans as described above (eta^2 value = 0.245),
- the profession of planners (eta^2 value = 0.14) and
- landscape plans as basis for comprehensive plans as described above (eta^2 value = 0.11).

The importance of the other tested factors, for example the size or population density of the community is not very high (eta^2 value ≤ 0.7) or the effect is not significant (p value > 0.05).

Landscape Planning as a Tool for Sustainable Development

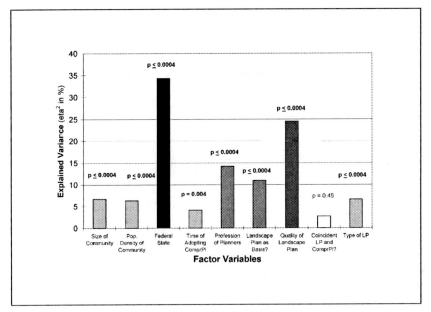

Fig. 7: Effect of variables on the consideration of nature conservation objectives in comprehensive plans (analysis of variance)

By means of factor analysis, three main factors have been generated from the data (Gruehn, 1998):
- the implementation willingness of the communities towards nature conservation and landscape management objectives,
- the sensitivity of the communities in perception of environmental problems and
- graphical and design competences of planners (fig. 8).

The effect of these factor variables on the consideration of nature conservation objectives in comprehensive plans is documented in figure 8.

With an eta^2 value = 0.51 the implementation willingness turns out to be a fundamental factor on the consideration of nature conservation objectives. By the way, this factor is significantly pushed by the quality of landscape plans (Gruehn, 1998).

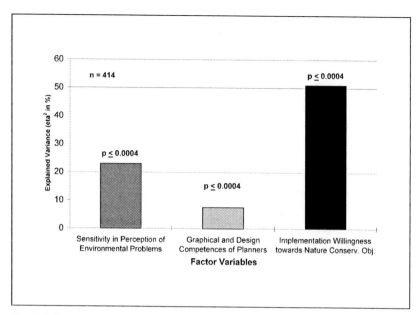

Fig. 8: Effect of variables extracted by factor analysis on the consideration of nature conservation objectives in comprehensive plans in Germany (analysis of variance)

Overall, landscape planning has significant effects on the decision making process in comprehensive planning in Germany.

REFERENCES

1. GfU, 1999, Landschaftsplan der Gemeinde Nennhausen (Landkreis Havelland), Berlin.
2. Gruehn, D., 1998, Die Berücksichtigung der Belange von Naturschutz und Landschaftspflege in der vorbereitenden Bauleitplanung – Ein Beitrag zur theoretischen Fundierung und methodischen Operationalisierung von Wirksamkeitskontrollen, Europäische Hochschulschriften 42 (22), P. Lang, Frankfurt 510 pp.
3. Gruehn, D. and Kenneweg, H., 1998, Berücksichtigung der Belange von Naturschutz und Landschaftspflege in der Flächennutzungsplanung, Angewandte Landschaftsökologie 17, Landwirtschaftsverlag, Münster, 492 pp.
4. Gruehn, D. and Kenneweg, H., 2002a, Kritische Evaluation der Landschaftsplanung im Rahmen der Bauleitplanung in Rheinland-Pfalz. Abschlussbericht zum gleichnamigen Forschungsbericht;
   http://www.naturschutz.rlp.de/lapla/eval_landschaftsplanung.pdf.
5. Gruehn, D. and Kenneweg, H., 2002b, Wirksamkeit der örtlichen Landschaftsplanung im Kontext zur Agrarfachplanung, BfN-Skripten 59, Bonn-Bad Godesberg, 156 pp.
6. Krönert, R., Steinhardt, U. and Volk, M., 2001, Landscape Balance and Landscape Assessment, Springer, Berlin, Heidelberg, New York, 304 pp.

7. Wende, W., 2000, Sicherung der Qualität von Umweltverträglichkeitsstudien durch die Landschaftsrahmenplanung, in: Naturschutz und Landschaftsplanung – Moderne Technologien, Methoden und Verfahrensweisen. D. Gruehn, A. Herberg and C. Roesrath, ed., Mensch und Buch, pp. 289–298.

THE RUSSIAN SCHOOL OF LANDSCAPE PLANNING

A. N. ANTIPOV & YU. M. SEMENOV
Institut of Geography SB RAS, Russia

Abstract: The ideology and number of methods of applied landscape research have been thoroughly developed in Russia rather long ago. But in fact the principles and methods of landscape-ecological concepts become real highly formally and inadequately effectively. In 1992, the Agreement on Cooperation in the field of Environmental Protection was signed between the Russian Federation and the Federal Republic of Germany. Within the framework of the Russian-German cooperation on landscape-planning, landscape-planning activities on "Ecologically Oriented Land Use Planning in the Baikal Region" were carried out by the Institute of Geography of the Siberian Branch of the Russian Academy of Sciences with active advisory support from the Federal Agency for Nature Conservation. It can be started that on the example of the Baikal Region the Russian-German cooperation, has set the stage of an extensive implementation of landscape-planning tools as a basis for sustainable territorial development.

Keywords: Landscape planning, the principles of geoecological and socio-functional landscape analysis, methodological approaches, sustainable development, geosystems, Baikal Region, Pribaikalsky National Park, Russian-German cooperation.

The key prerequisites to the proper choice of a man-environment interaction strategy include a correct assessment of the natural-resources potential involving a comprehensive appraisal of geosystems and tendencies for them to be transformed under the influence of both natural and anthropogenic factors. One possibility of solving the problem of optimizing land use and territorial planning of sustainable regional development could be through landscape planning enjoying wide use in Germany and in a number of European Union countries.

In the first place, LP is a set of methodological tools used to construct the spatial organization of the society's activity in particular landscapes providing for sustainable nature management and preservation of the basic functions of these landscapes as a life-support system; and, secondly, it is a communicative process enabling the interests of natural resources users and nature-management problems to be identified, conflicts settled, and an agreed-upon plan of action and measures to be worked out, which would involve all entities of conservancy and economic activity on planning territories [1].

Notwithstanding the fact that the LP concept made its appearance in the Russian geographic literature comparatively recently (let along its practical implementation in nature management in Russia), it has relatively long been customary in this country. Thus, as early as the end of the 19th century, a Special Expedition of Russia's Forestry Department was organized, with the principal objective (as opined by its head V.V. Dokuchayev), to work out integral, strictly systematic and consistent measures directed toward improving the natural conditions for agriculture, with water management put in good order. Practical efforts were preceded by a detailed inventory of natural conditions through pilot studies in experimental areas: geodetic, geological, soil and climatic surveys. Also widely known is the research done by the Resettlement Bureau-organized expeditions in Siberia and the Far East, involving fundamental cartographic engineering on component assessment of the natural conditions over the territories to be settled by peasants arriving from the European part of Russia, and by special expeditions with the task of substantiating the location of the USSR's productive forces.

Unfortunately, the possibilities of exploiting landscape-based assessment and land-use planning are limited by the lack of areal landscape survey and relevant services. Furthermore, such efforts, as well as the official practice of ecological examination of the substantiation for designated strips of land and the creation of various production facilities in Russia, do not constitute a planned, continuous process of ecological accompaniment of the project – from its conception through implemen-tation to subsequent supervision, nor do they imply a legislative finalization of assessment/examination results and remain virtually unused but for the particular projects of economic activity [1].

Doubtless the LG being used in Germany offers in this respect distinct advantages over the approaches outlined above. LP is implemented at different scales and is consistent in spatial levels with general territorial planning. The most important principle implies the vertical continuity of goals concurrent with identification of tasks, activities and programmes when passing from the top planning level (landscape programmes) to a lower one (landscape plans and lower).

Within the former USSR, projects similar to landscape planning were implemented predominantly in the Baltic republics: Lithuania and Estonia. In the Russian Federation there is essentially no practical experience of such efforts. Meanwhile, as pointed out above, it is in Russia that the ideology and a number of methods of applied landscape research have been fairly thoroughly developed, and a number of publications gave an outline of the principles of geoecological and socio-functional landscape analysis as elaborated by geographers – landscape scientists. In the 1980s, a form of regional ecological analysis, known as "Territorial Complex Schemes of Nature Conservation", was also gathering momentum. Organizationally and with regards to information content, however, landscape planning matches best the district planning widely known and widespread in Russia, the principles and methods of which rely, to a certain extent, on landscape-ecological concepts, although it is these principles that become real projects highly formally and inadequately effectively.

In 1992, the agreement on cooperation in the field of environmental protection was signed between the Russian Federation and the Federal Republic of Germany. Pursuant to the agreement, on the initiative of the German Agency for Nature Conservation (Germany) the Irkutsk Oblast Administration and the German Agency for Technical Cooperation (GTZ GmbH) entered into an agreement in 1994 to implement a joint Russian-German project, and during 1994–1998 on the territory of Irkutsk oblast (in the Irkutsk and Olkhon districts) within the framework of the Russian-German cooperation on landscape planning, landscape-planning activities on "Ecologically Oriented Land Use Planning in the Baikal Region" were carried out by the Institute of Geography of the Siberian Branch of the Russian Academy of Sciences (Irkutsk) in collaboration with the Institute of Geography of the Russian Academy of Sciences (Moscow), Irkutsk Oblast Administration, the Federal Agency for Technical Cooperation (Eschborn), and the "Ecology + Environment" planning group (Hannover).

With Russia's natural and socio-economic conditions fundamentally differing from those in Germany, it is not possible to automatically extend the German landscape-planning experience to Russia's territory, and this required rather serious efforts in order to perfect the planning methodologies. In Germany, where native natural complexes are actually non-existent, whereas the level of social relations in the realm of nature management is enormously high, the prime objective of planning is to preserve the natural bases for human life activity, even at the sacrifice of the interests of some groups of land users. In the Baikal region, however, with its currently conspicuous degradation trends in the socio-economic structure, the goal of planning looks more complicated and involves a need to respect the interests of local residents, while securing guarantees for their natural environment preservation.

Certain updates were made to the methodological approaches as such. In Germany, territorial and landscape planning implies two independent processes that are integrated at a certain stage in one form or another. In Russia, the current socio-economic problems are so serious that, unless they are taken into account, the realization of nature-conservation measures is becoming highly questionable. These realities are taken into consideration through a more extensive involvement of socio-economic factors of territorial development in the planning procedure; hence instead of landscape plans focusing mainly on nature conservation, it was necessary in our conditions to produce ecologically oriented land use plans, with a significant emphasis on socio-economic aspects of nature management.

To date, such issues as the typology and legal status of territorial development plans have not been properly worked out either theoretically or in practice in the Russian Federation, which is also true for the planning procedure as such. In Russian legislation, this sphere of legal regulation is characterized by numerous lacunas and a lack of an integrated approach. The new Urban Planning Code of the Russian Federation [2] though not yet having definitely solved the problem has to a certain extent straightened out the system of urban-development plans and specified the planning document titles. One can speak of some analogies in the hierarchy of planning documents as defined in the Urban Planning Code of the RF. Meanwhile, given the spatial dimensions of planning objects, certain corrections should be made to the succeeding scales of territorial analysis and generalizations.

In elaborating the planning methodology, the authors proceeded from the provision that LP is a tool for organizing the society's ecologically expedient activities, and its prime objective is not only to secure guarantees for the natural potential's long-term efficiency but also to secure guarantees of the local residents' rights to a decent life [3, 4].

LP is implemented as an hierarchical system in which assessments, planning provisions and prescriptions of all levels are not contradictory to each other but complementary instead, combining with each other on the "inclusion of counterflows" principle where framework recommendations (suggestions "from higher-up") serve as guidelines for more detailed instructions at lower planning levels, while being themselves formed under the influence of suggestions "from lower-down" *(table 1)*.

Table1: Levels and scales of urban and landscape planning

Administrative level	Urban planning levels	Landscape planning	Landscape planning scale
RF entity Group of entities	Consolidated scheme of urban planning, territorial complex scheme	Landscape planning	From 1:1,000,000 to 1:200,000
Municipal okrug, district, group of districts	Territorial complex scheme of urban planning	Landscape framework plan	From 1:200,000 to 1:50,000
Territory of local self-government, of a major city, specially protected territory	Territorial complex system of urban planning	Landscape plan	From 1:50,000 to 1:25,000
Human settlement, city community, SPT section, part of a territory under local self-government	Master plans	"Green" plan	From 1:25,000 to 1:5,000

The LP procedure has a type character and, landscape plans are normally drawn up in five main stages (*Fig. 1*):
- Inventory: collecting and summarizing all available information about the territory's natural environment, socio-economic conditions, structure and

The Russian School of Landscape Planning

land use characteristics, as well as identifying major nature-management conflicts within the context of analyzing the territory's imminent ecological problems;
- Assessing the natural conditions and potential of the planning territory in terms of their significance and sensitivity, as well as appraising the land use patterns;
- Elaborating the sectoral target concepts of natural resources utilization for particular natural components;
- Elaborating the integrated target concept of territory utilization; and
- Working out the program for major directions of action and measures.

Inventory

Collecting available data on natural environment health and social-economic conditions in the territory	Revealing major conflicts and problems in nature management sphere and social-economic conditions	Identifying natural environment components to be analyzed and high-priority natural component development functions

Estimation

Elaborating criteria for estimation scale of sensitivity and particular natural component values	Territory zoning as per value and sensitivity of specific natural components to realize goal component function

Elaborating component goals of territorial development

Developing a concept to integrate estimation criteria into territorial component development goals	Territory zoning according to component goals

Integrated concept of territorial development goals

Developing a concept for integrating component goals of territorial development	Territory zoning according to integrated goals

Concept of major direction for actions and priority measures to be taken

Integrating land-use conflicts with target concept	Defining measure types and their zoning	Defining measures by economic sectors

Fig. 1: Landscape frame plan development stages (detailed)

To implement the project, two model land areas were selected: the River Goloustnaya basin with the adjacent Lake Baikal's shore area, and the Olkhon administrative district ("rayon") *(figure 2)*. The R. Goloustnaya basin is a lowest-level territory of local (rural) self-government. The Olkhon district is a higher-level of local (district) self-government where its natural conditions, land use patterns, and the set of ecological and socio-economic problems are more varied.

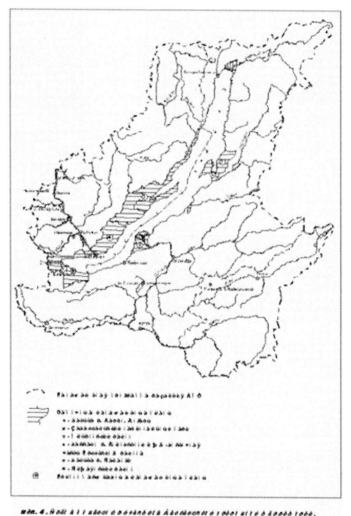

Fig. 2: River Goloustnaya basin, Olkhon administrative district

Ecologically oriented land use plans for the R. Goloustnaya basin[5] and Olkhon districts[6] were framework plans. Their main purpose was to pinpoint the principal guidelines for sustainable socio-economic development of the planning territories, with due regard for the sufficiently complete and stringent restrictions as stipulated by the law "On Lake Baikal Protection"[7].

Within the context of developing and implementing the project decisions of the framework plan for ecologically oriented land use in the Olkhon district, a package of cartographic documents was drawn up for functional zoning of agricultural lands within the Pribaikalsky Natural Park, the landscape plan was prepared for the area prioritized for recreational utilization along Lake Baikal's Maloye More (Smaller Sea) shore, and practical measures were started to improve the socio-economic and ecological situation, as well as a number of demonstration activities.

In subsequent years, the Institute of Geography SB RAS, with active consultative and financial support from the Federal Agency for Nature Conservation, Germany (Professor, Director A. Winkelbrandt, and Dr. H. Schmauder) continued the development of the methodological apparatus, primarily through the extensive use of landscape planning methods in different directions of nature conservation on the territory of the Baikal region. A scheme of ecological zoning of Baikal's natural territory at a scale of 1:1,000,000 was developed for the first time in Russia[8] to become a first document in the enforcement of the Russian Federation Law "On Lake Baikal Protection", Also developed were landscape framework plans for the Slyudyanka district[9] and for the southern part of the Irkutsk district, R. Selenga basin[10], and the Zabaikalsky Natural Park, as well as large-scale landscape plans for the settlement of Listvyanka and the town of Baikalsk[11] (see Fig. 2). This research provided a basis for developing the principles of landscape planning and the concept of its development in Russia, and for working out methodological recommendations for landscape planning[1,12].

The specific feature of a landscape programme as the upper hierarchical level in the landscape planning system implies identifying the principal functional (target) zones of utilization of the entire planning territory, with due regard for preservation of Lake Baikal's unique ecological system and prevention of any adverse effects on its health from economic and other activity. Its development is based on analyzing the natural-spatial structure, anthropogenic loads on geosystems, and existing and potential limitations on economic activity[8]. The prime objective of the landscape programme for Lake Baikal's natural territory is to identify the types of ecological territories that to a certain extent regulate the economic activity. This could only be achieved by simultaneously solving two problems: preservation of Lake Baikal's natural territory as a World Natural Heritage site, and ensuring sustainable socio-economic development of this territory, without infringing the rights and freedoms of the people living on this territory.

Thus the programme relies on the territorially differentiated approach to identifying the types of ecological territories (zones). They constitute geographic systems having a differing environment-forming significance and resistance to anthropogenic stress. They are differently affected by objects of economic activity that modify and transform the natural geographic systems or separate landscape components.

Three ecological zones were identified on Lake Baikal's natural (LBN) territory (*Fig. 3*): the Central Zone, including Lake Baikal with islands, the water-protection zone adjacent to Lake Baikal, and specially protected natural territories adjacent to lake Baikal; the Buffer Zone, the territory beyond the central ecological zone, including Lake Baikal's catchment area within the Russian Federation, and the Zone of Atmospheric Influence, the territory beyond Lake Baikal's catchment area within the Russian Federation up to 200 km in width to the west and north-west.

With large-scale LP at a community level, its objectives are generally not to solve general problems of territorial development but address specific high-priority issues defined in the context of nature conservation-related requirements by framework guidelines of higher-level plans.

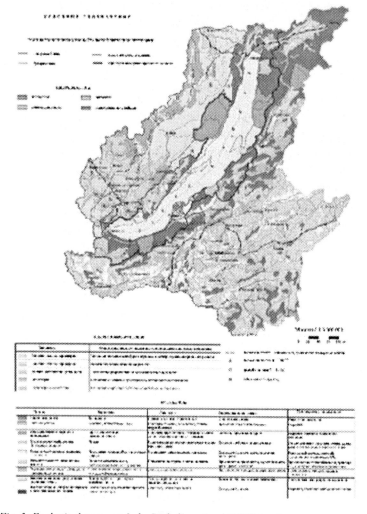

Fig. 3: Ecological zones on Lake Baikal's natural (LBN) territory

A vital aspect of large-scale LP is the conjugate analysis of problems in the spheres of policy, social welfare, the economy, and ecology as such – with regards to the chief aim to develop a particular territory, defined in the process of framework planning and subsequent analysis of the aims tree. Each of the above spheres is subdivided into specialized elements. Thus, the economic sphere is further broken down into the processing industry, forestry, water industry, transport, power engineering, and commerce.

To take into account the peculiarities of natural-ecological conditions in the context of large-scale LP, and assess and elaborate the goals of development, the most significant natural components are selected. Soils, species and biotopes, landscapes, and their recreational potential more often than not, serve as such on the scale involved.

The Institute of Geography has gained experience in using LP to solve industries-specific problems: land management, water-protection zoning, urban planning, and assessment of the environmental impact from industrial projects under development. It is at a large-scale level that LP is practicable in solving industries-specific tasks.

As pointed out above, functional zoning of agricultural lands was carried out for the Olkhon district, Irkutsk oblast. Using LP methods in land management allows a correct assessment of the potential of lands and functional zoning thereof by establishing a strict line of demarcation between different-purpose lands, based on estimating their quality, sensitivity to anthropogenic stress and current utilization, and determining the goal functions for further development. The scheme of functional zoning of lands was made at a scale of 1:25,000, as it is basic to land management documents.

Basic to the procedure of functional zoning is the map of actual land use which is drawn to show the territorial and natural-economic structure of agrarian land use, the distribution of land reserves between land users, its industrial specialization, the distribution of human settlements and production facilities (livestock-breeding farms), and the ethnic composition of population. The map is divided into portions covering individual cadastral zones, integral territories under the jurisdiction of a particular body of local self-government, and including lands of all categories (lands for agricultural use, settlements, industry, power engineering, transport, communication, radio and TV broadcasting, informatics, support of space activities, defense, security, other special-purpose lands, specially protected territories and objects, forest reserves, and water resources), and land users, irrespective of their departmental jurisdiction. The contents of each map of actual land use, depicting the agroterritory under the jurisdiction of a particular rural or village administration is reflected in a separate, individualized legend.

For the natural-ecological conditions to be included in the territorial analysis, and in the assessment and elaboration of the goals of development using the LP methodology in order to justify land organization, the most significant natural environments are the species and biotopes, soils, landscapes, and the recreational potential. The functional zoning scheme of agricultural lands in the Olkhon district, Irkutsk oblast, within the Pribaikalsky National Park, combines the zoning of the types of territorial development targets and of the territory's economic functions.

The current urban development activities are pursued with due regard for the ecological and natural features of settlement territories and aim at the attainment of favorable living condition, regulation of nature management, and environmental protection. In so doing, the problem of sustainable development of human settlements as well as territories between them is dealt with, including a minimization of harmful effects from economic and other activity on the natural environment, and its use in the interests of the present and future generations. The use of the LP approaches is the most beneficial for the achievement of the general objectives so formulated. This may be exemplified by the landscape planning effort done for the town of Baikalsk[11].

The first stage of planning involved a water-protective and integrated functional ecological zoning of the territory of Baikalsk and its surroundings, using assessment and analytical 1:10,000 and 1:25,000-scale maps, with the town and its environment treated as an integral territory. This was followed by the definition of the goals of territorial development for the zones identified, using the LP technology and bearing in mind the existing normative-legal restrictions imposed by federal and regional legislation.

The block of cartographic documents, thus prepared, included 18 maps. The resulting material for analyzing the problems of the town's development consisted of synthesized maps of territorial development goals for surface water and the biota, as well as two integral maps of an even higher level of generalization: integrated goals of territorial development, and integral zoning as per the kinds of major measures to be taken. To realize the identified categories of integrated goals of territorial development, the focus of action and measures was suggested. For the entire territory, 10 areas were identified, which were characterized by their development goals, with a general description of their natural structure and its human-induced modification (geographic localization). For these areas, actions and concrete measures were defined, the realization of which is indispensable for achieving the goals of territorial development of separate areas.

Thus it can be stated that the Russian-German cooperation, with active support from the authoritative Federal Agency for Nature Conservation, Germany, has set the stage of an extensive implementation of landscape planning tools as a basis for sustainable territorial development. With proper legal finalization of the procedure and results of such activities, this country's territorial problems could be resolved, with preservation of its unique natural resources and with no damage to its socio-economic status.

REFERENCES

1. Landscape Planning: Main Principles, Methods, European and Russian Experience. Bonn-Moscow-Irkutsk, 2002.
2. Urban Planning Code of the Russian Federation. Rissiyskaya gazets, 1998, No. 73.
3. Antipov A.N., Kravchenko V.V. and Semenov Yu.M. Landscape planning in the Baikal region. Geografiya i prirod. Resursy, 1997, No. 4.
4. Antipov A., Hoppenstedt A., Kravchenko V., and Semenov Ju. Oekologisch orientierte Landnutzungsplanung in der Baikal-Region. Garten und Landschaft. Muenchen, 1997, Heft 9.
5. Ecologically Oriented Land Use Planning in the Baikal Region. The R. Goloustnaya basin: a 1:200,000-Scale Framework Plan of Ecologically Oriented Land Use. Irkutsk – Hannover, 1997.
6. Ecologically Oriented Land Use Planning in the Baikal Region. Olkhon District: a 1:200,000-Scale Framework Plan for Ecologically Oriented Land Use. Irkutsk, 1998.
7. Federal Law of the Russian Federation (1 May 1999 No. 94-FZ) "On Lake Baikal Protection". Rossiyskaya gazeta, 1999, No. 36.
8. Ecologically Oriented Land Use Planning. Ecological Zoning of the Lake Baikal Natural Territory. Irkutsk, 2002.
9. Ecologically Oriented Land Use Planning in the Baikal Region. Slyudyanka District. Irkutsk, 2002.
10. Ecologically Oriented Land Use Planning. The R. Selenga Basin. Irkutsk, 2002.
11. Ecologically Oriented Land Use Planning. Territorial Development of the Town of Baikalsk and Its Natural Zone. Irkutsk, 2003.
12. Handbook on Landscape Planning. In 2 vol.: Vol. I. Principles of Landscape Planning and the Concept of Its Development in Russia; vol. II. Methodological Recommendations for Landscape Planning. Moscow: State Center for Ecological Programmes, 2000–2001.

CHAPTER 6

CHALLENGES AND THREATS FOR ENVIRONMENTAL STABILITY IN CENTRAL ASIA

HISTORICAL EXPERIENCE AND ESTIMATION OF MODERN LAND TENURE OF THE INNER ASIA

A. TULOKHONOV
Baikal Institute of Nature Management of the Siberian Branch of the Russian Academy of Sciences, Russia

Abstract: The modern condition of land resources efficiency and their economic use is the result of long interaction of society and nature with the purpose of a useful product reception (food stuffs, clothes, raw material for the industry, medicines, receptions of energy, etc.). On the example of using the steppe ecological system of the Inner Asia it is considered more than a centenary history of agrarian nature management of the indigenous population in conditions of a private property and primitive agriculture, nomadic cattle breeding at the end of 19 and the beginning of 20 centuries, during a socialist collective planned economy (1917-1990), and after disintegration of the USSR in conditions of transition to market relations. It is proved that the agriculture was more effective in the conditions of the private property even at absence of the industrial technologies. The structure of agriculture was corresponded to natural and climatic conditions as much as possible. In the steppe areas of China and Mongolia there were no special changes for this period, therefore here was kept adaptive ecologically safe land tenure and traditional culture of the indigenous peoples.

Keywords: Inner Asia, steppe areas, sustainable farming, agriculture development, historical and geographical approaches, Baikal region, cattle breeding, animal genetic fund, preservation of traditional culture.

1. INTRODUCTION

The Inner Asia represents a unique combination of arid ecological systems. They are bordered by mountain ranges that were generated in conditions of arid climate removed from sea water areas as much as possible. As consequence of moisture lack there was a specific standard of farming and ethnic groups with a nomadic way of life. Therefore the steppe landscapes of the Inner Asia are characterized by a moisture lack and low efficiency. The farming is poorly developed here and the agriculture is directed on the nomadic image as much as possible appropriate to the natural conditions. Thus the culture of the indigenous peoples occupying the vast region in the centre of the Asian continent is inseparably linked with the biotic environment. Destruction of these connections has a pernicious effect on both sides. To optimize this mutual relation it is offered to make the retrospective analysis of the agrarian nature management for 100 years and analyses the experience of animal husbandry and farming at the different hierarchical levels and to restore genetic fund of the native animals. An ultimate goal of the researches of our institute is the creation of the system of adaptive nature management as much as possible appropriate to the environment potential.

2. FEATURES OF THE INNER ASIA NATURAL CONDITIONS

From our point of view the Inner Asia represents east ending of the Great steppe and territorially includes traditionally cattle breeding regions of the Southern Siberia, Mongolia and Northwest China (Gumilev L.N., 1990). These mainly steppe territories are removed on thousand kilometers from sea water areas and limited by mountain ranges. From this territory the drain of the rivers in Pacific, Arctic oceans and non-flowing hollows of the big lakes is carried out. Thus the river Selenga, running into the lake Baikal and further proceeding by the rivers Angara and Yenisei forms river system in the extent more than 7 000 km - the longest in the world.

Such geographical position predetermines a moisture lack and sharply continental features of the climate. Historically this territory is occupied with mainly Mongolian peoples with typical nomadic way of life. Despite the external world globalization the way of life of the local population has not undergone essential changes here. There are no large centres of industrial pollution.

Therefore the resource potential of this territory is of interest, except the huge prospects concerning mineral raw material, and first of all as the largest source of cattle-breeding non-polluting production. For example, only in Mongolia the livestock of domestic animals makes more than 30 million heads at the population about 2 million persons.

Thus contrary to the economically advanced states the agricultural manufacture does not use industrial technologies here and it is more nearer to traditional pasturable forms of cattle breeding. By this means, the low cost price of cattle-breeding production and its high ecological characteristics are achieved. Essentially important to note that the main characteristic of such housekeeping is the adaptability or «fitted» to the environmental natural ecological system when the principle - « the greatest result with the least expenditure» is the chief.

3. EVOLUTION OF THE SIBERIAN FARMING IN 19 AND 20 CENTURIES

The example of such adaptive managing is the structure of the agriculture of Siberia on a boundary of 19 and 20 centuries when construction of the Transsiberian railway main and the intensive development of this territory began. On the official data 30 % of an arable land in Transbaikalia are damaged by wind or water erosion of soil and have already lost a part of the fertility and about 100 thousand hectares of the lands represent moving sand in an extreme degree of degradation (Tulokhonov A., 1989).

Such deterioration of efficiency of land resources in the Baikal region can be considered as appropriate consequence of irrational agricultural nature management. Sharp increase of economic activities of the person finally pursued the purpose of the efficiency rising of the agricultural complex through the intensive use of land resources. However, despite the big volumes of the capital investments the received economic parameters were appeared much below expected.

The given situation has developed in many respects owing to extremely centralized planning obviously underestimating specificity of the natural conditions of the Baikal region in the common natural and economic complex of Siberia and the country. The similar conclusion follows from the retrospective analysis of the structure of the agricultural production in Transbaikalia at the end of the 18 century in comparison with the other Siberian regions (provinces) existing at that time (Tobolsk, Tomsk, Yenisei, Irkutsk and Transbaikalian). Essentially to note that in the same latitude direction there is the east transfer of the atmospheric masses and gradual reduction of their humidity to a minimum in Transbaikalia located in the centre of the Asian continent.

Such historical and geographical approaches to an estimation of agrarian nature management is explained that in conditions of the private economy removed from commodity markets of the Siberian surburbs the agricultural production was adapted to a local natural conditions as much as possible, i.e. the ecology and economy mutually supplemented each other. First of all the distinct cattle-breeding specialization of Transbaikalia is attracted the attention. This specialization is expressed not only a maximum quantity of cattle per capita, but also the minimal sown area. Thus the grain productivity in Transbaikalia was conceded only to parameters of Tomsk province. In average year on climatic conditions the gross yield of grain was exceeded sowing norms, almost in 5 times, for rye the productivity was made 31.2 poods (pood – 16 kilograms) from state dessiatina (1.09 hectares), for wheat-29.9 poods from hectares, for oats-33.8 poods, for buckwheat - 27.9 poods, for barley - 32.9 poods, for winter rye - 33.4 poods.

The other feature of Transbaikalia agriculture of that time consists in specificity of structure of the sown areas. This region was sharply distinguished from other areas of Siberia. In Transbaikalia the basic area of crops was occupied by the rye, and least of all was remained at winter rye. The mentioned cultures are somewhat antipodes on the aspirations to climatic conditions.

Therefore it is not casual that in Transbaikalia testing deficiency of a moisture ensuring and sharp temperature differences obviously prevails summer rye, in structure of crops grain as unpretentious culture. At the same time the low temperatures and small capacity of a snow cover limited crops of winter rye. The rye crops are rather limited in Transbaikalia. Moreover it is uneasy to notice consecutive reduction from the Tobolsk area to Transbaikalian province the sown area of winter rye and wheat in a direction from the west to the east and the opposite tendency for rye (*table 1*) considering distribution of crops of these cultures to Siberia.

Table 1: The influence of the natural conditions at the structure of the agricultural production of Siberia on the boundary of 19 and 20 centuries from data

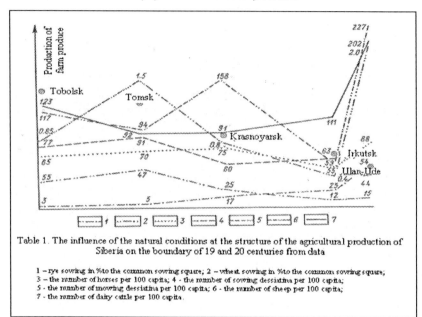

Table 1. The influence of the natural conditions at the structure of the agricultural production of Siberia on the boundary of 19 and 20 centuries from data

1 – rye sowing in % to the common sowing square; 2 – wheat sowing in % to the common sowing square; 3 – the number of horses per 100 capita; 4 - the number of sowing dessiatina per 100 capita; 5 - the number of mowing dessiatina per 100 capita; 6 - the number of sheep per 100 capita; 7 - the number of dairy cattle per 100 capita.

The marked feature is the reflection of the region climatic conditions. It is known that the prevailing western moisture transfer of air masses in Siberia determines the increase of continentality and common aridity of a climate to the east, aside Transbaikalia which is characterized in this respect by the conditions least favourable for the agriculture. From considered cultures is most adapted rye, to such natural conditions occupying the most part of crops at the end of the last century and winter rye grew only in separate Pribajkal'je areas. Gradually the areas of wheat crops are reduced to the east; wheat is enough exacting to external conditions.

It follows that in the farming structure of Siberia developing under spontaneous influence of private manufacture and from the environment such adaptation gave a unique opportunity of conducting a profitable economy. It is necessary to note that the similar structure of crops was kept long time. For example in 1926 in Buryatia rye was occupied 64.8 % of the sown area of the main crops, wheat – 15.2 %, oats – 8.3 % and winter rye – 6.3 % and then was destroyed during the Great Patriotic war.

Estimating a level of the agriculture development in Transbaikalia on the boundary of 19 and 20 centuries, it is necessary to allocate the levels of its specialization. They are basically distinguished from other Siberian provinces.

At the first level of specialization the priority belongs to cattle breeding above farming. At the second level among plant-growing production the leading part belongs to rye and the smaller part to wheat. At a level of the structure estimation of cattle breeding the basic attention was given to dairy cattle and horse-breeding at the subordinated role of sheep breeding.

As the result of such "adaptive" conducting the agriculture, its parameters of grain productivity and meat manufacture were comparable to the modern data even without taking into account those capital investments which were allocated to the agriculture according to plan. As practice shows, the breach of these principles in a zone of the risky farming of Transbaikalia frequently has negative results.

Attempt of extensive development of farming is one of the examples of irrational nature management during virgin lands ploughing. Stating a retrospective estimation to this phenomenon it is necessary to tell that the virgin and the fallow lands assimilation has not influenced on grain manufacture. Gross yield of grain during all statistical reporting is determined not by the sown area, but it is connected to favourable weather conditions in the greater degree.

Such instability of the level of the agricultural production is reflection of the increasing exhaustion of biological efficiency of natural landscapes. In these conditions, despite of significant capital investments in the agriculture, its efficiency become more and more dependent from the natural conditions. In Transbaikalia the maximal grain yield completely coincide with a maximum quantity of precipitation in the summer period. But in droughty years there is a sharp reduction of the productivity. Though such abnormal changes it was not observed before.

Analyzing the process of virgin and fallow lands assimilation in 1954-1959, it is necessary to note that 90 % made the ploughed pastures, fallow lands, haymakings, i.e. the best lands. Negative consequences of virgin lands assimilation it was aggravated by that circumstance that they have coincided in time with the period of sharp increase of a sheep-breeding livestock. In other words the reduction of the area of pastures was accompanied by the growth of a cattle livestock and first of all of sheep. It is natural that such negative phenomena should remain without ecological and hence economic consequences.

The increase of specific load of cattle on the staying pastures has strengthened their degradation processes, having created similarity of a vicious circle from problems of fulfilling the planned targets both on farming and on cattle breeding. These lacks of extensive methods of managing were showed especially vividly in the second half of 50th years when the agriculture of Buryatia starts to feel constant deficiency of natural forages and this tendency grows in due course.

In these conditions the decision of economic tasks will not always be coordinated to norms of rational nature management that more and more complicates the ecological situation in the all Baikal region. We inevitably come to a conclusion about a necessity of wide use of ecologically correct forms of nature management realizing special value of nature protection actions at the considered territory.

Intensive economic assimilation of new areas of Siberia puts forward the development of effective nature protection measures in a number line of paramount tasks. In this respect completely special place occupies the basin of the lake Baikal (the Baikal region) within the limits of which the certain development was received by the industry and the agriculture. A number of the governmental documents undertake various measures on reduction of negative influence of human activity by a nature of region. Thus the basic attention was given to prevention of industrial pollution of environment which is mainly located in the lower current of the river Selenga and in separate places of the coast of the lake Baikal.

On this background the agrarian nature management covers considerably the big area though this influence on the nature of region externally is not so appreciable. Probably, therefore ecological consequences of the agricultural production have not derived a due estimation till now and available development are devoted to the decision of the private questions of soil and water resources protection. Finally underestimation of such difficult phenomenon as agricultural nature management inevitably reduces efficiency of separate nature protection measures.

For industrial production the decision of environmental problems is reduced to reduction of harmful waste products first of all but the agriculture in region should develop by the intensification and simple restrictive reforms are unreal here. Besides, its scales and structure are determined by a wide complex of the natural and anthropogenic factors, frequently described by difficult internal and external interrelations.

Not casually the ameliorative works which have been carried out in a number of regions of Siberia have not resulted in increasing of fertility of the land; moreover they have worsened its condition. Despite of various measures on development of sheep breeding, production of mutton was reduced from 7.9 thousand tons in 1970 up to 5.5 thousand tons in 1982 only in Buryatia. It is possible to give the other examples when the decision of separate economic tasks demanded large material inputs has not given expected result. The similar negative moments in many respects are determined by the infringement of traditional connections of the agriculture with the natural environment. The structure of cattle breeding of the Siberian provinces of that time *(table. 1)* is even more specific.

Table 2: Territorial accommodations of domestic cattle in Siberia in the end of XIX-XX centuries, heads on 100 persons of the population

Province	Horses	Dairy cattle	Sheep and goats	Pigs	All cattle
Tobolsk	65,0	123,1	117,0	47,7	352,8
Tomsk	70,5	91,8	94,1	35,4	291,8
Yenisei	75,4	91,2	158,8	40,3	365,6
Irkutsk	55,5	111,4	59,9	43,3	270,1
Transbaikalia	87,9	202,0	227,0	48,6	565,5

4. ADAPTIVE CATTLE BREEDING - AS A BASIS OF THE SUSTAINABLE FARMING OF THE INNER ASIA

As we have noted earlier the agriculture of Transbaikalia took completely special place in a common natural and economy complex of Siberia (*tab. 2*). Recalculation of quantity of a cattle livestock on 100 persons of the population shows an obvious priority of cattle-breeding specialization of Transbaikalian province. Conceding in the population to the majority of the Siberian provinces Transbaikalia surpassed them in all kinds of domestic animals in calculation on 100 persons of the population and 1,5-2 times by quantity of all cattle.

Within the limits of Transbaikalia 1/5 of all meadows of Siberia settled down here. That quite provided a cattle livestock with necessary forages, and the population with livestock products.

However already in those times the factors limiting extensive development of transbaikalian cattle breeding, were appeared such as erosion, repasture of separate pastures, etc. In this connection some conclusions of the Transbaikalian commission about improvement of the cattle breeding are very actual today. These conclusions were appeared at the end of the 19 century: it is necessary to consider that the insufficient provision of dry forage is the source of the two weakest parties of the transbaikalian cattle breeding. And it in turn occurs that the transbaikalian owner pursues the purpose that in his economy there was cattle as much as possible not paying attention on quality and efficiency of the cattle and not being conformed with available quantity of a dry forage; he is convinced that his cattle breeding economy was bringing little profit because of insufficient quantity of cattle, he tries to increase the quantity, but at the same time loses in a quality. The stock of a dry forage remains and if it grows a little, but not proportionally to the growth of the number of cattle » (Materials.., 1898).

Growth of a cattle livestock was accompanied by reduction of pastures and haymakings everywhere. And if the minimal area of pasture of cattle in Buryatia in 1898 made about 4 million dessiatina, then in 1980 the area of pastures did not exceed 1.3 million hectares from which the significant part was eroded and in many respects has lost the initial bioresource potential. This was promoted by sharp increase of a sheep livestock, being the most powerful pasturable "press" in comparison with other kinds of domestic animals. For less than 100 years there was almost two-repeated increase of a cattle livestock and almost thrice-repeated reduction of pastures. Hence specific loading on pastures has increased in 6 times on minimal estimation.

In 1960 on 100 hectares of agricultural lands in Western Transbaikalia were 58 sheep, in East Transbaikalia - 54 sheep, while as a whole in Russia - 28 sheep. In this respect the specific sheep livestock of Transbaikalia came nearer to a parameter of the Northern Caucasus (61 sheep on 100 hectares of land) - primordial sheep-breeding area of the country. In 1974 the parameters of Buryatia surpassed the data in Russia almost in 3 times, Irkutsk area - in 4,1 times and the Chita area - on 15,7 % on a density of sheep livestock on 100 hectares agricultural lands. Meanwhile the efficiency of the natural fodder lands Buryatia appreciablly conceded those in the mentioned regions (Bazarov D., 1985).

Finally such unlimited loading on the natural environment had negative effect on economic parameters of the agriculture. Despite of increase of capital expenses the production of meat remains at a level 50-60 thousand tons for a long time and the production of mutton is reduced even.

Thus it is possible to assume that approximately up to the middle of the 50th years the scales of agricultural nature management were limited mainly to socio-economic factors (lack of manpower, absence of powerful engineering etc.). At the same time the natural forage reserve had the certain reserve for the extensive development of the agricultural complex and was balanced with the existing level of cattle breeding in the certain degree. Then this rather stable system of agrarian nature management was broken by ploughing up of virgin lands, first of all of natural pastures and increase of a sheep livestock which quantity has achieved almost 2 millions of sheep in 1980. From the point of view of a complex estimation of bioefficiency of used natural landscapes such livestock can not be provided with forages any more within the limits of the given region and at the attempt of its preservation for the long period inevitably will cause irreversible changes in all natural and economy complex of the basin of the lake Baikal.

Finally such "scissors" between natural opportunities of the natural environment and agrarian nature management can not exist long. And sooner or later there should be a returning to ecological balance between consumers and bioefficiency of landscapes. Each reduction of number of a sheep livestock, except the military years, can be considered as the aspiration of the natural and economy complex to return to sustainable condition provided with necessary forages and considered in the system analysis as the phenomenon of homeostasis. Therefore the increased case of domestic animals in connection with the lack of the forages, noticed in last years, indirectly testifies to the unsustainable position of all agricultural complex of the region.

The extensive way of the development of the agriculture has exhausted its opportunities and has entered the ecological contradiction with opportunities of natural bioresources. To some extent it is possible to compensate the reduction of natural fertility of the lands due to additional capital investments on land improvement, selection of cattle, etc. (Namzhilova L., Tulokhonov A., 2001).

However, in the conditions of the Baikal region these measures on the intensification of the agricultural production, despite of huge capital investments, have not brought essential results till now. Even if it is assumed that due to increase of efficiency of these measures we can increase total parameters of agricultural complex. It is hardly possible to keep duration and profitability of such conducting of the branch. Other limiting factor of the development of the agriculture of Transbaikalia is its position in the basin of the lake Baikal with its limited mode of nature management. Therefore discrepancy of modern structure of agricultural complex to natural climatic conditions can be considered as one of the main reasons of its low efficiency and of growing negative influence on the ecology of the Baikal region.

In other words by the end of the 20 century as a result of centralization and state regulation of the agricultural production with introduction of agrarian technologies alien for region, traditional land tenure of steppe territories of Siberia and Inner Asia was completely destroyed. The development of the agricultural production of that time was not provided with bioresource potential of the natural environment. The proceeding from real conditions the further growth of the agricultural production is possible only for the account of larger material inputs providing full requirement of the agriculture in imported forages, fertilizers, etc., or due to change of structure and technology of the agricultural production according to the formed ecological situation and opportunities of the natural environment.

Having assumed that the first way in modern conditions is unreal we should come to a conclusion about a necessity of the basic change of the structure of the agrarian complex of the region. This structure should be adapted to the bioresources of the region with a special mode of the adaptive nature management.

From this point of view the most optimal structure of the agriculture with prevalence of cattle-breeding branches and in farming specializing on grain crops, less exacting to favourable weather conditions and also on the fodder grasses adapted to the local natural conditions as much as possible is represented to us. As the recommendation it is possible to consider the returning to the former structure of the agriculture is expedient when more than half of the sown area was occupied by rye.

Thus, it is possible to approve that adaptability of the agrarian nature management in the conditions of the market relations is the major condition of the efficiency of the agricultural production and preservation steppe ecological systems of Siberia and Inner Asia. Only structural reorganization of the agrarian complex does not solve all problems. It is necessary the creation of a modern infrastructure and highly effective genetic fund of domestic animals.

5. RESTORATION OF THE NATIVE KINDS OF DOMESTIC ANIMALS GENETIC FUND

One more tendency of our researches is directed on the restoration of the appropriate strains structure of a livestock of domestic animals. During the planned socialist economy traditional kinds of cattle were completely replaced with breeds of higher efficiency. However service demanded the big expenses due to the state grants. It is necessary to note that despite of external similarity the native kinds of animals differed by adaptation to an extreme environment, practically not demanding a special care (Atlas of migratory mammals, 1999).

With this purpose in our institute is created the experimental economy which includes camels, yaks, khainaks, Buryat sheep and horses. The majority of these animals is revealed and brought from other areas of the Inner Asia. Only the Buryat cow has not found till now. However its morphological indications are found out in dairy cattle in the east areas of the Inner Mongolia. On precomputations the unit of traditional kinds of cattle at pasturable feeding is more effective in 8-9 times than the stalled keeping.

One of the major problems of land tenure optimization of the Inner Asia is the organization of an experimental economy on preservation and restoration of the genetic fund of the native animals of this territory. It will allow farms to survive in the market conditions and to lower a loading on land resources and to restore the efficiency of pasturable ecological system. The pedigree cattle-breeding will allow receiving an additional source of incomes for the local population.

6. PRESERVATION OF THE CULTURE, THE ECONOMY AND THE ENVIRONMENT OF THE INNER ASIA

Now our institute continues the researches on optimization of farming and pasturable cattle breeding already at a level of model areas and parts at the territory of Russia (Tuva, Republic of Buryatia, Chita area), Mongolia and China (the Inner Mongolia, Sintzyan-Uigur autonomous areas).

The local objects with municipal government are carried out in these territories where field routing researches were fulfilled. These researches had the type of seasonal nomads of different domestic animals kinds, social and economic consequences of nomadic management. (Culture and Environmental ..., 1996). Transformation of the economy and the culture of the peoples with mongolian language in conditions of the integral socialist form socialist of ownership in three countries is investigated. It is established that the condition of the environment is different despite of united national features. The national communities have undergone the greatest transformation in Russia. The ecological systems became degraded. They underwent by intensive repasture and the agrarian complex does not correspond to the principles of rational nature management.

In Mongolia and China the traditional managing has not undergone basic changes and consequently the efficiency of the steppe biocoenosis is much above in comparison with the Russian agricultural enterprises where repeatedly carried out the experiments with collectivization, ploughing of virgin lands, crops of corn and peas (the Economy, the culture and the environment in the areas of the Inner Asia, 2001).

It is essentially important to realize that rational nature management is the indissoluble symbiosis of natural and social component of the united ecological systems where is determining the behaviour of the person and scales of his influence on the environment. Thus the destruction of the natural balance as a result of the intensive operation of land resources is inevitably negatively reflected on social development of the society. Quite clearly that it is difficult to return to the last way of life and nomadic traditions at the present stage It is more important to use advantages and comfort of the civilization in interests of those who keeps the bases of the adaptive land tenure (Namzhilova L., Tulokhonov A., 2000).

7. CONCLUSION

Thus it is expedient to allocate the following accents in the perfection of the land tenure of the Inner Asia.
1. The way to increase the efficiency of existing structure of agricultural complex consists in studying the experience of adaptive nature management of Siberia in the pre-revolutionary past.
2. It is necessary to restore the pedigree structure of native agricultural animals, as the most adapted to severe climatic conditions of the Inner Asia (sheep, horses, camels, dairy cattle, yaks, etc.).
3. First of all the rational land tenure and its efficiency is determined by a way of life and the culture of the society owning land resources. Only the free owner at support of the state can use the main value - the land in his interests and in the future generation interests.
4. Steppes of the Inner Asia are one of the few places on a planet where the society and the nature still keep integrity of structural communications. The nomadic economy corresponds to the natural conditions of the region as much as possible. The production of nomadic cattle breeding allows receiving non-polluting production, as a strategic resource for the globalized world.

Information on the Author

The professor Arnold Kirillovich Tulokhonov - the director of the Baikal Institute of Nature Management of the Siberian branch of the Russian Academy of Sciences, the professor of the Buryat State University (Ulan-Ude). The basic state documents on protection and rational use of natural resources of the lake Baikal basin are developed under his leadership. Scientific interests - especially protected natural territories, geoinformation systems, ecological education and training, traditional nature management, ecological conflicts. He is the member of the Russian Academy of Sciences since 2003 and since 1985. He graduated from Irkutsk University in 1971.

REFERENCES

1. Gumilev L.N. (1990). Ethnic genesis and biosphere of the Earth. Leningrad, 528p. (In Russian).
2. Tulokhonov A.K. (1990). Historical and geographical aspects of the connection of the agriculture of the Baikal region with the natural environment. News of the Academy of Sciences, series of geography. 38-45p. (In Russian).
3. Materials of Kulomzin commission for research of landed property and land tenure in Transbaikalian area (1898.) Saint Petersburg. (In Russian).
4. Bazarov D.B., Mihajlov I.I., Fadeeva N.V. (1985). Southern Transbaikalia - large base of the Siberian sheep breeding. Pribaikalje and Transbaikalia. Moscow, 443-456p. (In Russian).
5. Namzhilova L.G., Tulokhonov A.K. (2000). Evolution of the agrarian nature management in Transbaikalia. Novosibirsk, 200p. (In Russian).
6. Atlas of migratory ammals (1999.) Novosibirsk. – 283p. (In Russian and English).
7. The culture and Environmental in the Jnner Asia, vol. 1, 2, White Horse Press, Cambridge, UK.
8. Economy, culture and environment in areas of the Inner Asia (2001.) Novosibirsk. 279p. (In Russian).

NO MAN'S LAND

Environment Influences in Central Asian Security

P. H. LIOTTA
*Executive Director, Pell Center for International Relations and Public Policy
Newport, Rhode Island USA*

Abstract: While environmental and human security remain both evolving and contested concepts in numerous theoretical debates, there should be little doubt that the *vulnerability* aspects that these security issues involve present serious long-term challenges to the success and stability of Central Asia. Aside from offering a general approach to the meaning of environmental and human security, this article also argues that there are crucial differences between *threats* and *vulnerabilities*, distinguishes between the two, and suggests why recognizing that difference has important implications for policy decisions in Central Asian security. Additionally, this article offers several theoretical models that have been proposed in recent research and considers their relevance to the region. Specifically, this review addresses what have been argued as "trigger mechanisms" that can unleash violent conflict, create socio-economic disparity, and induce long-term insecurity.

Keywords: environmental security, human security, theoretical models for scarcity and violence, stability models, Central Asia, transboundary resources.

1. INTRODUCTION

In what should be considered an act of great significance, the Nobel Peace Prize Committee awarded its prestigious recognition to an environmentalist, Wangari Maathai of Kenya, on October 8th, 2004. Acknowledged for her contributions to sustainable development, democracy and peace, the Nobel Committee specifically cited:

Peace on earth depends on our ability to secure our living environment. Maathai stands at the front of the fight to promote ecologically viable social, economic and cultural development in Kenya and in Africa. She has taken a holistic approach to sustainable development that embraces democracy, human rights and women's rights in particular. She thinks globally and acts locally.[1]

Maathai's example sets a positive note for the consideration of environmental influences in Central Asia and Mongolia. Although this is the first time since 1970 (when Norman Borlaug was recognized for his contributions to the "Green Revolution") that an environmentalist has been recognized with the Peace Prize, her award suggests that environmental conditions can be a cause as much for co-operation as for conflict.

Indeed, while environmental and human security remain both evolving and contested concepts in numerous theoretical debates, there should be little doubt that the *vulnerability* aspects that these security issues involve present serious long-term challenges to the success and stability of Central Asia. Aside from offering a general approach to the meaning of environmental and human security, this article also argues that there are crucial differences between *threats* and *vulnerabilities*, distinguishes between the two, and suggests why recognizing that difference has important implications for policy decisions in Central Asian security. Additionally, this article offers several theoretical models that have been proposed in recent research and considers their relevance to the region. Specifically, this review addresses what have been argued as "trigger mechanisms" that can unleash violent conflict, create socioeconomic disparity, and induce long-term insecurity. It is paramount that we recognize that not all security issues involve "threats"; rather, the notion of vulnerabilities is as serious to some peoples—and some regions—as the familiar "threat" metaphor of armies massing at the borders, or barbarians at the gates. Those who form policy and make critical decisions on behalf of states and of peoples must, ever increasingly, focus on aspects of traditional "national security," in which military forces will likely continue to play a preëminent role, as well as human security, in which "nontraditional" security issues predominate—in which other approaches will take center stage. If such a supposition proves true, and in a future where both "hard" and "soft" security will matter, those involved in policy decisions (and those affected by such decisions) will increasingly need to focus on aspects of *both* threats and vulnerabilities. There is a crucial need, then, to recognize the difference between these two categories.

A threat is *identifiable, often immediate, and requires an understandable response*. Military force, for example, has traditionally been sized against threats: to defend a state against external aggression, to protect vital national interests, and enhance state security. (The size of the U.S. and USSR nuclear arsenals during the Cold War made perhaps more sense than today because the perceived threat of global holocaust in the context of a bipolar, ideological struggle was far greater then.) A threat, in short, is either *clearly visible or commonly acknowledged.*

A vulnerability is often only *an indicator, often not clearly identifiable, often linked to a complex interdependence among related issues, and does not always suggest a correct or even adequate response*. While disease, hunger, unemployment, crime, social conflict, criminality, narco-trafficking, political repression, and environmental hazards are at least somewhat related issues and do impact security of states and individuals, the best response to these related issues, in terms of security, is not at all clear. While Canada, for example, has emphasized the relevance of human and environmental security to "high politics," and attempted to restructure its armed forces to meet these challenges, the relevance of military state-centered forces to address or "solve" non-state-centered issues is questionable.

Further, a vulnerability (unlike a threat) is *not clearly perceived, often not well understood, and almost always a source of contention among conflicting views*. Compounding the problem, the *time* element in the perception of vulnerability must be recognized. Some suggest that the core identity in a security response to issues involving human or environmental security is that of recognizing a condition of *extreme vulnerability*.

Extreme vulnerability can arise from living under conditions of severe economic depravation, to victims of natural disasters, and to those who are caught in the midst of war and internal conflicts. Long-term human *development* attempts thus make little to no sense and offer no direct help. The situation here, to be blunt, is not one of sustainability but of rescue.

R. H. Tawney, describing rural China in 1931, described the extreme vulnerability among peasants through a powerful image: "There are districts in which the position of the rural population is that of a man standing permanently up to the neck in the water, so that even a ripple is sufficient to drown him".[2] In such instances, the need for intervention is immediate.

But there also cases of long-term vulnerability in which the best response is uncertain. I have termed these problematic vulnerabilities—which are most difficult for policy analysts and decision makers, often driven by crisis response rather than the needs of long-term strategic planning—*creeping vulnerability*. Given the uncertainty, the complexity, and the sheer non-liner unpredictability of creeping vulnerabilities, the frequent—and classic—mistake of the decision maker is to respond with the intuitive response—the "gut reaction." Thus, the intuitive response to situations of clear ambiguity is, classically, to *do nothing at all*.

To be clear here: avoiding disastrous long-term impacts of such vulnerabilities (which can evolve over decades) requires strategic planning, strategic investment, and strategic attention. To date, states and international institutions seem woefully unprepared for such strategic necessities. Moreover, environmental and human security, since they are contentious issues, often fall victim to the *do nothing* response because of their vulnerability-based conditions in which the clearly identifiable cause and the desired prevented effect are often ambiguous.

In essence, we have moved from the dynamic of the traditional *security dilemma* to encompass issues in the twenty-first century that will include as well a new *human dilemma* in specific geographic locations that require sustainable development and long-term investment strategies. Plausible vulnerability scenarios might reasonably include:

- different levels of population growth in various regions, particularly between the "developed" and the "emerging" world—to incorporate

disproportionate population growth—youth bulges—and unprecedented levels of urbanization unseen in human history;
- the outbreak and the rapid spread of disease among specific "target" populations (such as HIV/AIDS) as well as the spread of new strains of emerging contagions such as SARS.
- significant climate change due to increased temperatures, decline in precipitation, and rising sea levels;
- the scarcity of water and other natural resources in specific regions for drinking and irrigation, and the compounding growth among populations dependent on transboundary water resources;
- the decline in food production and the need to increase imported goods;
- progressing soil erosion and desertification;
- increased urbanization and pollution in "megacities" (populations of ten million or more) around the globe, with the recognition that in the Lagos-Cairo-Karachi-Jakarta arc over the next two decades most will migrate to urban environments that lack the infrastructure to support rapid, concentrated population growth.
- the need to develop warning systems for natural disasters and environmental impacts—from earthquakes to land erosion.

These emerging vulnerabilities will not mitigate or replace more traditional hard security dilemmas. Rather, we will see the continued reality of threat-based conditions contend with the rise of various vulnerability-based urgencies. Paradoxically, creeping vulnerabilities will likely receive the least attention, even as their interdependent complexities grow increasingly difficult to address over time.

Admittedly, my suppositions above that insist on a distinction between threat and vulnerability become somewhat suspect in the so-called "Age of Terror." While no one doubts that certain states and actors are under "threat" from al Qaeda and Jemaah Islamiyah, the shadowy nature of such loosely grouped networks defies the traditional sense of threat. Loose terrorists "networks" often display the following characteristics: the facility to operate effectively as a lateral (and noncentralized) network, the ability to learn, the capacity to anticipate, and the capability to "self-organize" or reconstitute after they have been struck.[3] As such, these networks operate on the faultline between threat and vulnerability, and too narrow a focus on either "threat" or "vulnerability" will only lead to frustration—and failure.

2. NEW AND OLD SECURITY

In the classical sense, security—from the Latin *securitas*—refers to tranquility and freedom from care, or what Cicero termed the absence of anxiety upon which the fulfilled life depends. Notably, numerous governmental and international reports that focus on the terms "freedom from fear" and "freedom from want" emphasize a pluralist notion that security is a basic, and elemental, need.

Yet in the once widely accepted realist understanding, the state was the sole guarantor of this anxiety absence: security extended downwards from nations to individuals; conversely, the stable state extended upwards in its relations to influence the security of the international system. Individual security, stemming from the liberal thought of the Enlightenment, was also considered both a unique and collective good.[2] Moreover, despite the abundance of theoretical and conceptual approaches in recent history, the right of states to protect themselves under the rubric of "national security" and through traditional instruments of power (political, economic, and especially military) has never been directly, or sufficiently, challenged. The *responsibility*, however, for the guarantee of the individual good—under any security rubric—has never been obvious.

It does seem significant that aspects of "non-traditional" security issues that have long plagued the so-called "developing" world—issues that include environmental degradation, resource scarcity, epidemiology, transnational issues of criminality and terrorism—can increasingly affect the policy decisions and future choices for powerful states and world leaders as well. As disparate as these "non-traditional" issues may be, the "developed" world is now confronted with similar, human-centered vulnerabilities that had often been present previously only for developing regions. The implications of this changing security environment for the analyst and policy maker are therefore potentially tremendous.

The future will require decision makers in to focus on a broad—and broadening—understanding of the meaning of security. The 1994 *United Nations Development Programme* (UNDP) report, for example, attempted to recognize a conceptual shift that needed to take place:

The concept of security has for too long been interpreted narrowly: as security of territory from external aggression, or as protection of national interests in foreign policy or as global security from the threat of nuclear holocaust. It has been related to nation-states more than people. . . . Forgotten were the legitimate concerns of ordinary people who sought security in their daily lives. For many of them, security symbolized protection from the threat of disease, hunger, unemployment, crime [or terrorism], social conflict, political repression and environmental hazards. With the dark shadows of the Cold War receding, one can see that many conflicts are within nations rather than between nations.[4]

In 2003, the UN Commission on Human Security expanded this concept to include protection for peoples suffering through violent conflict, for those who are on the move whether out of migration or in refugee status, for those in post-conflict situations, and for protecting and improving conditions of poverty, health, and knowledge.[5]

Essentially, states and regions, in a globalized context, can no longer afford to solely emphasize national security issues without recognizing that abstract concepts such as values, norms, and expectations also influence both choice and outcome. Societies, whether in the emerging world or in the "developed world" (admittedly, a rather arrogant term), are increasingly witnessing an unfolding tension: While there are expectations that states have for protecting their citizens, citizens increasingly hold their states accountable.

Yet within in the debate on human security, there has emerged an increased focus on the rights of the individual. This debate has led to intriguing possibilities and, most definitely, uncertain outcomes. It remains unclear, however, whether an ethical and collective policy to support human security will be the focus of most states in the future—or whether any such policy could logically be de-linked from a systematic association with power, and powerful states.

Traditional and analytic bases	Security type	Specific security focus	Specific security concerns	Specific security hazards (threats/vulnerabilities)
Non-traditional	Environmental security	The ecosystem	Global sustainability	Mankind through resource depletion, scarcity, war and ecological destruction
Traditional, realist-based	National security	The state	• Sovereignty • Territorial integrity	Challenges from other states (and 'stateless' actors)
Traditional and non-traditional, realist and liberal-based	"Embedded" security	• Nations • Societal groups • Class economic • Political action committees/ interest groups	• Identity inclusion • Morality values of conflict • Quality of life • Wealth distribution • Political cohesion	• States • Nations • Migrants • Alien culture
Non-traditional, liberal Marxist-based	Human security	• Individuals • Mankind • Human rights • Rule of law • Development	• Survival • Human progress • Identity and governance	• The state itself • Globalization • Natural catastrophe and change

Fig. 1: Alternate "Security" Concepts

Increasingly, therefore, decision makers and policy analysts will need to focus on multiple referents as objects of security. Despite the "clean" distinctions made in Figure 1, there are increasing levels of "danger" in too closely following the precepts of one security concept at the expense of another.

Some brief explanation of the concepts used in Figure 1 might prove useful. In essence, the distinctions move from a "top-down" global emphasis to a "bottom-up" individual focus.

- *Environmental security* emphasizes the sustained viability of the ecosystem, while recognizing that the ecosystem itself is perhaps the ultimate weapon of mass destruction. In 1666 in Shensi province, for example, tectonic plates shifted and by the time they settled back into place, 800,000 Chinese were dead. Roughly 73,500 years ago, a volcanic eruption in what is today Sumatra was so violent that ash circled the earth for several years, photosynthesis essentially stopped, and DNA samples suggest that the precursors to what is the human race amounted to only several thousand survivors worldwide. Yet from an alternative point of view, mankind itself

is the ultimate threat to the ecosystem. Thus, from a radically extreme perspective, elimination of humanity proves the ultimate guarantee of the ecosystem's survival.
- *National security* represents the traditional understanding of security, to include the protection of territory and citizens from external threats—from other states, and, more recently, "stateless actors. Hyper-emphasis on state security, especially in the emergence of "homeland security," impacts the two following concepts of security, especially regarding the practice of individual liberties and the freedom to participate openly in civil society.
- *Embedded security* is not synonymous with the more commonly used term "societal security." Rather, embedded security is somewhat symbiotically (perhaps parasitically) linked to other security concepts. It often represents the narrow interests of specific communities, nations, or political action groups within a state. In its extreme form, it can lead to social stratification, the fracturing of "common" interests, and xenophobia. Samuel P. Huntington, all normative judgments aside, focused on such embedded security groups in his 2004 work, *Who Are We? Challenges to America's National Identity*.
- *Human security*, other than a common agreement on the focus on the individual, is still an emerging concept. In the September 1999 issue of *Security Dialogue,* Astri Suhrke pointed to a fundamental bifurcation that human security as conceptual approach and policy principle continues to suffer from: Is it related more to long-term "human development," such as was suggested in the 1994 United Nations *Human Development Report*, or (as a security issue) does it constitute a principle of intervention during immediate crisis, such as Rwanda in 1994 or Kosovo in 1999 or even Iraq in 2003?[6] The answer to either question is "Nes"—a little bit of "no" and a little "yes."

Thus, while some (including this author in the past) have argued that there may be a growing convergence between what was traditionally called "national security" and the still developing concept of "human security," there appears to be an even more powerful counterargument in which the opposite trend is apparent.[7] In interventions as disparate as Somalia in 1993 to Liberia (at various stages of disintegration) to the Balkans and Iraq in 2003, there seems to have emerged an overt increase in American and British hegemonic behavior, accompanied by an uneven commitment to issues involving human security. While Prime Minister Tony Blair could speak of "universal values" and President G. W. Bush proclaimed that "Freedom is the non-negotiable demand of human dignity," foreign policy choices regarding intervention were almost exclusively made when such choices satisfied the narrow, selfish and direct "national security interests" of more powerful states.[8] If such choices also satisfied certain aspects of human security, then all the better.

Yet, as the blatant international failure in 1994 in Rwanda to do anything illustrates—other than a collective international decision to do *nothing*—human security is hardly proving to be the trump card of choice in decisions by states to intervene in the affairs of other states, to include violating traditionally respected rights of sovereignty. In other words, taken to extreme forms, both human security and national security can be conceptually approached as antagonistic rather than convergent identities. Each, in its exclusive recognition, remains problematic.

So-called ethical practice in foreign policy that acts on the behalf of individual citizens, for *any* state—not just the United States—is most accommodated (and accommodating) when such actions "mesh," however, with achieving the ends of more traditional national interests of the more powerful state. As a basis for international action, we have yet to achieve a consensus on what constitutes "international interests," who should support it, and who should uphold it.

Undoubtedly, increasing numbers now speak out on behalf of what the International Commission on Intervention and State Sovereignty has termed the "responsibility to protect": the responsibility of some agency or state (whether it be a superpower such as the United States or an institution such as the United Nations) to enforce the principle of security that sovereign states owe to their citizens. But there is a dark side of this proposition, of course, is that the "responsibility to protect" also means the "right to intervene." In the topology of power, dominant states will intervene at the time and place of their choosing.

My argument, therefore, in examining the unfolding security realities that have emerged in recent history, is to focus on multi-level, multi-referent aspects of security.

3. DEFINING ENVIRONMENTAL SECURITY

Regarding Central Asia, it may be appropriate to consider how theoretical approaches to "non-traditional" (or non-military) security issues have been considered elsewhere in the recent past. To do this correctly requires recognizing some fundamental shifts in security definition—particularly with the emergence of human and environmental security themes in the 1990s.

In terms of precise categorization, there are critical differences between human and environmental security. In the broadest sense, environmental security considers issues of environmental degradation, deprivation, and resource scarcity; by contrast, human security examines the impact of systems and processes on the individual, while recognizing basic concerns for human life and valuing human dignity. Yet as numerous examples illustrate, complex interactions within various environments often place stress on the security of the individual. Thus, environmental and human security often co-exist in a complicated interdependence best conceptually considered as "extended security."

Policy makers would be wise to recognize this conceptual approach. Yet for research to be relevant to policy makers, it should almost always contextualize significance within a specific human- and regional-oriented perspective. To be blunt, there is a specific and pragmatic reason for emphasizing these issues in terms of security: doing so makes the topic both accessible for decision makers and provides a basis for determining present and future policy. Ole Wæver, one of the earlier influences (along with Barry Buzan) in promulgating the "new security agenda" reflects a certain skepticism, nonetheless, about the ability to influence policy through reframing the understanding of security:

A security issue demands urgent treatment: it is treated in terms of threat/defence, where the threat is external to ourselves and the defence often a technical fix... traditionally the state gets a strong say when something is about security. To turn new issues (such as the environment) into "security" issues might therefore mean a short time gain of attention, but comes at a long-term price of less democracy, more technocracy, more state and a metaphorical militarisation [sic] of issues. For this reason, environmental activists and not least environmental intellectuals who originally were attracted to the idea of "environmental security" have largely stepped down. Security is about survival... The invocation of security has been the key to legitimizing the use of force, and more generally opening the way for the state to mobilize or to take special force... Security is the move that takes politics beyond the established rules of the game.[9]

There are, however, any number of overextending assumptions in the above reference. Above all, is the assumption that security is an extreme term that can only be couched in terms of threat, and that the state—as political monolith—can only respond to with the use of force. Security is far richer in contextual meaning than such a stratified identity seems to allow.

Security is a basis for both policy *response* and long-term *planning*. Further, the use of force—particularly military force—is often an ineffectual and irrelevant response to the "new security agenda." Thus, the argument that "environmental security" is simply a mask for military intervention is argument that is, at best, thin. What *is* true is that the understandings of, and definitions for, environmental security range so broadly that its meaning takes on something for everyone—and perhaps, ultimately, nothing for no one.

For the specific relevance of the term "environmental security" applied to the Central Asia, the broadest relevant definition should be, and should remain, an understanding that environmental security centers on a focus that seeks the best effective response to changing environmental conditions that have the potential to reduce stability, affect peaceful relationships, and—if left unchecked—could contribute to the outbreak of conflict. Perhaps ironically, the best overall definition for environmental relevant to the Central Asian space was written two decade ago, when Norman Myers argued that:

National security is not just about fighting forces and weaponry. It relates to watersheds, croplands, forests, genetic resources, climate and other factors that rarely figure in the minds of military experts and political leaders, but increasingly deserve, in their collectivity, to rank alongside military approaches as crucial in a nation's security.[10]

In contrast to Wæver's suspicious pessimism regarding the true political motives for the environmental security agenda, and in support of Myers' above ideas on the need to rethink—and re-conceptualize—security, one would hope that both military and political leaders will come to widely recognize the validity of environmental security for strategy and policy initiatives. Based on the tensions, disagreements, and uncertainties regarding environmental influences in Central Asia, however, it is not clear—and, sadly, not likely, that this recognition will forestall some potentially disastrous outcomes in the region.

4. SUGGESTIONS FOR FUTURE RESEARCH

While acknowledging that such forecasts are *not* strict predictions—but rather *projections*—from reasonable indicators of current and recent trends and effects, there can be little doubt that significant change may well occur in the Central Asian context regarding environmental influences. Such change will directly affect the security calculus of the entire region. As such, a number of general conclusions and concerns could be raised about the shifting landscape of Central Asia and the critical uncertainties that will inevitably emerge:

1. The environment forms part of a larger network of cultural, political, and economic linkages in Central Asia. To date, little multilateral agreement has been reached that will solve the wider negative consequences of collapsing "extended security," exemplified by the potential impending water crisis in the region.
2. Unless significant agreements and institutional agreements are reached, there will equally be little incentive to establish or sustain early-warning and conflict prevention centers for the Central Asia.
3. The specific relevance of the term "environmental security" to Central Asia should be framed as an understanding that environmental security centers on a focus that seeks the best effective response to changing environmental conditions that have the potential to reduce stability, affect peaceful relationships, and—if left unchecked—could contribute to the outbreak of conflict.
4. A number of *uncertain vulnerabilities* must be vividly presented to policymakers as having serious long-term consequence, to include their eventual emergence as *threats*. These vulnerabilities in Central Asia may include—but are certainly not limited to—disease, hunger, unemployment, crime, social conflict, terrorism, narco-trafficking, political repression, and environmental hazards. Vulnerabilities, if left unchecked over time, *become* threats.
5. Military and political leaders should recognize the validity of environmental security for strategy and policy initiatives. The application of these particular arguments for geo-specific research has *not* reached a point of precise refinement.Models, developed to date, remain insufficient

6. Further research must concentrate on obtaining reliable and broad indicators of environmental change measured over longer periods of time. Although recent research suggests vastly increased upswings in ecological/natural disasters in Central Asia, more data is required.
7. We must recognize the relevance—and the danger—of adding the term "security" to either environmental or human-centered concerns. To be blunt, there is a specific and pragmatic reason for emphasizing these issues in terms of security: doing so makes the topic both accessible for decision makers and provides a basis for determining present and future policy.
8. Solving environmental issues in Central Asia may help solve socio-economic issues such as the transition to democratic, or stable, governance in each state and mitigating the causes and effects of transnational terrorism in the region.

Ultimately, the inconclusive and sometimes contradictory results of various models and frameworks relevant to an examination of "extended security" in Central Asia leave us in a state of uncertain certainty. Environmental change is occurring now that clearly will affect the security calculus, but we simply do not know enough or have available data for definitive proof. Further, while we may recognize vulnerabilities, or aspects of what may be vulnerabilities, we simply do not know which will emerge as threats.

Thus, researchers who might desire a specific quantitative outcome in studying and applying focus to geo-specific regions, such as the Central Asian space, are still unable to "bound" expected outcomes and prevent potential future negative security events. Such modeling could be inherently dangerous, however, particularly if used as the exclusive basis for foreign policy decisions to intervene or abstain in attempting to control the boundaries of a collapsing state or region. Models, driven by quantitative outcome answers, may ignore the value of qualitative process examinations that help frame more appropriate questions. Further, predictive modeling may lead as well to determining "permanent failure states": those states that, based on quantitative analysis alone, would appear to have no possibility of social, political, or economic recovery.

Ultimately, the complexity of human interaction with and response to complex environmental influence may well lie beyond any viable or accurate modeling attempt. Admittedly, this is a contentious conclusion, but probably an honest one as well. Reasonable strategies for regional security and ecological stability in Central Asia thus need to balance preoccupation with rationality with the recognition that policymakers almost always lack clairvoyance, suffer from cultural blinders, and are often driven by contingency responses. Most often, these crucial actors enter into a kind of psychological—and sometimes physical—St. Vitus dance until exhaustion or resolution set in.

At best, we should hope for the participation of multiple decision makers and for the desire for all within the region to involve. Assessing the future, we can only know that dynamic change is coming—but how, in what way, and whom this change will most directly affect remains yet one more uncertain certainty we must consider in defining both the critical geography and the geostrategic no man's land of Central Asia.

REFERENCES

1. http://nobelprize.org/peace/laureates/2004/press.html
2. James C., Scott, *The Moral Economy of the Peasant* (New Haven, CT: Yale University Press: 1977) pg. 1.
3. For an extended discussion of this phenomenon, see P. H. Liotta. "Chaos as Strategy," *Parameter* (Summer 2002), pp. 47-56.
4. < http://carlisle-www.army.mil/usawc/Parameters/ 02summer/liotta.htm>
5. Adam Smith, for example, in *The Theory of Moral Sentiments,* mentions only the security of the sovereign, who possesses a standing army to protect him against popular discontent, and is thus "secure" and able to allow his subject the liberty of political "remonstrance." By contrast, M. J. de Condorcet's argument, in the late eighteenth century, suggested that the economic security of individuals was an essential condition for political society; fear—and the fear of fear—were for Condorcet the enemies of liberal politics. These distinctions are ably considered in Emma Rothschild's "What is Security? The Quest for World Order", *Dædulus: The Journal of the American Academy of Arts and Sciences,* Volume 124, Number 3 (June 1995), available at http://web.lexis-nexis.com; also see: Emma Rothschild, "Economic Security and Social Security", paper presented to the UNRISD Conference on Rethinking Social Development, Center for History and Economics, Cambridge, Massachusetts, 1995.
6. Commission on Human Security, "Protecting and Empowering People." <http:www.humansecurity-chs.org/finalreport/outline.html> (19 June 2003)
7. See Suhrke's useful discussion in "Human Security and the Interests of the State," *Security Dialogue,* Vol. 30, No. 3, September 1999, 270-271.
8. For one perspective that suggested that a convergence between human and national security was occurring in the period after September 11, 2001, see P. H. Liotta, "Boomerang Effect: The Convergence of National and Human Security," *Security Dialogue*, Volume 33, Number 4, pp. 473–488, as well as the responses of: Brooke A. Smith-Windsor, "Terrorism, Individual Security, and the Role of the Military: A Reply to Liotta," *Security Dialogue*, Volume 33, Number 4, pp. 489–494; P. H. Liotta, "Converging Interests and Agendas: The Boomerang Returns," *Security Dialogue*, Volume 33, Number 4, pp. 495–498.
9. For the most clarifying statement of the "Blair Doctrine," see "The Doctrine of the International Community," delivered to the Chicago Press Club during the U.S.-led NATO intervention in Kosovo in 1999. http://www.number-10.gov.uk/output/Page1297.asp; the Bush position is expressed in the *National Security Strategy of the United States of America,* 17 September 2002, most specifically in the "Introduction" and "II. Champion Aspirations for Human Dignity." http://www.whitehouse.gov/nsc/nss.html

10. Quoted in Thomas Scheetz, "The Limits to 'Environmental Security' as a Role for the Armed Forces"; paper provided by the author. Wæver's original remarks, titled "Security Agendas Old and New, and How to Survive Them," were prepared for the workshop on "The Traditional and New Security Agenda: Influence for the Third World," Universidad Torcuato Di Tella, Buenos Aires, Argentina, 11-12 September 2000.
11. Norman Myers, "The Environmental Dimension to Security Issues, *The Environmentalist,* 1986, 251.

PHYSICAL - GEOGRAPHICAL CHARACTERISTICS OF THE ALTAI REGION

DR. D. ENKHTAIVAN
Institut of Geography, Mongolian Academy of Sciences, Mongolia

The Altai Mountains are characterized by their unique natural scenery and numerous cultural - archeological monuments. The interests in biodiversity conservation and sustainable development of at least four countries, i.e. China, Kazakhstan, Mongolia and Russia, are focused on that region. Whereas the contiguous territories of the four countries are similar in their demographic structure, ethnic composition, climatic conditions and socio-economic characteristics, they differ in terms of people's philosophy, legal frameworks, institutional structures, economic situation and activities in nature protection.

The Altai Mountains are situated in the center of Eurasia at the frontier of two major natural zones of the northern hemisphere, i.e. the humid boreal and the arid desert-steppe zone. The mountain system forms the most elevated part of the North-Asian continental watershed separating the river runoffs of the Arctic Ocean basin and the Central-Asian internal drainage basins. The Altai is characterized by complicated geological and relief structures, various landscapes from taiga to deserts as well as diverse vegetation and wildlife.

The Altai Mountains form an entire geographical system, which can be subdivided into the Siberian, the Mongolian and the Gobi Altai. The study is focused on the Siberian and the Mongolian Altai. In the northwest, the Altai borders on the South-Siberian steppes, in the north the Zhoria Mountains connect the Altai with the Kuznetsk Alatau, in the northeast the Tuva basin is located between the Altai and the Sayan mountains. In the southwest of the Altai there are the Kazakh lowlands and the Irtysh River with the Zaysan Lake, which is the border to the Tarbagatai Mountains. In the south, the Altai passes into the Dzhungaria and the Gobi desert and in the east it passes into the Big Lakes Lowlands and the steppes of the Outer Mongolia. The Central-Altai forms a huge rock massif of 4000 m height. The mountain ridges spread into west, northwest and north direction originating from the Tabyn-Bogdo-Ola massif in the central mountain area. The relief is dominated by typical alpine formations such as hillside peaks, sharp crests and deep glacial cirques.

The Altai are old folded mountains. The current mountain scenery was formed in the Tertiary (Miocene, Pliocene), when the Altai was lifted again as a whole block. Thus, diverse sedimentary, abyssal and metamorphic rocks from paleozoic era to Holocene could develope. The oldest strata are from Upper paleozoic era, Cambrian and Ordovician age. Rocks from Devonian, Carbon and lower Perm age are only locally spread. The sediments of Tertiary are often found in inter-mountainous lowlands where they form strata of several hundred meters. Quaternary sediments are widely spread except on peaks and upper watersheds. During the Pleistocene, vast areas of the Mongolian Altai were covered by huge glaciers reaching a thickness of up to 500 m. As a result, typical glacial landforms as U-shaped valleys and cirques with lakes and moraines can be found throughout the Altai. The altitude of the nival belt rises from 2300 m in the west up to 3200 m in the arid southeast. During the Pleistocene, the nival zone stretched approx. 1000 to 1200 m below the present level. There are about 70 large glaciers in the Altai, covering an area of approximately 600 sq. km (Walter & Breckle, 1994). Permafrost relief with stone polygons, solifluction and thermokarst lowlands are extensively spread in the region.

The basins of interstitial ground water are situated in places of metamorphic, sedimentary and magmatic rocks of Paleozoic periods; in the inter-mountainous artesian basins the main water-bearing horizons were formed in Quaternary, Mesozoic, upper and mid Paleozoic deposits. Flows of mineral water are observed in the basins as well.

The location of the Altai region in the center of Eurasia and the great distance to the oceans determine its climatic conditions to a great extent. In the West, the climate has a slightly oceanic character and in the East it is sharp continental. The humid western (Atlantic) air masses bring plentiful precipitation to the western and central part of the mountains concentrated on the summer months. The quantity of rainfall increases from south to north and from west to east. The annual precipitation in the north and west mountain ridges adds up to 2500 mm, in the southern Altai up to 1200-1500 mm. In the Mongolian Altai the precipitation is below 160 mm per year. The winters are mostly poor in snow. The thickest snow cover can be found in North-Eastern and North-Western Altai. There, the snow cover reaches 3 m on the windward slopes whereas it is of only 10-35 cm in the Central Altai valleys.

It lasts more than 1900 hours per year and in alpine lowlands even 2600 hours. The winters are long and harsh and the vegetation period (with a daily temperature of above 5°C) amounts to 130-190 days. The temperature regime depends on the height above sea level, ridges stretch, relief position and slopes exposition. Because of the inversion of temperature in lowlands, early or late frost occurs frequently. In winter, the temperature in lowlands can be up to 20 degrees colder than at the slopes. There are big differences in temperature between winter and summer, day and night. The annual average temperature is 3.8°C at the Zaysan Lake in the semi-desert South, -2°C at an altitude of 1000 m and in the inter-mountainous basins it declines to -7.2°C (e.g. Chuyskaya lowland). The average temperatures in January/July makes up –21/+16°C at 1000 m height and –32/+10 to 12°C at a height of 2000 m.

The Altai Mountains are the origin of the most important water resources of the wider region. Particularly the big glaciers hold very large water reserves. The main aggregations of glaciers are the following: on the

Katunsky ridge (Sapozhnikova, Gr. Berelsky, Gebler, Rodzevich glaciers), on the North- Chuisky ridge (Gr. Maashei, Left and Right Aktru glaciers), on the South-Chuysky ridge (Sofiisky, Gr. Taldurinsky, etc.) and on the Tabyn-Bogdo-Ola mountain massif. In the Mongolian part of the Altai Mountains there are 250 small and big glaciers covering 514 sq. km. The Altai Mountains define the flow of the largest Siberian river, the Ob, which is formed by the major water-ways Bija and Katun. The main river of the Southern Altai is the Irtysh, which comes from the Mongolian Altai. It forms the border of the South Altai in China and Kazakhstan and runs into the river Ob in Russia. The Irtysh has large tributaries (the rivers Bukhtarma, Ulba, Uba) and a number of small tributaries. The Khovd watershed is located in western Mongolia's Altai Mountains, bordering China on the west and Russia to the north. With an area of 58,000 sq. km and an elevation gradient of 3270 meters, it is the largest and most important watershed in Western Mongolia. Originating in the glaciated peaks of the Altai, the Khovd River flows 516 km east to Lake Khar Us Nuur and two other. Feasibility Analysis for a Transboundary Biosphere Territory in the Altai Mountains inter-connected lakes surrounded by semi-arid desert steppe. The water from the Khovd River is essential to the vitality of these aquatic ecosystems. The rivers in the Altai Mountains are to a large extend fed by the thawing of snow and glaciers. About 7000 lakes are found in the Russian Altai, about 1000 lakes in the Kazakh Altai and some thousand lakes more in China and Mongolia. The lakes differ in origin (tectonic, glacial and karst), size and water composition. Other large lakes are the Uureg Nuur (238 sq. km), Achit Nuur in the Eastern Altai, the Khoton-Nuur (50 sq. km), Khurgan- Nuur (23 sq. km) and Dayan-Nuur (67 sq. km) in the Mongolian Altai. These lakes are of glacial origin.

The soil layer depends on the geographical location, i.e. altitude and climatic conditions. The highest parts of the mountains are zones of permanent snow, glaciers and rocks. Other soil types, such as light chestnut, brown desert soils as well as solonchak, solonets and meadow-salt soils are characteristic for desert-steppes and desert zones, e.g. of the Zaysan lowland. Mountainous-tundra and mountainous-meadow soils are 30-40 cm thick and rich of peat. The chernozems of the plains have thick humus horizons from 40 up to 120 cm and are therefore the best arable lands. Podsolized and leached chernozem up to 600cm/m. Typical, mainly ordinary and south chernozems, chestnut soil 900-1100m. Dark chestnut, mainly chestnut and light chestnut 1300-2500m.

NEW CHALLENGE OF THE MODERN CIVILIZATION AND GLOBAL ECONOMY IN THE NORTH OF THE WESTERN SIBERIA

G.N. GREBENUK & F.N. RJANSKY
The Humanities University, Faculty of Natural and Exact Sciences, Nizhnevartovsk, Russia

Abstract: In our opinion the beginning of the 21st century is marked by three most important and indirectly connected changes which are capable to influence essentially the destinies of mankind as a whole, and in particular the part of it which lives in the circumpolar zone of the Northern hemisphere. First of all it is global changes of climate, the reduction, and probably exhaustion of stocks of oil and gas on the earth, and also changes in mass consciousness of millions and millions of people on the planet.
In the report we would like to discuss problems and facts connected with changes of climate in Western Siberia, including its northern part.

Keywords: Western Siberia, The North, changes of climate, oil and gas, the migration of population, the problems of infrastructure, ecology.

The reconstruction of climate and natural environment of Holocene and Pleistocene of Siberia (1) has shown a cyclic and also naturally repeating character of these changes. (*Picture 1*) Our researches (2) have determined a natural hierarchical fractal figure of existential changes. From here, the representation about the discrepancy of the warming processes - cold snaps and dryness - humidity appears: there is no such cold snap within the bounds of which there would be no less long and less deep warming, and there are no such warmings which would not be interrupted by series of fine cold snaps. Therefore it is important to be defined with the tendency and to construct a right trend.

The analysis of the last thousand years of the sub Atlantic period of the climato-stratigrafical scheme of the late after-glacial age and

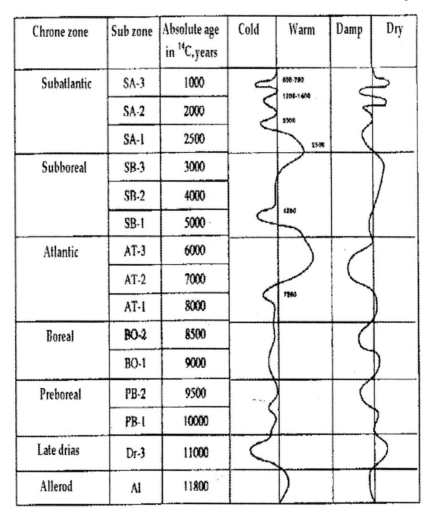

Picture 1: The climatostratigraphical scheme of the late after-glacial age and Holocene of the Western Siberia (by Volkova and others, 1989)

Holocene of Western Siberia (3) allows marking out three stages of a cold - heat and dryness - humidity:
- the first (initial) - warm - damp,
- the second (VI-VII centuries) named « a small glacial age » - cold - dry,
- the third (the end XX and transition to XXI century) - transition to the warm and damp period.

During warming in the central part of Western Siberia cedar and cedar-pine woods, with participation of deciduous ones on valleys of the rivers were widely distributed. Cold snaps led to pauperization of structure of arboreal vegetation and to occurrence of sparse growth of birch and birch-pine with shrub birches. Approximately three hundred years ago began the present stage, which is characterized by the common warming and the aridization of climate

(4). E.D.Lapshina's researches (5) have shown, that in present time in the south of a wood zone of Western Siberia reduction of водности of bogs is observed and it is expressed in settling of peatbogs opened earlier. In opinion of this author «development of a modern marsh cover in a taiga zosne of Western Siberia occurred against background of dynamically changing climatic conditions of the natural environment. It appreciably explains heterogeneity of their flora, a variety of vegetative communities and landscape structure which combine attributes of various bioclimatic epochs of Holocene, and in some cases the attributes have been inherited from more ancient, before- Holocene stages of development of a vegetative cover of West Siberian plain (with 25)».

Russians will penetrate to the north of Ob pool at the end of a climatic optimum (7-11 centuries). At 12-13 centuries from River Pechora on its inflow (River Us (Ус)) «through-a-stone way» has been laid, which was used for leaving to the river Sob (the left inflow of Ob in its bottom current (6).

Active development of Siberia by Russians has coincided with the next small glacial age 1550-1850 years. The period was characterized by fall of temperature on the average in 1.2-2°C (7). The cold snap was accompanied by increase of the extreme natural phenomena: Flooding, droughts during warm seasons, early colds in the autumn, severe winters, returns of colds in the spring and long rainy periods.

According to meteorologists, during the period between 1880 and 1940 the mid-annual temperature raised in 0.25°C, later - since 1940-1970 it fell in 0.2°, and since 1970 to1980 year it raised in 0.3°. Rise in temperature continued in 80th years too. In connection of changing temperature the humidity changed too. It is necessary to note, that warming till 1940 was observed only in northern hemisphere, but last decades rises in temperature and changes of humidity are fixed all over the globe. Probably these phenomena are natural and cyclic; these are original circulations of different periodicity and extent in time and space.

There is a base to think (2), that changes of temperature and the humidity, begun in 80th years and captured globe have 324-years periodicity, as well as « the small glacial age » of 16-19 centuries, the time of arrival and development Siberia and the Far East by Russians and by analogy they can be named «a small climatic optimum ». However, many foreign and Russian authors consider (8), that at this time warming will be serious and changes will be much faster than before. The reason is, that in an atmosphere dioxide of carbon and other gases are intensively collected from burning oil, coal and gasoline. They act like glass in a hotbed: it passes solar beams through, but detains an output of heat. It has led to warming of an atmosphere more then usually, that is known now under the name "hotbed effect". Having broken a natural course of change of a climate of a planet, this dangerous effect has changed such crucially important variables as deposits, the wind, and a layer of clouds, oceanic currents and the sizes of polar ice caps.

Tool gaugings of a near ground atmosphere of air have been conducted since 1860. According to the World meteorological organization, global increase of mid-annual temperature for the period since 1860 to 2000 has reached 1°C with following dynamics. Since 1860 to 1935 the temperature has increased on 0.4°C; then, after a short break, coincided with the Second World War, global increase of mid-annual temperature has made 0.8°C. In 1989 The British meteorological management has made a conclusion, that this year was the warmest on a planet during previous hundred years. The six of ten warmest years of the 20th century fell on 80th years. In 1989 in Moscow for the first time during 110 years of supervision January 1989 has broken all records of heat.

In 90th years warming proceeded and 1998 has already began the warmest year during 20 century. According to some information the mid-annual temperature was higher on +1.5°C.

In 1979 in Geneva the First World conference on a climate has taken place in which academician E. K. Fedorov has expressed opinion, that in the future changes of a climate will be inevitable and that it will become evident during the nearest decades. In November 1988, the first session of intergovernmental group on climatic changes has passed under aegis of the World meteorological organization in Geneva. During it three working groups have been created, one of which should be engaged in a scientific estimation of climatic changes and the available scientific information; the second should generate the strategy of possible actions. It was entrusted to third group to predict influence of climatic changes on an environment and their social and economic consequences.

The commission of the Bundestag (Germany) had stated opinion about questions of protection of an atmosphere in 1989; they said that the nearest decade (90th years) because of activity of man, the average air temperatures would raise on 1.5 – 4.5°C. As a result of warming, the level of World Ocean can rise on 1.5 meters, which is fraught with unpredictable consequences. But even the Program of the United Nations on the environment rather pessimistically estimated an opportunity of quick changing of the situation. We now know how to remove dioxide of sulfur from emissions, but we have not learned to neutralize dioxide of carbon effectively. Certainly the decision of a problem can promote reduction of cutting down of woods, but even fast restoration of a part of them will not solve it completely, and in 2040 inhabitants of the Earth will be already about 10 billion.

Warming of a climate conducts to thawing of polar ice and increase of a level of World Ocean. According to forecasts by 2050 possible increase of a level of World Ocean will make 150см. (By the end of 21century it will make 3-4м.); and then extensive areas of oceanic and sea coasts, with set of cities, appear under water. Already now the temperature of sub glacial water in area of North Pole has grown almost on 2°C, owing to this thawing of ice from below has began. Feature of a situation is those, that the nearest millennium some regions with unstable humidifying become drier, that will lead to degradation of the grounds and loss of crops. And damp areas will become even damper. In the North of Western Siberia winters will be damp and warm, and summer - hotter and droughty.

Following the changes of a climate there inevitably will come changes in position of natural zones (*Picture 2*).

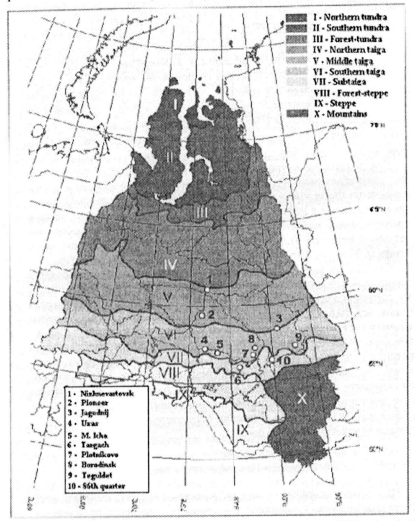

Picture 2: Bioclimatic zones and arrangement of key sites (by Papshina, 2003)

We can judge about the events, which would happen, by paleographical facts of the Atlantic period (8000 - 4500 years), which are characterized by the maximal development of wood vegetation. Northern timberline reached a sea coastal line. In the beginning of the Atlantic period (8000 years ago) in the center of Western Siberia it was birch-pine woods with a fur-tree. About 7500 years ago, after a short cold snap in country between Ob and Irtish woods with birch and fur-tree and a fir, and also with an elm on valleys of the rivers, began to prevail. On sandy ground of river valleys and hollows of an ancient drain were generated intrazonal pine woods (« tape pine forests »), which since then have not undergone essential changes (9). The maximal distributions dark coniferous trees (a cedar, a fir, a fur-tree) and deciduous trees (a linden, an oak, an elm) breeds of trees fall on the end of the Atlantic period (6000-4500 years). Exactly this interval is counted a climatic optimum of Holocene in Siberia. It is characterized by the highest July temperatures (tVII-+ 22o ... +23°C, tI-16 ...-21°C); the annual amount of deposits reached 600-650мм (3). Mid-annual temperatures were positive and reached +1.2 ... +1.5°C. In the second phase of warming in conditions of increasing deficiency of a moisture displacement of southern border of a wood zone to the north reached 500-600км is marked. But it was not face-to-face, but connected with internal reorganization and changes in structure of vegetative communities, that has led to formation on a place of a modern southern taiga hemiboreal birch and pine-birch woods and wood meadows – elanies. The period of the most active development of process of forming the bogs is also connected with the Atlantic optimum. On watersheds, high terraces and in valleys of an ancient drain of a bog enter into transitive, and later into a riding stage of development.

During the same period on a place of Sahara there was a savanna with groves of acacias and the abounding in water of the rivers; Central Asia was the same. The level of Aral Sea was on a mark 72м.

It is necessary to note, that now in Western Siberia the vegetative cover and the inhibited organic mass in peat adjournment are in a condition of unstable balance, which will be easily disturbed at any moment. In conditions of prospective warming, drying up of peat adjournment with their subsequent ignition together with a wood cover on waterless valleys and emission in an atmosphere of catastrophic amount of carbon with unpredictable consequences can take place. Professor Horst Shtrung (Germany) and biologist L. Agafonov, who during last five years investigated together the North of Western Siberia in it's part near Ural, have established that all karstic lakes in forest-tundra and northern taiga increase every year because of constant thawing of frozen ground on 5-6см. As a whole for last 20 years the temperature of multifrozen breeds close to surfaces in the North of Siberia has increased on 1оC. The result of thawing of a long-term frozen ground can be gaping of grounds (грунтов), lowering of the bases of buildings and capital constructions, failures and breakages of oil and gas pipelines.

Since the second half of the 19th century vigorous process of receding of glaciers of Altai and movement of wood vegetation upwards on slopes also began. In 90th years and in the beginning of 21st century mountain glaciers recede not only on Altai but also on the Himalayas. According to climbers and old residents, for example, glacier Ngozumba in area of file Annapurna intensively melts and recedes for a year on 100-200м.

Warming already affects in reduction of structure of ice in Ob and Irtish pool; on Irtish the period of navigation has increased for a week.

A position of biota becomes essentially difficult. The received results (10) allow making a conclusion that natural and anthropogenic-technogenic (антропогенно-техногенный) tendencies have different orientations in modern fast changes of the natural environment and a climate of Western Siberia. In view of an estimation of modern anthropogenic, which intensively and deeply influences structural reorganization of ground and sea ecosystems of region, biota gets in catastrophically narrowed area of crossing different in orientations limiting factors. These authors bring an attention to the question: «whether there are any natural mechanisms of adaptation to modern global changes of the natural environment and a climate for biogeocoenose of a high level of organization; and if there are no such ones, whether enough time was released to develop and use such mechanisms? (With 14) ».

In the rivers of North of Ob pool there were essential changes in volumes of catch and in a variety of breeds of fishes. We shall tell in the middle current of Ob the livestock of muksun was sharply reduced at its high catches down the current, but in area of Pravdinsk bream and a pike perch has much appeared. Also last decade sharp distribution of cases of opisthorchis (about which earlier in the North did not hear) became new.

In connection with sharp changes in water-meadows of the rivers of Ob pool the mass output of a brown bear to watersheds is marked, where the network of technical devices of an oil-and-gas complex, there were also cases of a unmotivated attack on workers of crafts is placed, fatal outcomes. In Far North inadequate reduction of a livestock of a polar bear is faster than reduction of an ice field of Arctic Ocean is fixed. There is an opinion demanding the deep analysis, that it is connected with weakening of immunity to natural pathologies because of an overstrain of immune system of a polar bear in conditions of strengthening of contrast of weather and ice-landscape conditions in places of its dwelling. Probably, the mechanism of mass and fast extinction of mammoth's faunae during an epoch of the end of a glacial age was those.

The migration of the most mobile part of biota to the North is also fixed. Last six years in Hantomansijsk autonomous region for the first time vaccination against tick's encephalitis began to be carried out, in 2004 only in Nizhnevartovsk region 250 stings of ticks are fixed. Thus the amount of stings annually grows on 15-25 %. Last stings are fixed now in second half of September. In August 2004 stings of ticks for the first time are fixed in the center of Yamal-Nenets autonomous region.

Mass cases of migration of large animals (lynxes, a wild boar and a golden eagle) and birds from deciduous and forest-steppe zones of the south of Western Siberia to the north are also marked.

Changes are observed also in structure of the vegetation, testifying the beginning of displacement of vegetative zones, occurrence, and strengthening or better growth of forest-steppe kinds in structure of taiga's landscapes. These changes are actively used in changes to more southern park and garden cultures in northern cities and personal plots surrounding them.

It is impossible to exclude, that fast thawing of glaciers of Mountain Altai has provoked there earthquakes of August, 2003 which in 17 minutes were reflected in Nizhnevartovsk as 3-4 points which in it's turn, have caused intensive excitement such as "tsunami or session" in Ob, and also electromagnetic fluctuations which have suspended the movement of herd of muksun, going in Nizhnevartovsk region. Changes in landscapes were intensified, especially in the sites near the water-meadow, connected with more intensive washouts. There are attributes of more intensive deformations connected with greater mobility of geomorphologic elements than earlier.

We offered for consideration the hypothetical mechanism with powerful ingression on the Bottom and Average Ob caused by concurrence of a cycle of flooding on the river Ob with two or three week rise of coastal waters of ocean because of fast thawing of ices during spring-and-summer time. Such mechanism in Caspian Sea has taken place in 2004 on the river Sulac. Rise of the river Ob on 16 meters instead of usual 10-11 meters will lead to flooding of Nizhnevartovsk, and on 19 meters will lead to flooding of the basic oil-and-gas areas and all the highways of region, going lengthways the river Ob.

The social aspect of warming and the processes of drainage forest fires, landslips and flooding in the south of Russia, in Central Asia, Kazakhstan and on Caucasus connected with it has already caused a sharp immigration stream to the north, the significant part of which is an abundance of labor and requires serious regulation. However this theme is for the special work.

Here there are real consequences because of imposing of "hotbed effect" on cyclic warming were supposed by us (at the end of 80th years) by 2050 [2].

1) *Canada.* Reduction of amount of deposits would lead to poor harvests on the fertile grounds of Ontario.
2) *The USA*, Colorado. Falling of a water level would disturb an agricultural production, water supply and manufacture of the electric power in 8 states, including California.
3) *The USA the Middle West.* Summer seasons would be drier and hotter, that would lead harmful consequences for an agriculture.
4) *Newfoundland and New Scotland.* A plenty of icebergs becomes a real danger for navigation.
5) *The USA. Great Lakes.* The most intense waterway in the world would become free from ice of 11 months in a year, but because of lower water level a cost of ship transportations would considerably increase, and manufacture of the electric power would decrease.
6) *The Western Europe.* In this area the climate does not worsen, because "hotbed effect" does not effect on Gulf Stream.
7) *Greenland.* The part of an ice cover would thaw here, and as a result the sea level would rise on 20-40см.

8) *Russia. The European part; Central Asia; Kazakhstan and Western Siberia of Russia.* The season of growth of agricultural crops would increase for 40 days, but drier climate would demand expensive land improvement.
9) *The Arctic zone.* Ports in Siberia, on Alaska, Bering sea and islands of the Canadian archipelago would be free from ice for the whole year, increasing the volume of a mercantile shipping.
10) *The south of the Far East of Russia, East China.* Productivity would increase because of a plenty of rains.
11) *India and Bangladesh.* Both countries would be exposed to stronger typhoons and flooding.
12) *Equatorial Africa.* The damp Torrid Zone would move to the north and would bring moisture to Chad, Sudan and Ethiopia dried up in present times.
13) *Antarctica.* Because of a plenty of snowfalls and cold rains the thickness of the Antarctic ice cap would increase, and it would partly detain the rising of sea level because of "hotbed effect".

Last researches of an ice core from Antarctica allow supposing, as one of variants, the next cycle of a cold snap, but it's hierarchical level and scale is not clear and demands longer analysis.

Rations and roles of a natural cycle of warming - cold snaps and the "anthropogenic" factor are also very important. Influence, which activity of the person renders on processes of a hydrological cycle, already for a long time involves scientists and practical worker [14]. In 30 years of 18 centuries, strong shallowing of many European rivers have already started to explain a cutting down of woods in their pools. In 40 years of 19 centuries in Petersburg Academy of Sciences the commission that included academicians K.F. Ber, P.I. Keppen, etc. was created; their objects included the studying consequences of cutting down of woods in pool of Volga [15]. In a scientific seal of the European countries and Russia the performance «About the loss of water in springs, in the small and big rivers of the cultural countries at lump sum strengthening of high waters» of Austrian engineer G. Vensa in 1873 caused very wide discussion. [16]. He explains the decrease in a water level in the rivers observed in those years by destruction of woods and drainage of bogs. The Viennese, Copenhagen and Petersburg Academies of Sciences studied G. Vensa's works. The special commission of the Petersburg Academy of Sciences has come to a conclusion, that the shallowing of the rivers is the result of reduction of deposits, and in relation to the Russian regions this natural mechanism has been actually proved [17].

So, a number of significant events of catastrophic character of last time have emphasized growing importance of long-term trends of functioning of geosystems. In our opinion, natural cycles of change of statuses in a history of the Earth concern to a number of the processes having material effect on functioning of geosystems at a regional level. Prominent feature of course of such processes is the geophysical complex of a lot of natural cyclical variations to which swings over of dryness - humidity with economically appreciable droughts and flooding, earthquakes the sharp atmospheric phenomena (tornados, hurricanes, typhoons, etc.), long severe winters and so forth concern,

In territory of Western Siberia from northern seas up to southern mountains all listed displays of the specified processes which combinations carry both regional, and a local kind of territorial distribution are observed. Use of such approach allows developing of typologization of areas and zones of regions in conformity with the characteristics described above. For the subsequent analysis correlation of received typologization with the available facts of natural and economic division into districts is necessary. The basic essence of operation here is consists in definition of time of a natural and economic cycle developed at present time in region [18].

Changes concern two aspects. On the one hand, development of those directions of economic activities for which the optimum complex of natural and cyclic conditions is created is necessary.

In result we mean priorities of regional development for the intermediate term and long-term periods including changes in investment policy, in formation of the credit mechanism, the budget of region. In conditions of development of market relations these factors receive the accelerated development and will be shown much faster in practice of managing. The variation of organizational structures and administrative interactions of economic units of region will be other important point of influence of the mentioned change of natural factors. We mean here the formation of floppy structures from the small and average enterprises, capable to realize those or other advantages for the considered period. As consequence of such formation will be development of processes of cooperation by manufacture, financings and other fields of activity.

On the other hand, the development of those directions for which unfavorable conditions are formed is necessary. As a whole it is necessary to note the important feature of display of the specified processes. First of all it is possible character of their display, also in tendencies in intermediate term and long-term aspects that conducts to necessity of the certain updating carrying out of a policy of regional development. The theory of reliability of functioning of complex systems, in our opinion, is responded with this. The payoff function for plans for development of region with allowance for influences of long-term trends of functioning of complex systems is under construction on the basis of these positions. Such approach provides a scientific basis for production of a problem about reservation of means in regional economy.

The analytical researches, which were carried out, have allowed revealing new aspects in formation of a control system of wildlife management in region on long prospect.

REFERENCES

1. Проблемы реконструкции климата и природной среды голоцена и плейстоцена Сибири. - Новосибирск: Изд-во Ин-та археологии и этнографии СО РАН, 2000.- Вып.2.-472с.
2. Рянский Ф.Н. Фрактальная теория пространственно-временных размерностей: естественные предпосылки и общественные последствия.-Биробиджан: Изд-во ИКАРП ДВО РАН, 1992.-50с.
3. Волкова В.С., Бахарева В.А., Левина Т.П. Растительность и климат голоцена Западной Сибири// Палеоклиматы позднеледниковья и голоцена. М.: Наука, 1989.-С.90-96.
4. Жуков В.М. Климат и процесс болотообразования // Научные предпосылки освоения болот Западной Сибири. М.: Наука, 1977.-С.13-29.
5. Лапшина Е.Д. Флора болот юго-востока Западной Сибири. - Томск: Изд-во Том.ун-та, 2003.-296с.
6. Кузнецова Ф.С. История Сибири. Ч I: Учебное пособ.-2-е изд.-Новосибирск: ИНФОЛИО – пресс, 1999.-256с.
7. Жилина Т.Н. Малая ледниковая эпоха (1550-1850г.г.) как причина риска развития сельского хозяйства Западной Сибири // Экологический риск/ Мат. втор. Всероссийской конфер.-Иркутск: Изд-во Института географии СО РАН, 2001.- С.9-11.
8. Антропогенные изменения климата / ред.М.И. Будыко, Л.: Гидрометеоиздат, 1987.-404с.
9. Гребенюк Г.Н. Типы кедровых лесов бассейна реки Вах (правобережье Средней Оби): Монография. - Новосибирск: ЦСБС СО РАН, 2004.-165с.
10. Бляхарчух Т.А. История растительности юго-востока Западной Сибири в голоцене по данным ботанического и спорово-пыльцового анализа торфа // Сиб. Экол. Журн.-2000. Т.7, №5.-С.659-668.
11. Титов Ю.В., Овечкина Е.С. Растительность поймы реки Вах. –Нижневартовск: Изд-во Нижневарт. пед. инст-та, 2000.-123с.
12. Зольников И.Д., С.А. Гуськов, А.Н. Дмитриев, А.Е. Богуславский, В.А. Баландис. Исследование динамики экологических обстановок Западной Сибири с учетом изучения палеоэкологических аналогов современных глобальных изменений природной среды и климата // Экологич. Риск / мат. втор. Всероссийской конфер.- Иркутск: Изд-во Института географии СО РАН, 2001.-С.11-14.
13. Коркин С.Е. Природные опасности долинных ландшафтов Среднего Приобья / Автореф. канд. дисс.-Барнаул: ИВЭП СО РАН, 2004.-22с.
14. Ткачев Б.П., Булатов В.И. Малые реки: Современное состояние и экологические проблемы. Аналит.обзор/ГПНТБ СО РАН.- Новосибирск, 2002.-114с.
15. Арманд А.Д. и др. Механизм устойчивости геосистем.- М.:Наука, 1992.-207с.
16. Рахманов В.В. Водорегулирующая роль лесов// Тр. Гидрометеоцентра СССР.- 1975.-Вып.153.-192с.
17. Маслов Б.С., Минаев И.В. Мелиорация и охрана природы. - М.: Россельхозиздат, 1985.-271с.
18. Рянский Ф.Н., Юсупов В.Р. Региональное районирование долгосрочной динамики геосистем и его использование в хозяйственной практике // Состояние природной среды Зейско-Буреинской равнины и сопредельных территорий. Перспективы ее использования и охрана. - Благовещенск, 1991- С.41-43.

SUMMARY OF OTHER WORKSHOP LECTURES

by N. DOBRETSOV

Paleoclimate during desertification and modern tendencies

Nikolai Dobretsov, Russia

Much importance in the formation of the global climatic system is attached to Asia, a large land with elevated and severe topography that has a pronounced effect on planetary air circulation. The history of modern climate and environment formation in this territory during Cainozoe is observed on intercontinental profiles. One of the fullest profiles among the Pleistocene continental sediments are loess-soil profiles, and the reflected record of climate events shows the cyclic variations in environment and climate during this period of time. The data on 100 loess-soil profiles of West Siberia were analyzed, and the revision of the already published materials was conducted. The comparison of the structure of full loess-soil sequence in Siberia within Bruenes period (0-780 th. years) with the structure of continuous global climate records makes it possible not only to reveal the frequency, but to show the correlation between climate and environment changes on this territory during Pleistocene and the global ones.

The comparison of loess record in West Siberia and continuous global records of climate and environment changes demonstrate their good correlation. It is indicative of the global character of loess record and confirms the global scale, code identity and the uniform mechanism of climate fluctuations as in the middle altitudes of Central Asia as on the whole planet. Some distinctions in the structure of time series of climatic sequences obtained from various materials reflect the different response of the system to orbital forcing. Good correlation between temperature fluctuations and changes in the intensity of air circulation reflected in the loess-soil and other global climate records show the relationship of full air circulation and the change of its main factor, i.e. temperature difference in low latitudes and pole.

Interdisciplinary approach to the study of paleoclimate and introduction of new analysis techniques such as synchrotron emission allow one to decode the continuous records of climate fluctuations. It makes it possible to predict for sure the desertification processes in specific geographical areas as well as to forecast the single changes in ecosystems' state, for instance, under permafrost degradation.

Tendencies of Land Use in Large Regions of the World (by the Example of Ukraine)

Leonid Rudenko, Ukraine

Land use, as defined in the paper, is the influence of man and society on nature components, which is revealed in peculiarities of different resources use for the purpose of the society activity support, and is grounded on ecological capacity of the territory estimation and conservation of natural properties of nature components, landscape and biological diversity. Irrational approaches to land use in Ukraine during the 20th century have resulted that at the beginning of the 21st century Ukraine's economy was one of the most nature-consuming in the world. The power intensity of the national GDP was 14,31 times more than world average one, electric power intensity – 8,8, water-consuming – 2,83 times more. The atmospheric emission of carbon dioxide per unit of GDP was 15,25 times more then in the world on average. The main conclusion from the analysis of land use in Ukraine is lack of understanding by most of the authority representatives the importance of natural properties of nature components conservation and their value as the basis for goods and services production as well as providing proper living conditions of population.

Land relationships and land use issues in Russia: return to the traditions and moving forward

B.I. Kochurov & M.N. Baulina, Russia

We interpret the analysis of a complicated land use issue in Russia as a social order of the state that is consistently introducing decentralization of management in the Russian economy and renders many natural resources management functions to the lowest (local) management level.

All measures that are to be carried out according to the main legislation acts – Land Code of the Russian Federation and Federal Law "On General principles of Local Government in the Russian Federation" - for example by a municipal body as a subject of land use management and protection, require serious scientific and methodical justification.

For this purpose it is necessary from the positions of the modern science and current practice to perform a retrospective analysis of all consequences of socialistic transformations of nature with regard to land use: ploughing up of virgin lands in Siberia and Kazakhstan, "total" digression of grazing lands, "large-scale" land-reclamation, transfer of northern rivers, elimination of "unpromising" villages, etc. Finally it is necessary to reveal the reasons of the unsuccessful attempt to establish "socialistic land management" and to develop its theoretical backgrounds.

It has been stated that land relationships make a triad – three main groups of land relationships: the economic, legislative and territorial (land management) relationships. This definition connects the economics of land use with land legislation and the theory of territorial location of various economic activities (agriculture, cattle breeding, recreation, etc.).

Use of land based on deliberate needs of the society, regulated by legislative grounds and organized within a territory makes the process of land use.

Regulations concerning local land relationships are most developed. Most weak chain in the triad of the land relationships is land use organization that is characterized by deep stagnation within last 80 years.

Formation of land use organization in Russia should be referred to Stolypin's reform. In the soviet times the land use organization officers, while making local zoning and giving backdate justifications to "historical transformations of rural settlements", distorted the data on actual status of lands in the state statistics (which is going on at present) and issued numerous works on the success of collective farms.

Thus the territorial organization of land use in Russia at all levels is to be developed from zero point using observations of the currently developing land use and recording the local factors impacting the process.

Problems of Land Use and Biodiversity Conservation in the contiguous territories of Western Mongolia

Navaanzoch Kh. Tsedev, Mongolia

1. Prompt natural resources development both in West Mongolia and throughout the whole Mongolia territory is observed at present. This has led to some problems like progressively growing soil deterioration, pasture emaciation and onset of desertification. On the whole the land is used for natural cattle pastures, growing cereals and vegetables and the extraction of natural resources like coals, clay and nonferrous metals.

Stock raising branch still remains the main branch of economy since infrastructure and industry are poorly developed here. According to the data of the Animal Husbandry Institute, in the second half of the 80s about a third of all pasture areas of our country was already in a degraded state. Today, the present situation is being aggravated intensively. Not only soil is being degraded but also the whole complex of the landscape has been destroyed. The same situation exists in the contiguous country, namely the Republic of Altai.

The load on the earth exceeds permissible ones that contribute to desertification, impacts on the soil, plants, grass and other complexes of the pasture landscape. For example: capacitance to pasture at the territory around Khar-Us reserve has been estimated, i.e. annually it can feed 158,2 – 301,4 thousands of conventional sheep heads. Now at the territory about half million of heads of cattle is concentrated and excess in capacitance to pasture of 200-300 thousands of conventional heads of sheep is marked due to the following anthropogenic impacts on the ecosystem of Khar-Us reserve; a). Its territory's use as pastures and hayfield; b). Its use for private purposes (timber of poplar, birch and elm is used for building purposes, bushes- for heating).

These two types of land use prevail in this territory. Besides this, the areas near cities, towns, centers of soums, summer and winter nomad camps, the watering places for cattle and wells are subjected to degradation and pollution as well. Also lots of branching roads in the countryside has appeared. The anthropogenic impacts let to disturbance of nature balance and the appearance of movable sand that accelerate desertification processes.

Some researchers consider that the basic method for realization of nature management conservation problem can be landscape planning accompanied with creation of a series of different scale maps of some areas. Other researchers think that the most effective method for combating against desertification is biological land reclamation; planting wild grass, in particular, Agropyron Sibirica, that is important as hay grass and as a pasture plant as well.

2. Western Mongolia is distinguished by rich biodiversity. Many specific biological species are endemic ones. 26 kinds of vertebrates living in Mongolian Altai are included in the Red Book of Mongolia.
Five kinds of them such as: Tenches, Fairtail Sea Eagles, Surviving Sea Gulls, otters, Snow Leopards are included in the Red Book of the International Union of Nature Protection.

In order to protect biological diversity in Mongolia we ought to intensify nature protection activities in the special protected areas and fix strict custom control over export of valuable, rare, endangered, endemic and other kinds of animals; as well as to create international transboundary special protected areas in the contiguous regions.

Ecological safety and land use problems

Yuri Vinokurov, Russia

Altai takes an important place on the physical-geographic and social-political map of Central Asia. Altai serves as a regulator and re-distributor of air and water masses being on the top of the great watershed that isolates river basins of Arctic Ocean from internal-drainage areas of the Asian mainland. Altai is situated at the joint of natural zones and serves as a biosphere knot: the place of meeting and interlacing of north taiga, steppe and semi-desert landscapes. A great variety of natural environments and thus generated landscape complexes provide for valuable biodiversity and availability of endemic and relic species of flora and fauna of the world importance.

Various natural-climatic conditions, complicated geological structure, flora and fauna richness predetermine availability of natural resources, i.e. mineral raw, water, agroclimatic, biological, etc.

Social and political significance of Altai is defined by the fact that it is located in the periphery (interlink of boundaries) of four countries and characterized by different population, cultures and religions, types and traditions of nature management.

Ecological safety of Altai is formed to a great extent outside its ecosystem boundaries. Transboundary transfers of air masses from industrially polluted territories and their accumulation in the steppe territories inside the mountain hollows effect greatly on its state.

Extensive land use typical for Altai threatens ecological safety not only of Altai but the Euro-Asian continent as a whole.

Agrarian nature management with prevalence of distant pasture cattle breeding is characteristic for all four countries of the region. Agriculture takes place in the intermountain hollows on steppe landscapes. Until the present time forest exploitation was widely spread. Also bee keeping, wild plants gathering and different trades are popular here. In East Kazakhstan and Russia mining is typical due to availability of rich mineral-raw resources of Altai. Raw material processing is performed in Kazakhstan. In Russia only primary processing with the following export of processed and enriched raw to metallurgic works of Kuzbass, the Urals and other regions is made. Joint development of minerals has been started at the boarder of Russia and Mongolia, Russia and Kazakhstan.

Land use under severe and often extreme natural conditions (high mountains, semi-desert landscapes, permafrost, significant fluctuations of annual and daily temperatures, strong winds, thin fertile soil layer, short vegetation period, etc) resulted in revealing serious ecological and social-economic problems including the ones of transboundary character. These problems are the following: pollution of soil cover and lost of fertility; high and unsystematic load on pastures and their digression; climate change; aggravation of biodiversity and degradation of flora and fauna habitat; loss of markets and lack of solvent demand for local production; poaching and smuggling export of valuable natural raw.

The solution of many problems is possible solely at the international level with involvement of all countries situated in Altai as well as international organizations, in particular, within the framework of Transboundary Biosphere Territory of Altai and International Coordination Committee "Our common home-Altai".

Eco-Environmental Situation and Protection Countermeasures in Altai Mountain Area of China

LIU Wenjiang, China

Altai Mountains is one of the important areas in Central Asia shared by four countries. The mountains have well-preserved Europe-Siberian flora and fauna and typical mountain forest – river – lake ecosystem, which have great implication for globe biodiversity and natural reserve. Given the great potential of development and aspiration for local poverty reduction, the southern Altai area within China has been recently emphasized and accelerated its development by various governments in terms of water, mineral and tourism resources, and resource-based international cooperation. In the forthcoming development activities, the achievement of coordinated resources development and environmental protection should be recognized. In consideration of the protection of the Altai's specific biodiversity and eco-environment, the paper first focuses on the description of biodiversity, status of conservation and natural reserve, analyses of ecosystem fragility and national implication of biodiversity conservation in southern Altai mountain area, based on the updated research results. In particular, the paper emphasizes the analysis on the scope, structure and function of the well-protected Kanas Natural Reserve established in 1980 and upgraded to one of the national natural reserves in 1986.

The potential resources development and its possible impact on the eco-environment are discussed. Water and mineral resources are the basic means in supporting the local social and economic development. Some approaches and policies are suggested to enforce their good management. Moreover, the tourism is recently envisioned to be the most important and potential sector for improving the local economy and reducing the poverty. The paper analyzes the position and role of Altai Mountains that have become one of the top tourism areas in Xinjiang and defined as the most potential sector in the cooperation with neighboring countries, especially Lake Kanas and its surroundings. Given such great potential, plans for tourism and eco-environment protection should be formulated in terms of environmentally sound management, coordinated policies and actions for tourism development among the countries concerned. In particular, the paper suggests extend the World Heritage Golden Altai Mountain by adding the Kanas Natural Reserve and other high-quality natural reserves within Altai Mountains, in order to raise its value and enforce the eco-environmental protection under the Man and Biosphere Reserve Program. In addition, the paper highlights the indigene participation of various organizations in tourism and capacity building to promote eco-tourism.

Thirdly, the paper recognizes the important role of the advanced technology of remote sensing and GIS to monitor the dynamic change of ecosystem and render possible assessment, especially in assessing the possible effect of the degraded frigid-temperate ecosystems due to human activities and even the global climate change. On the international level, it is necessary to establish a joint monitoring mechanism to exchange and share the information and knowledge.

At last the paper proposes the suggestions leading to sound eco-environmental protection and sustainable development. It concludes that the work in the following aspects should be enforced domestically and internationally. In order to achieve the sustainable development in Altai Mountains the following is necessary: (1) to incorporate eco-environment protection plan into the local social-economic development plans; (2) to establish the natural reserves, enforce environment-friendly water and mineral resources development and promote eco-tourism; (3) to establish the monitoring stations to monitor and assess the ecosystem and related resources changes; (4) to enforce the joint eco-environment protection actions and research projects; (5) to implement the joint environmental impact assessment in the international ports establishment, boundary trading and transnational tourism development; (6) to establish an effective institutional mechanism for effective cooperation and management through the participation of all stakeholders.

WWF approach to the establishing of Transboundary Biosphere Territory (TBT) «Altai»

Alexander Bondarev, Russia

In 2000 WWF experts have developed comprehensive scheme of protected areas (PA) - for Altai-Sayan Ecoregion (ASER).

TBT "Altai" is situated in the center of the ecoregion and it is no wonder, that the experts have paid special attention to it. Experts supposed to establish a network of PA with different status within the discussed territory. Among them "Uyngur" - a cluster of Biosphere Reserve "Katunskiy", National Park "Sailugemskiy", natural parks «Central Altai" and "Telengitskiy". The total area of proposed PA's make about 1.6 million ha, and 1.7 million ha including the territory of "Katunskiy" Reserve.

In 2000 WWF began implementation of the ECONET scheme for ASER. National Park "Katon-Karagai" was established in Kazakhstan with the territory of 643,000 ha in total and Natural Park "Argut" was created in Altai Republic (Russia) on the territory of 34,000 ha. Also project drafts for "Uyngur" and Natural Park "Ukok" were developed. WWF supported the project of ten-times enlargement of the territory of Natural Park "Chui-Ozy" in Altai Republic. Functional zoning were developed and implemented in National Park "Sielkhem" (Mongolia).

Taking into consideration importance of TBT "Altai" for biological, historical and cultural diversity, for traditional nature use and tourism, latter can be determined as a strategy for the development of the region, the most appropriate solution is the establishing of International Park of Peace within the borders of four countries: Russia, Mongolia, Kazakhstan and China.

All three countries, except Russia, have PA, which can be considered as a basis for the establishing of such a territory. These PAs are: National Park "Katon-Karagai" in Kazakhstan (0.6 million ha), National Parks "Altai Tavan Bogd" & "Sielkhem" in Mongolia (0.8 million ha), Khanassi Reserve (0.1 million ha), which one though has a status of reserve, but is closer to national park, due to engagement in tourism development, in China.

Thus logical solution for Russia would be creation of National Park "Altaiskiy" on the territory of Kosh-Agach, Ongydai and Ust-Koksa municipals with the area of 1.7 million ha. The most critical issue is the establishing of a park without exception territory from current land users.

Landscape planning should precede park establishing for identification borders and functional zoning.

Total area of proposed International Park can compound 3.2 million ha within the borders of existing and planned PAs. In order to attain integrity of the territory, the area will increase to approximately 5 million ha.

Such a model meets legal requirements of four countries. All national clusters of the future park will have similar objectives. The park will stimulate development of international tourism, cooperation and exchange; highlight attractiveness for national and international donors. Territories around the Park will benefit as buffer zones of national clusters mainly from the development of tourism infrastructure.

Comprehensive inner zoning of national portions of the Park will meet the concept of Seville Strategy for Biosphere Reserves.

Transboundary cooperation of the mountain communities of Altai and Mongolia

Klimova Vera, Kazakhstan

Harmonious development always stipulates the unity of spiritual and of material values. This postulate is right as for personal relations as for Trans boundary issues as well.

In our opinion the shortest and most effective way to establish the transboundary cooperation is in joint actions, through realization of NGOs projects and the efforts of local people of the Altai- Sayany region. This national democracy can be possibly achieved because life conditions, problems of mountain inhabitants in different countries are similar in Altai. Actually, Altai is our common home. Valuable international experience in collaboration of the different organizations dealing with mountain conservation has been already gained. For example, the Alliance of Mountain Organizations in Central Asia that was founded as an association of non-governmental organizations. The major goal of the Alliance is promotion and support of sustainable development in mountainous villages, improvement of life quality in local communities. The main resource of all activities is local people from villages and auls. These communities solve such issues as knowledge and experience exchange between mountainous communities of Central Asian countries, distribution of successful projects at national and regional levels, training and enhancing the qualification of mountain communities inhabitants through organizing seminars, trainings, foundation of craftsmanship centers, e.g. the ones dealing with making felt carpets, leather goods, processing of milk and producing of cheese, etc. The important element of activities is representation and protection of mutual interests of mountainous communities, support of constant links with partners and people responsible for decision making at all levels. It would be reasonable to broaden the informational coverage of the Alliance's activity and issue of common edition in four languages.

So, taking into consideration the world experience gained by regions, achievement of the goal on sustainable development in mountainous territories and also improvement of transboundary cooperation it seems possible to create the Altai Alliance of mountain communities.

It should be noted that through social organizations of mountain communities, the local people acquire the skills of self-governing and would take responsibility for solving many social issues. Besides, people would participate in decision-making of many vital social and ecological problems.

Since Altai region plays an important role in many countries, we can attract the attention of sponsors to Altai region's problems and financing different social and ecological projects through the Alliance. It is obvious that the soul of Altai is not only its nature, but also the people living there.

Enormous potential of transboundary cooperation is in the development of ecological tourism and involvement of local communities into this activity. First and foremost it concerns "the green tourism", in development of which local population is encouraged to participate. Here is possible to found the Altai-Sayany Association "Green house". It implies creation of common website, exchange of experience, arrangement of common seminars and trainings as well as transboundary tours.

Thus we could always live at peace and cooperate successfully in our common home Altai.

LIST OF WORKSHOP PARTICIPANTS

Directors:

Prof. Hartmut VOGTMANN
President of the Federal Agency for Nature Conservation,
Konstantinstr. 110, 53173 Bonn,
Phone: +49 228 8491 210
Fax: +49 228 8491 250
Email : Hartmut.Vogtmann@BFN.de
GERMANY

Prof. Nikolai DOBRETSOV
President of the Siberian Branch of Russian Academy of Sciences, prosp. Akademika Lavrentieva, 17,
Novosibirsk, 630090
Phone: +007 (3832) 30 05 67
Email: dobr@sbras.nsc.ru
RUSSIA

Lecturers or Key Speakers

Alexander ANTIPOV
Vice-director,
Institute of Geography SB RAS,
1, Ulan-Batorskaya Str.
664033 Irkutsk
Phone : (3952) 427820
Email : antipov@irigs.irk.ru
RUSSIA

Alexander BONDAREV
UNDP/GEF Project Manager
Phone: 83912279491
Fax: 83912276692
Email:
Alexander.bondarev@undp.ru
Abondarev@wwf.ru
RUSSIA

Igor BONDYREV
Professor, Head of Foreign Affairs Department,
Vakhushti Bagrationi Institute of Geography,
1, M. Aleksidze St., Block 8
0193 Tbilisi
Phone: (995-32) 334935
Fax: (995-32) 331417
Email : bond@gw.acnet.ge
REPUBLIC OF GEORGIA

Prof. Dr. Bernd CYFFKA
Institute of Geography
Goldschmidtstrasse, 5
D-37077 Goettingen
Phone: +49-551-398066
Fax: +49-551-398008
Email : Bcyffka@gwdg.de
GERMANY

A.A. CHIBILIOV
Director, Institute of Steppe
11, Pioneer St. Orenburg, 46000C
Phone (3532) 77-432
E-mail: steppe@elay.esoo.ru
RUSSIA

Dangaa ENKHTAIVAN
Dr., Head of Laboratory of Physical geography
Institute of Geography MAS,
P.O.B-361, Ulaanbaatar 210620
Phone.: 976-11-353461
Fax: 976-11-350472
Mobile: 99265560
Email: Taivan_geo@yahoo.com
MONGOLIA

Galina GREBENYUK
Dean of Department Natural & Precision Sciences
State Pedagogical Institute
56, Lenin str.
Nizhnevartovsk
Phone.: 12-90-24
Fax: 43-65-86
RUSSIA

Dr. Dietwald GRUEHN
Head of the Environmental Planning Department
Austrian Research Centers - systems research GmbH
Tech Gate Vienna,
Donau-City-Str. 1
A-1220 Vienna
Phone: ++43 (0)50 550 4581/4580
Fax: ++43 (0)50 550 4599
Mobile: ++43 (0)664 8251186
E-mail: dietwald.gruehn@arcs.ac.at
AUSTRIA

Ph. D. Mikhail KHANKASAYEV
Associate Director,
Institute for International Cooperative Environmental Research
Florida State University
2035 East Paul Dirac Drive,
226 HMB
Tallahassee, Florida 32310-3700
Phone: 850/644-5524
Fax: 850/ 574-6704
Email :
mkhail.khankhasayev@fsu.edu
USA

Natalia KHARLAMOVA
Geographical Department
Altai State University,
61, Lenina Avenue,
656049, Barnaul
Phone: (3852) 683115, 248999, 364448,
Email: harlamova@alt.ru
RUSSIA

Martin KLIMANEK
Mendel University of Agriculture and Forestry, Faculty of Forestry and Wood Technology
Institute of Geoinformation Technologies
Zemedelska 3, 613 00 Brno,
Phone: +420 545 134 017
Fax: +420 545 211 422
Email: klimanek@mendelu.cz
CZECH REPUBLIK

List of Workshop Participants

Boris KOCHUROV
Institute of Geography,
RAS, Moscow
Phone: 442-64-59
Email : kerwit@online.ru
inecol@rc.msu.ru
RUSSIA

Jaroomir KOLEJKA
Assistant Professor, Candidate of science
Mendel University of Agriculture and Forestry, Brno
Faculty of Forestry and Wood Technology,
Institute of Geoinformation Technologies
Zemedelska 3, CZ-61300 Brno
Phone: +420 545 134 018,
Fax: +420 545 211 422,
Email: kolejka@mendelu.cz
CZECH REPUBLIK

Igor KOROPACHINSKY
Academician, Advisor of RAS
Central Siberian Botanical Gardens
Novosibirsk
Vize - president Russian MAB
Phone: (3832) 30-41-01
Fax: (3832) 30-19-86
Email: root@hotgard.nsc.ru
RUSSIA

Bella KRASNOYAROVA
Head of Laboratory of Nature Management,
IWEP SB RAS
1, Molodyozhnaya St.
Barnaul 656038, Russia
Phone: (3852) 666506
Email: bella@iwep.asu.ru
RUSSIA

Dr. Peter H. LIOTTA
Executive Director
Pell Center for International Relations and Public Policy
100 Ochre Point Avenue
Newport, Rhode Island 02840
Phone: 001 401 341 2371
Fax: 001 401 341 2974
Email : peter.liotta@salve.edu
phliotta@cox.net
USA

Vladimir LOGINOV
Director, Institute for Environment and Natural Resources Use
Academician of National Academy of Sciences,
10 Staroborisovskiy Trakt,
BY-220114, Minsk
Phone: (375) 17264-26-32
Fax: (375) 172642413
Email: ipnrue@ns.ecology.ac.by
BELARUS

Anatoly MANDYCH
Research leader, Candidate of sciences
Institute of Geography, Russian Academy of Sciences,
Staromonetny per., 29
109017 Moscow,
Phone: 095 198-0291
Fax: 095 959-033
Email: amandych@mtu-net.ru
RUSSIA

Prof. Dr.Angelika MEIER-PLOEGER
Department of Organic Food Quality and Food Culture,
Faculty of Organic Agricultural Sciences, University of Kassel
Nordbahnhofstr. 1a
37213 Witzenhausen,
Phone: +810 495 542 981 712
Email: amp@uni-kassel.de
GERMANY

Dr. Kholnazar MUKHABBATOV
Head of Department of Regional Economy,
Institute of Economical Researches
Academy of Sciences,
Dushanbe 734025
Phone: 8-10-992372323948(home)
Email : Region_ek@rambler.ru
TAJIKISTAN

NAVAANZOCH X. Tsedev
Vice-Director for Science and Foreign Relation
Hovd State University
MONGOLIA

Adyal OIDOV
Mongolian Academia of Sciences,
Ulaan-Baatar
Scientific secretary of Society sciences
MAS
Professor, Sc. D
Phone: 265001
MONGOLIA

Dr. Petranka PIPEVA
Deputy of Dean
University "Prof. Dr. Assen Zlatarov"
1, Prof. Yakimov Blvd.
8010 Bourgas,
Phone: +35956858243
        628622
Fax: 858243
Email: ppipeva@abv.bg
       anamariana@abv.bg
BULGARIA

Dr. Doris POKORNY
Research Coordinator,
Rhon biosphere reserve
Oberwaldbehrunger Str. 4
D- 97656 Oberelsbach
Phone: ++49-[0] 9774-91020
Fax: -9102-21
Email : Doris.pokorny@brrhoenbayern.de
GERMANY

Victor REVYAKIN
Dean of Geographical Department
Altai State University,
66 Dimitrova Str.,
656099, Barnaul
Phone: (3852) 249989
       (3852) 238760
Email: decanat@geo.dcn-asu.ru
RUSSIA

Felix RJANSKI
Head of Department
State Pedagogical Institute
56, Lenin str.
Nizhnevartovsk
Phone: (3466) 147750
Email: ecologis@nptus.ru
RUSSIA

Elena RODINA
Kyrgyz-Russian Slavic University
Chief of Sustainable Development
of Environment Department
44 Kievskaya str.
Bishkek 720000
Phone: (0996312) 43-13-82
Fax: 22-27-76
     22-64-92
Email: ccproject@istc.kg
KYRGYZSTAN

Leonid RUDENKO
Corresponding member,
Director, Institute of Geography,
National Academy of Sciences of
the Republic of Ukraine
Phone: (380) 442346193
       (380) 444243195
       8-044 234-61-93
Fax: (380) 4424323
     8-044-234-32-30
Email: geograf@kiev.ldc.net
       geo-ius@lds.net
UKRAINE

List of Workshop Participants

David SMITH
United Nations Environment
Programme (UNEP)
38870 P.O. Box
30552 Nairobi
Phone: +254-20 624059
Fax +254-20 623861/622788
Email: David.Smith@unep.org
KENYA

Arnold TULOKHONOV
Corresponding member,
Baikal Institute for Nature
Management SB RAS,
6, Sakhyanova St., Ulan-Ude,
670047 Republic of Buryatia,
Email: bip@bsc.buryatia.ru
RUSSIA

Yury VINOKUROV
Director of Institute for Water and
Environmental Problems SB RAS
1, Molodezhnaya St.
Barnaul, 656038
Phone: (3852) 666055
Fax: (3852) 240396
Email: iwep@iwep.asu.ru
RUSSIA

Liu WENJIANG
Xinjiang Institute of Ecology and
Geography, CAS
Director, Department of International
Cooperation,
Phone: 86-991-7885304
Fax: 86-991-7885300
Email: wjliu@ms.xjb.ac.cn
CHINA

Alexander YEMANOV
Director of Geophysical Service
SB RAS
630090, Novosibirsk, Academician V.
A. Koptyuga avenue, 3
Phone: (3832) 33-27-08
Fax: (3832) 30-12-61
Email : emanov@gs.nsc.ru
          asomse@gs.nsc.ru
RUSSIA

Prof. JIEBIN ZHANG
Director of Integrated Water
Management Project,
Xinjiang Institute of Ecology and
Geography
Chinese Academy of Sciences
South Beijing Road 40-3, Urumqi,
Xinjiang 830011
Phone: 0086-991-7885379
Fax: 0086-991-7885320
Email: zhangjb@ms.xjb.ac.cn
CHINA

Other Participants

Andrei ARKHIPOV
Ministry of Economics of Republic of Altai
Gorno-Altaisk
Phone: (38822)-2-25-82
Fax: (38822)-2-31-30
Email: arhipov@economy.gorny.ru
RUSSIA

Yury BOBROVKIN
Chief scientific employee of
IWEP SB RAS
Barnaul
RUSSIA

Lidia BUREAKOVA
Head of Laboratory of Soil science & Agricultural chemistry
Altai State University, Barnaul
Tel.: 628 451
RUSSIA

Valery ERMIKOV
Head of UONI SB RAS
Novosibirsk
Phone: 383-230 3619
Fax: 30 20 95
Email: ermikov@sbras.nsc.ru
RUSSIA

Prof. Jean KLERKX
International Bureau for Environmental Studies
Brussels
Tel.: +322 767 7409
Fax: +322 306 4853
Email: jklerkx@pibes.be
BELGIUM

Sergei NOZHKIN
Department of Foreign Connections,
Administration of Altai Region
Barnaul,
RUSSIA

Alexander PALKIN
Director the Center of Science & Technologies Republic Altai
Gorno-Altaisk
Phone: (38822)-2-71-21
Email: Altai-nauka@mail.gorny.ru
RUSSIA

Heinrich SCHMAUDER
Senior civil servant,
Department II 1.3,
Federal Agency for Nature Conservation
Konstantinstr. 110, 53179 Bonn
Phone: +49 228 8491 241
Fax: +49 228 8491 245
Email : Schmauderh@BFN.de
GERMANY

Svetlana SURAZAKOVA
Director of GAF
IWEP SB RAS
Gorno-Altaisk
Phone: (38822) 2-67-86
Fax: (38822) 2-67-86
Email: spsuraz@rambler.ru
RUSSIA

Raisa VOROBYEVA
Director AF FSUP SRISSV "Progress"
Phone: 240203
Fax: 240203
Email: Af_progr@ab.ru
RUSSIA

Sergei ZINOVIEV
Advisor of Head of Republic Altai
Government of Republic Altai,
Gorno-Altaisk
Phone: (38822)-2-44-57
Fax: (38822)-9-51-21
Email: root@apra.ru
RUSSIA

List of Workshop Participants

Organizing Committee

Hartmut Vogtmann (Germany)
President of the Federal Agency for
Nature Conservation
Konstantinstr. 110D-53179 Bonn
Tel. +49-228-8491-210
Fax: +49-228-8491-250
vogtmannh@bfn.de

Heinrich Schmauder (Germany)
Federal Agency for Nature
Conservation
Konstantinstr. 110D-53179 Bonn
Tel. +49-228-8491-210
Fax: +49-228-8491-250
Schmauderh@bfn.de

Nikolai Dobretsov (Russia)
President of Siberian Branch of
Russian Academy of Sciences
17, Prospect Lavrentieva,
Novosibirsk, 630090
Phone/Fax: (3832) 300567
dobr@sbras.nsc.ru

Ermikov Valery (Russia)
Presidium SB RAS
17, Prospect Lavrentieva,
Novosibirsk, 630090
Phone: (3832) 303619
ermikov@sbras.nsc.ru

Yury Vinokurov (Russia)
Director of Institute for Water and
Environmental Problems SB RAS
1, Molodezhnaya St., Barnaul, 656038
Phone: (3852) 666055
Fax: (3852) 240396
iwep@iwep.asu.ru

Bella Krasnoyarova (Russia)
Head of Laboratory of Nature
Management, SB RAS
tel: (3852) 666506
bella@iwep.asu.ru

Printed in the United Kingdom
by Lightning Source UK Ltd.
110331UKS00002B/53